ELEMENTARY ELECTROMAGNETIC THEORY

VOLUME 2

MAGNETIC FIELDS, SPECIAL RELATIVITY AND POTENTIAL THEORY

BOOKS BY DR. CHIRGWIN, PROFESSOR KILMISTER AND DR. PLUMPTON ON ELECTROMAGNETIC THEORY

ELEMENTARY ELECTROMAGNETIC THEORY

Volume 1 STEADY ELECTRIC FIELDS AND CURRENTS

Volume 2 MAGNETIC FIELDS, SPECIAL RELATIVITY AND POTENTIAL THEORY

Volume 3 MAXWELL'S EQUATIONS AND THEIR CONSEQUENCES

ELEMENTARY ELECTROMAGNETIC THEORY

IN THREE VOLUMES

VOLUME 2

MAGNETIC FIELDS, SPECIAL RELATIVITY AND POTENTIAL THEORY

B. H. CHIRGWIN

Queen Mary College, London

C. PLUMPTON

Queen Mary College, London

AND

C. W. KILMISTER

King's College, London

PERGAMON PRESS

OXFORD · NEW YORK · TORONTO

SYDNEY · BRAUNSCHWEIG

Pergamon Press Ltd., Headington Hill Hall, Oxford
Pergamon Press Inc., Maxwell House, Fairview Park, Elmsford, New York 10523
Pergamon of Canada Ltd., 207 Queen's Quay West, Toronto 1
Pergamon Press (Aust.) Pty. Ltd., 19a Boundary Street, Rushcutters Bay, N.S.W. 2011, Australia
Vieweg & Sohn GmbH, Burgplatz 1, Braunschweig

First edition 1972

Library of Congress Catalog Card No. 75-151747

Printed in Hungary

08 016599 0 (hard cover)

08 016600 8 (flexicover)

CONTENTS

PREFACE TO VOLUME 2 vii

6. Magnetostatics 197

- 6.1 Introduction 197
- 6.2 Origin of the magnetic field 198
- 6.3 The magnetic displacement vector 203
- 6.4 Boundary conditions 216
- 6.5 The magnetostatic scalar potential 221
- 6.6 Small magnets 224
- 6.7 Inductance 229

7. Magnetization 243

- 7.1 Permanent magnets 243
- 7.2 Permeable media 255
- 7.3 Induced magnetization 256
- 7.4 Hysteresis 258

8. Potential Problems 263

- 8.1 Fundamental theorems and techniques 263
- 8.2 The method of images 270
- 8.3 Three-dimensional problems with axial symmetry 284
- 8.4 Two-dimensional problems 298
- 8.5 Gravitation 316

9. Electromagnetic Induction and Magnetic Energy 332

- 9.1 Faraday's and Neumann's Law 332
- 9.2 Observers in motion: the Lorentz transformation 336
- 9.3 The Lorentz transformation of field quantities 351

9.4 The Lorentz force on a moving charge 356
9.5 Non-steady currents 365
9.6 Initial conditions: switching 374
9.7 Quasi-steady currents 387
9.8 Magnetic energy and forces 399
9.9 Dissipation of energy by quasi-steady currents 406

10. The Displacement Current and Maxwell's Equations 417

10.1 The displacement current 417
10.2 Electromagnetic radiation 419
10.3 Boundary conditions at discontinuities 421
10.4 Poynting's vector 425
10.5 Maxwell's stress tensor and electromagnetic momentum 430

ANSWERS TO THE EXERCISES 435

INDEX 439

PREFACE TO VOLUME 2

THIS is the second of three volumes intended to cover the electromagnetism and potential theory usually included in an undergraduate's course of study. These books are intended only as an introduction to electromagnetism and have been prompted by discussions with first-, second- and third-year undergraduates.

The general scheme of the volumes is to start with the simple case of steady fields and to develop the appropriate generalizations when this constraint is relaxed. Thus, in the first volume we start with the fields associated with stationary charges and relax the stationary condition to allow consideration of the flow of steady currents in closed circuits.

In the first volume we considered the experimental results which require mathematical explanation and discussion, in particular those referring to phenomena which suggest that the simple Newtonian concepts of space and time are not fully valid. Then we considered steady state fields and dealt next with electrostatics, including dielectrics, energy theorems, uniqueness theorems, and ended that volume with a chapter on the steady flow of electric currents. SI units were used throughout, although the older systems were briefly mentioned.

In this second volume we first consider the magnetic field of steady currents —magnetostatics. This is followed by a chapter on the methods of solving potential problems drawn from electrostatics, magnetism, current flow and gravitation. Relaxing the constraint of stationary steady currents, we are led to consider electromagnetic induction when the current strengths in closed circuits vary or when the circuits move. This leads to the necessity of considering the breakdown of Newtonian ideas and the introduction of special relativity. When we further relax the constraint of closed circuits and consider the motion of charges in open circuits we are led to introduce the displacement current because of the relativistic theory already set up, and so are led to Maxwell's equations.

In the third volume we consider the implications of Maxwell's equations such as electromagnetic radiation in simple cases, and deal further with the relation between Maxwell's equations and the Lorentz transformation.

We assume that the readers are conversant with the basic ideas of vector analysis, including vector integral theorems.

Our thanks are due to the University of Oxford, to the Syndics of the Cambridge University Press and to the Senate of the University of London for permission to use questions set in their various examinations.

CHAPTER 6

MAGNETOSTATICS

6.1 Introduction

The electrostatic field arises from stationary charges in equilibrium; in this chapter we consider the field produced when the charges move so that they form a steady current. This field is called a magnetic (or, more accurately, magnetostatic) field. We may notice here that, as soon as questions of motion are introduced, it is necessary to specify the frame of reference with respect to which the motion is measured. This brings in at once the difficulties regarding our concepts of space and time mentioned before. In this chapter we begin by setting up the theory of magnetostatics from an experimental point of view. Then we go on to consider our results from the point of view of transformations between observers in uniform relative motion. We cannot carry out the (superficially attractive) alternative task of using the transformations to deduce the results for charges in motion from those for charges at rest, since only certain *special* sets of moving charges (e.g. two parallel currents) can be produced in this way from stationary charges. Any argument purporting to be such a deduction must therefore contain concealed assumptions, whose justification must ultimately be experimental.

The results of Chapter 5 imply that such currents must flow in closed linear circuits, governed by Kirchhoff's Laws or more generally satisfy the condition div $\boldsymbol{J} = 0$ and Ohm's Law. The existence of the magnetic field is indicated by the force exerted on a conductor carrying a current when it is placed in the neighbourhood of another such current (cf. the forces between neighbouring charges in electrostatics).

Historically the investigation of magnetism arose from the work of Gilbert in the sixteenth century with lodestone (permanent magnets) in the earth's field, but the best theoretical development does not follow the historical development and we postpone consideration of permanent magnetism until after we have developed the magnetic properties of steady currents.

We recall the elementary qualitative results concerning the magnetic fields of currents. They show, for example, by means of "iron filings" that the magnetic field lines form circles centred on a long straight conductor, or that in a long solenoid the field lines are straight and directed parallel to the axis of the solenoid, so that the current in the solenoid circulates around the field lines. The essential feature of all these results is the "corkscrew" rule relating the directions of field and current.

In Vol. 1 we saw that an electrostatic field vector resembles the velocity pattern of a fluid diverging from its sources, in that case the electric charges. Elementary experiments with currents, such as Oersted's experiment, suggest that a magnetic field vector resembles the velocity of a fluid around its vortex filaments, the currents. This "corkscrew" relationship between the field vector and its sources, which are associated with a direction, is more difficult to visualize than the divergence relation, and gives rise to formulae involving vector products and differentiations by means of **curl** instead of products by scalars and differentiations by **gradient** and divergence. Nevertheless, apart from these basic differences there is a close analogy between the "field view" of magnetostatics and the corresponding "field view" of electrostatics as discussed in § 1.2.

6.2 Origin of the magnetic field

Corresponding to the point charge of electrostatics, the linear current is an origin of a magnetostatic field and can be approximated in practice by the use of a fine conducting wire. However, a linear current must form a closed circuit; by using some form of "return lead at infinity" we can consider a long (infinite) straight current. A small loop carrying a current is another convenient form of circuit, which can also be approximated in practice, but the commonest form of linear circuit is a wire loop of finite dimensions carrying a current.

We shall define a vector $\boldsymbol{B}$ which measures the strength of the magnetic field, and corresponds to the vector $\boldsymbol{E}$ of the electrostatic field, in terms of the force experienced by a "test circuit" (cf. test charge) carrying a small current. A long straight current situated in the neighbourhood of another current experiences a force at right angles to the conductor, either towards (currents parallel) or away from (currents antiparallel) the other conductor. A small coil experiences a couple and a force identical to those caused by the action of an electrostatic field on an electric dipole; in fact, two small coils interact just as do two electric dipoles. Since the small (plane) coil with a current circulating around its perimeter is the most convenient arrangement we use it as the starting point for our definitions.

We call a plane coil of small area α carrying a (small) current I around its perimeter *a magnetic particle* (or magnetic dipole) whose moment is defined by

$$\boldsymbol{m} = I\alpha\hat{\boldsymbol{e}} \tag{6.1}$$

where $\hat{\boldsymbol{e}}$ is a unit vector perpendicular to the area and related by the right-hand corkscrew relation to the circulation of I around the perimeter. This vector is given the name "electromagnetic" moment and is measured in A m^2, as is clear from the definition (6.1). A magnetic dipole experiences both a couple and a force when in the neighbourhood of a current; when this force and couple are reduced to a base at the centre of the area the couple $\boldsymbol{\gamma}$ is given, approximately, by

$$\boldsymbol{\gamma} = \boldsymbol{m} \times \boldsymbol{B} \tag{6.2}$$

or, more accurately, we *define* the vector $\boldsymbol{B}$ by

$$\hat{\boldsymbol{e}} \times \boldsymbol{B} = \lim_{|\boldsymbol{m}| \to 0} (\boldsymbol{\gamma}/|\boldsymbol{m}|). \tag{6.3}$$

The force on a magnetic dipole is (see § 2.10)

$$\boldsymbol{f} = \textbf{grad}\,(\boldsymbol{m} \cdot \boldsymbol{B}) \tag{6.4}$$

or, in cartesian components (with summation convention),

$$f_i = m_j B_{j,\,i} \tag{6.4a}$$

where we assume that $\boldsymbol{m}$ is constant and we denote the derivative of the component of a vector by an additional suffix, viz.

$$B_{j,\,i} = \frac{\partial B_j}{\partial x_i},$$

since the definition of $\boldsymbol{B}$ is associated with an area, and for historical reasons the name given to this quantity is *magnetic flux density*, and the unit is named *tesla* (symbol T).

In order to extend this definition to a finite circuit, not necessarily plane, we draw an arbitrary surface spanning the circuit C and cover it with a mesh of infinitesimal circuits each carrying the same current I, Fig. 6.1. The internal currents cancel out, leaving the current I around the perimeter. This device is equivalent to replacing the finite circuit by a *magnetic shell*, a layer of magnetic particles of uniform strength. The moment of an element of area $d\boldsymbol{S}$

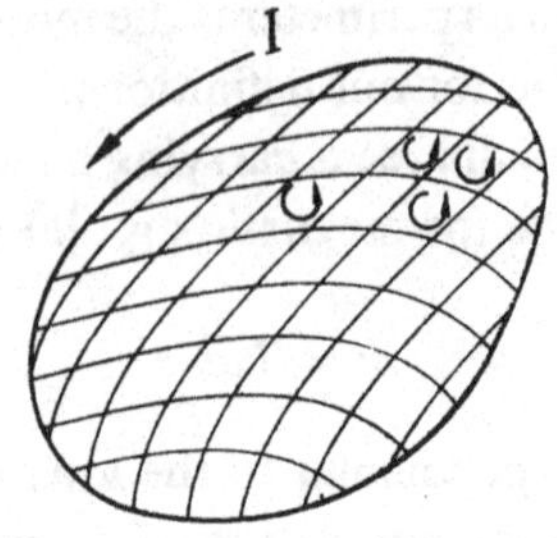

FIG. 6.1

of this surface Σ is $I\hat{\nu}\,dS$, where $\hat{\nu}$ is the unit normal. The forces (6.4a) acting on the elements, when integrated, give the resultant force on the circuit as

$$F_i = I\iint \nu_j B_{j,i}\, dS.$$

We now wish to simplify the expression for F_i. We start from Stokes's theorem expressed in terms of components

$$\oint_\Gamma \boldsymbol{a}\cdot d\boldsymbol{s} = \iint_\Sigma \textbf{curl}\, \boldsymbol{a}\cdot d\boldsymbol{S} \quad \text{or} \quad \oint_\Gamma a_i t_i\, ds = -\iint_\Sigma \varepsilon_{ijk} a_{j,k}\nu_i\, dS,$$

where t_i are the components of the unit tangent to the perimeter curve Γ of the open surface Σ whose element of area is $\nu_j\, dS$. (Here ε_{ijk} denotes the alternating symbol which is ± 1 according to whether ijk is an even or odd permutation of 1, 2, 3 and is zero when any two of i, j, k are equal.) This form of Stokes's theorem can be generalized to the case where a has any number of suffixes. Thus, for two suffixes,

$$\oint_\Gamma a_{ip} t_i\, ds = -\iint_\Sigma \varepsilon_{ijk} a_{jp,k}\nu_i\, dS.$$

In this formula we choose $a_{ip} = \varepsilon_{ipq} b_q$, where b_q are the components of some vector (field). Then,

$$\begin{aligned}\oint_\Gamma \varepsilon_{ipq} b_q t_i\, ds &= -\iint_\Sigma \varepsilon_{ijk}\varepsilon_{jpq} b_{q,k}\nu_i\, dS \\ &= -\iint_\Sigma (\delta_{kp}\delta_{iq} - \delta_{kq}\delta_{ip}) b_{q,k}\nu_i\, dS \\ &= -\iint_\Sigma (b_{i,p}\nu_i - b_{q,q}\nu_p)\, dS. \end{aligned} \tag{6.5}$$

With $\boldsymbol{b} = \boldsymbol{B}$ (or $b_i = B_i$) this gives in vector notation

$$\boldsymbol{F} = I\oint_\Gamma \hat{\boldsymbol{t}}\times \boldsymbol{B}\, ds + I\iint_\Sigma \nu\, \text{div}\, \boldsymbol{B}\, dS. \tag{6.6}$$

But the spanning surface Σ is arbitrary and the forces acting on a circuit must depend only on the perimeter of this surface. This, in conjunction with (6.5) or (6.6), implies that

$$\operatorname{div} \boldsymbol{B} = 0. \tag{6.7}$$

$$\therefore \quad \boldsymbol{F} = I \oint_{\Gamma} \hat{\boldsymbol{t}} \times \boldsymbol{B} \, \mathrm{d}s. \tag{6.8}$$

This may be thought of as a force produced by every vector element of the circuit I being acted on by a force

$$\mathrm{d}\boldsymbol{F} = I \, \mathbf{d}\boldsymbol{s} \times \boldsymbol{B}. \tag{6.8a}$$

It should be noted that we cannot *deduce* eqn. (6.8a) for an isolated piece of current since (6.8) has been proved only for *closed* circuits. Later (§ 9.4) we see that this result also follows from the assumption that the Lorentz force on a particle, which carries a charge q and moves with velocity $\boldsymbol{v}$, is $q\boldsymbol{v} \times \boldsymbol{B}$. [With this result, and the assumption that the magnetic field surrounding a long straight current is directed around circles centred on the current, we see that the force acting on one of two parallel straight conductors is directed in the plane containing the conductors.]

The relation (6.8a) is the same as Fleming's left-hand rule. Equations (6.8) and (6.8a) together with the definition (6.2) are the magnetic counterparts of those of § 2.4 giving the force on a charge and the definition of the field strength $\boldsymbol{E}$. The condition (6.7) is the magnetic counterpart of the equation $\mathbf{curl}\, \boldsymbol{E} = \boldsymbol{0}$. The latter relation was derived from energy considerations, but these are not applicable here; detailed reasons for this fact are given later. Briefly, the reason is that to maintain a current in a circuit against resistance requires the expenditure of energy from a battery, but when a circuit moves in a magnetic field electromagnetic induction (see Chapter 9) will affect this current and may require the battery to supply energy if the current I is to remain unaltered, as is necessary when we assume that $\boldsymbol{m}$ is constant.

The condition $\operatorname{div} \boldsymbol{B} = 0$ is very important and implies that the magnetic field lines are closed loops, e.g. the circles surrounding a long straight conductor. Further, this condition implies the existence of a vector field $\boldsymbol{A}$, such that

$$\boldsymbol{B} = \mathbf{curl}\, \boldsymbol{A}. \tag{6.9}$$

This vector is called the vector potential, and corresponds to the scalar potential V of electrostatics. It is not uniquely defined for we can add to $\boldsymbol{A}$ the gradient of an arbitrary scalar. Nor is there a "physical" definition of $\boldsymbol{A}$ in terms of work done to place a circuit in position in the field by bringing it from infinity. When the circuit is "brought from infinity" through the field, electromagnetic induction requires the battery maintaining the current to provide an unknown quantity of energy in addition to the work done by the agent moving the coil.

We obtain expressions for $\boldsymbol{A}$ in terms of the currents [cf. eqn. (6.19) for $\boldsymbol{A}$ with (2.15) for V] after we have introduced the magnetic form of the inverse square law.

Although we cannot separate the energy required to maintain a constant current from the energy required to move a circuit against the forces of the field, the concept of "potential energy" of a circuit in a field $\boldsymbol{B}$ can be used. Our formulae (6.2) and (6.4) were based on the assumption that the work done by the forces acting on the magnetic particle in a displacement of the particle was equal to the decrease in $-\boldsymbol{m}\cdot\boldsymbol{B}$, assuming $\boldsymbol{m}$ did not alter during the displacement. This result was derived from the behaviour of an electric dipole in an electrostatic field where the true potential energy was $-\boldsymbol{m}\cdot\boldsymbol{E}$ (in the electrostatic case no expenditure of energy was needed to maintain $\boldsymbol{m}$ constant). Replacing a finite linear circuit by the equivalent magnetic shell, we find that the "work function" for this circuit is

$$-I\iint \hat{\boldsymbol{\nu}}\cdot\boldsymbol{B}\,\mathrm{d}S = -NI, \tag{6.10}$$

where N stands for the flux of $\boldsymbol{B}$ through the circuit C. If the circuit moves so that each point has a velocity $\boldsymbol{v}$ which may vary from point to point we obtain, from the formula for differentiating an integral with a moving boundary,

$$\frac{\mathrm{d}N}{\mathrm{d}t} = \iint \hat{\boldsymbol{\nu}}\cdot\{\boldsymbol{v}\,\mathrm{div}\,\boldsymbol{B}+\mathbf{curl}\,[\boldsymbol{B}\times\boldsymbol{v}]\}\,\mathrm{d}S = \oint_C (\hat{\boldsymbol{t}}\times\boldsymbol{B})\cdot\boldsymbol{v}\,\mathrm{d}s,$$

where we have put $\mathrm{div}\,\boldsymbol{B} = 0$ and used Stokes's theorem.

Suppose now that a finite circuit is given an arbitrary velocity $\boldsymbol{v}$. Then the rate of working of the magnetic forces on the circuit is

$$-\frac{\mathrm{d}W}{\mathrm{d}t} = \oint \mathbf{d}\boldsymbol{F}\cdot\boldsymbol{v} = I\oint (\hat{\boldsymbol{t}}\times\boldsymbol{B})\cdot\boldsymbol{v}\,\mathrm{d}s = I\,\frac{\mathrm{d}N}{\mathrm{d}t}\,.$$

$$\therefore\ W = -IN+\text{constant}. \tag{6.11}$$

Hence in calculating the force on a circuit we may use $W = -IN$ as a force function, *assuming I remains constant*. Because the last condition may require expenditure of more energy we do not use the term "potential energy" for W. Since $B = \text{curl } A$ we obtain an alternative form

$$N = \iint \hat{\nu} \cdot B \, dS = \oint \hat{t} \cdot A \, ds. \tag{6.12}$$

The latter form depends only on the position of the perimeter, as we should expect, since the spanning surface is arbitrary.

Exercises 6.2

1. If $\hat{e}$ in eqn. (6.3) is chosen successively in the directions i, j, k, then

$$i \times B = F_1, \quad j \times B = F_2, \quad k \times B = F_3.$$

Prove that

$$B = \tfrac{1}{2}[F_1 \times i + F_2 \times j + F_3 \times k].$$

2. Use the summation convention and the alternating symbol to prove the following identities:

$$b \cdot (a \times \text{curl } c) = a \cdot [(b \cdot \nabla)c] - b \cdot [(a \cdot \nabla)c],$$
$$\text{curl } [(r \times e) \times B] = (r \times e) \text{ div } B + B \times e - [(r \times e) \cdot \nabla] B,$$

where r is the position vector and e is a constant vector, the other vectors being functions of position.

3. If the magnetic shell of Fig. 6.1 is regarded as a rigid body, the system of forces acting on it can be reduced to a force F and a couple $\Gamma(O)$ about the origin of coordinates. Show that F is given by eqn. (6.6), and that

$$\Gamma(O) = I \iint \{\hat{\nu} \times B + r \times [(\hat{\nu} \cdot \nabla) B] + r \times (\hat{\nu} \times \text{curl } B)\} \, dS.$$

By using the formulae of question 2 show that

$$\Gamma(O) = I \int r \times (\hat{t} \times B) \, ds + I \iint (r \times \hat{\nu}) \text{ div } B \, dS.$$

[This confirms the interpretation of eqn. (6.8) that there is a force $I(\hat{t} \times B)\, ds$ on each element of current.]

6.3 The magnetic displacement vector

The quantity B is the force aspect of the magnetic field; now we introduce the counterpart of the electrostatic vector D.

According to our analysis by the field outlook, at any point of a field there are two aspects, illustrated in the analogy of the stretched string by the tension in the string—here the vector B—and the "displacement" or strain in the string. It is this second aspect which we consider now; it will be denoted by a vector H. The name given to the quantity H is "magnetic field strength" (because in early discussions this vector rather than B was used to measure the field strength). The units, as we shall see, are amperes per metre (A m^{-1}).

Some difficulty arises in describing and visualizing the "corkscrew" relation which lies at the basis of the magnetic field.

Consider first a long solenoid. Effectively the current in this arrangement is a sheet of current—a surface current—which flows around the cylinder in a direction perpendicular to the generators. We can imagine the space inside the solenoid filled with narrow tubes all parallel to the axis and packed tightly like a bundle of sticks or a honeycomb, Fig. 6.2. Now if each of these

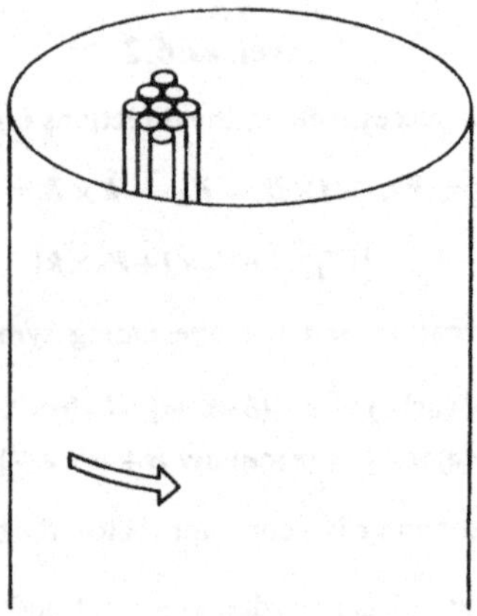

FIG. 6.2

narrow tubes had a current circulating around it equal in strength to that around the actual solenoid the currents would cancel out internally and just leave the actual current around the solenoid (cf. the mesh of small circuits used to replace a finite circuit in the previous section). In the field outlook we regard the actual currents (and charges) as the "end-effects" of the state of affairs in the medium. Here if we regard the space inside the solenoid as filled with this bundle of "capillary solenoids", and the rest of space as empty, the "end-effect" is just the actual current distribution found on the solenoid. If the solenoid is very long (infinite) there is no magnetic field outside corresponding to the absence of "capillaries" in our construction. It is this distribution of "capillaries" which is described by the vector $\boldsymbol{H}$. Before considering the general quantitative relation involved here, we examine the simple case of a long straight conductor.

We construct an arrangement of "capillaries" which takes the form of circles having the current as axis, the tubes being therefore closed, and they fill all space outside the wire. (See Fig. 6.3.) In effect, every line of the magnetic field runs along the middle of one of the capillary tubes. Now, if each of these capillaries has a surface current of strength $I/(2\pi r)$ amperes

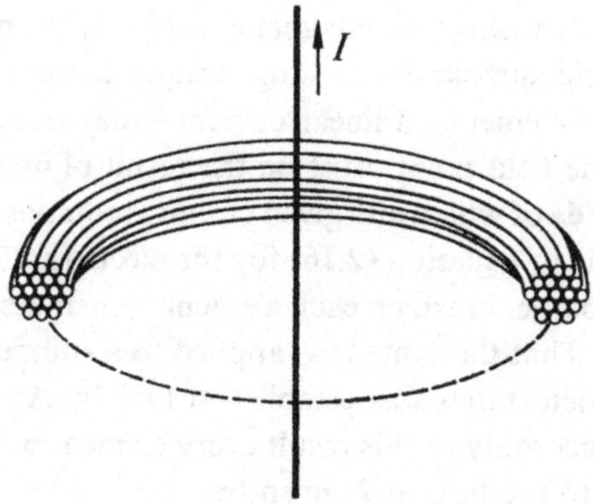

FIG. 6.3

per metre (A/m) circulating around it like the current around a very narrow solenoid, r being the radius of the circle forming the capillary, then the "end-effect" at the wire is a current of strength I along the wire, the currents cancelling out at all other points of space.

This circulation of surface currents around capillaries is the "strain" in the medium corresponding to the presence of a current in a conductor bounding the space. Since this circulation corresponds to a "vorticity" everywhere (borrowing a term from hydrodynamics), we denote it by a vector directed along the capillary of magnitude equal to the surface current and related to the sense of circulation around the capillary of magnitude equal to the surface current and related to the sense of circulation around the capillary by the right-hand corkscrew rule.

Using cylindrical coordinates (r, ψ, z) we can write down the expression for $\boldsymbol{H}$ when the current I flows along the polar axis, viz.

$$\boldsymbol{H} = \left\{0 \quad \frac{I}{2\pi r} \quad 0\right\}. \tag{6.13}$$

We obtain this value because the upward-flowing current on the inside of each capillary must add up around its whole circumference to I.

Example. Two equal and opposite current sheets of strengths $\pm K$ flow in infinite parallel planes separated by a distance d. The corresponding distribution of "capillaries" fills the space between the planes, the capillaries being directed at right angles to the direction of current flow and parallel to the planes. In this case $|\boldsymbol{H}| = K$ and is independent of the distance between the planes.

We now generalize this idea and investigate how the "capillaries" are arranged when they correspond to some arbitrary distribution of currents. In other words, we must find the relation between the vector $\boldsymbol{H}$ and the currents which are causing the magnetic field. By using the cylindrical symmetry of the field surrounding a long straight current we arrive at eqn. (6.13), but the basic element—a linear current—may occur in many shapes and the effect at one field point must be the result of integrating the effects of all the elements $I\,\mathbf{ds}$ of which any given circuit is composed. The expression (6.13) is very similar to equation (2.16) for the electrostatic displacement of an infinite line of charge, in which each element contributes according to the inverse square law. That the same law, applied to a source which is a vector $I\,\mathbf{ds}$, holds in magnetostatics was established first by Ampère and then by Biot and Savart. According to this result every element of the circuit Γ gives a contribution $\mathbf{d}\boldsymbol{H}$ to the field at P given by

$$\mathbf{d}\boldsymbol{H} = I\frac{\mathbf{ds}\times\boldsymbol{r}}{4\pi r^3}, \tag{6.14a}$$

see Fig. 6.4. It is impossible to have an isolated current element (currents

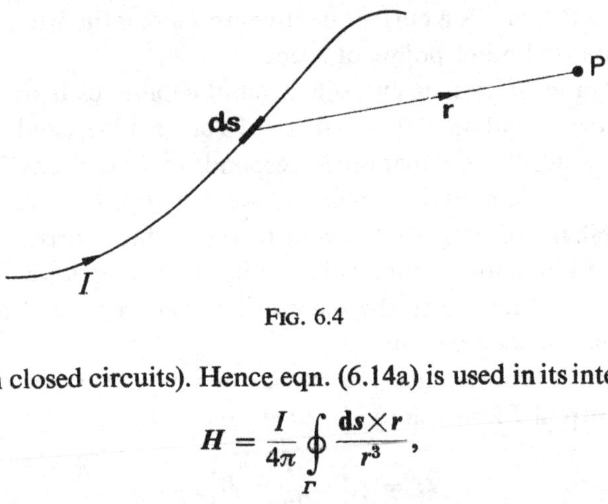

FIG. 6.4

must flow in closed circuits). Hence eqn. (6.14a) is used in its integrated form

$$\boldsymbol{H} = \frac{I}{4\pi}\oint_{\Gamma}\frac{\mathbf{ds}\times\boldsymbol{r}}{r^3}, \tag{6.14}$$

where $\boldsymbol{r}$ is the position vector from the element to the field point.

When the "medium" in which the linear currents are situated is a vacuum we assume, as in electrostatics, that the stress and strain vectors $\boldsymbol{B}$ and $\boldsymbol{H}$ are proportional and write

$$\boldsymbol{B} = \mu_0\boldsymbol{H} \tag{6.15}$$

(this corresponds to $\boldsymbol{E} = \boldsymbol{D}/\varepsilon_0$ in electrostatics). The quantity μ_0 is the *permeability* of free space.

As defined in eqn. (6.3), $\boldsymbol{B}$ has the units of force divided by current and area, i.e. N/Am², but this is not the name used. For historical reasons, the flux of $\boldsymbol{B}$ across an area, i.e. $\iint \boldsymbol{B}\cdot\mathbf{d}\boldsymbol{S}$, is measured in a unit called a weber and so $\boldsymbol{B}$ is measured in Wb/m² and is called "magnetic flux-density" rather than magnetic field. Our definition of $\boldsymbol{H}$ by means of surface current in the "capillaries" gives it the units of surface current strength, i.e. A/m. The definition of μ_0 in (6.15) therefore gives permeability the units of Wb/Am or henries per metre. The importance of the flux of $\boldsymbol{B}$ arises in connection with electromagnetic induction, where the quantities weber and henry occur.

Using eqn. (6.14) we can now write down the field vectors for the field arising from a (closed) linear circuit Γ as

$$\boldsymbol{B} = \mu_0\boldsymbol{H} = \frac{\mu_0 I}{4\pi}\oint_\Gamma \frac{\mathrm{d}\boldsymbol{s}\times\boldsymbol{r}}{r^3}. \tag{6.16}$$

This integral is a function of the coordinates of the field point P and, provided P never coincides with a point on the circuit,

$$\boldsymbol{B} = \frac{\mu_0 I}{4\pi}\oint_\Gamma \left\{\mathbf{grad}_P\left(\frac{1}{r}\right)\times\hat{\boldsymbol{t}}\right\}\mathrm{d}s,$$

where $\hat{\boldsymbol{t}}$ is the unit tangent and $\mathbf{grad}_P$ denotes differentiation w.r. to the coordinates of P. Since $\hat{\boldsymbol{t}}$ does not depend on the position of P we can write

$$\boldsymbol{B} = \mathbf{curl}_P\left\{\frac{\mu_0 I}{4\pi}\oint\frac{\hat{\boldsymbol{t}}}{r}\,\mathrm{d}s\right\} = \mathbf{curl}\,\boldsymbol{A}, \tag{6.17}$$

where

$$\boldsymbol{A} = \frac{\mu_0 I}{4\pi}\oint\frac{\hat{\boldsymbol{t}}}{r}\,\mathrm{d}s.$$

This is the magnetic (vector) potential corresponding to $V = Q/(4\pi\varepsilon_0 r)$ for the electrostatic potential of a point charge Q at the origin.

Example 1. Current I flows in a circular loop of wire of radius a. Prove that the magnetic field at a point on the axis of the loop, at a distance z from its plane, is directed along the axis and is of magnitude

$$\frac{Ia^2}{2(a^2+z^2)^{3/2}}.$$

A current I flows in a long uniform closely wound solenoid with n turns per unit length. Prove that the magnetic field at a point on the axis, inside the solenoid and well away from the ends, is directed along the axis and is approximately of magnitude nI. Show that at a point on the axis and in the plane of the end winding the magnitude of the field is approximately $nI/2$.

We use the axes shown in Fig. 6.5 (i), where P is at $\{0\ 0\ z\}$ and the element of arc is at $\{a\cos\theta\ \ a\sin\theta\ \ 0\}$.

$$\boldsymbol{r} = -\boldsymbol{i}a\cos\theta - \boldsymbol{j}a\sin\theta + \boldsymbol{k}z, \qquad \mathrm{d}\boldsymbol{s} = (-\boldsymbol{i}a\sin\theta + \boldsymbol{j}a\cos\theta)\,\mathrm{d}\theta;$$

$$\boldsymbol{H} = \frac{I}{4\pi}\oint\frac{\mathrm{d}\boldsymbol{s}\times\boldsymbol{r}}{r^3} = \frac{I}{4\pi}\int_0^{2\pi}\frac{\boldsymbol{i}za\cos\theta+\boldsymbol{j}za\sin\theta+a^2\boldsymbol{k}}{(a^2+z^2)^{3/2}}\,\mathrm{d}\theta = \frac{Ia^2\boldsymbol{k}}{2(a^2+z^2)^{3/2}}.$$

To find the field in the solenoid we choose P to be the origin of coordinates and consider an element of the solenoid at Q to be a circular current of strength $nI\,dz$ situated at a distance z from P, Fig. 6.5 (ii). The field at P is in the z-direction, and

$$\mathbf{H} = \mathbf{k}H = \frac{\mathbf{k}nIa^2}{2}\int_{-\infty}^{\infty}\frac{dz}{(a^2+z^2)^{3/2}} = \mathbf{k}nI,$$

the integral being evaluated by the substitution $z = a\tan\theta$. (This result reproduces the value obtained by considering the "bundle of capillaries".)

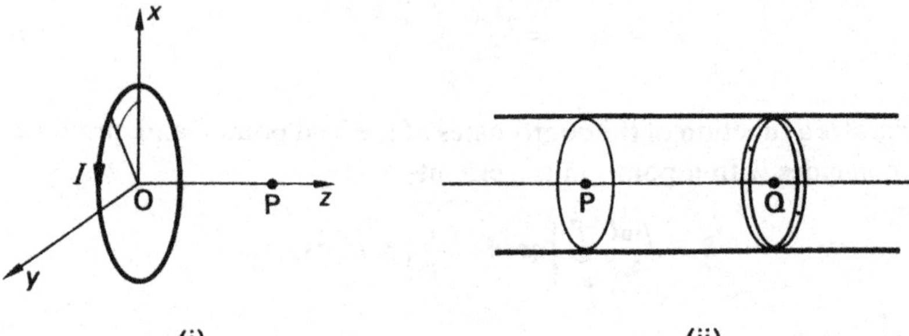

FIG. 6.5

When P is situated at the centre of the end we take the limits of integration to be 0 and ∞, so that, in this case,

$$H = \frac{nIa^2}{2}\int_{0}^{\infty}\frac{dz}{(a^2+z^2)^{3/2}} = \frac{1}{2}nI.$$

Example 2. Two equal currents flow in the same sense round two circular coils of radius a placed perpendicularly to the line joining their centres, which are at a distance $2c$ apart. Show that, in order to make the magnetic field as nearly as possible uniform at points on the axis near the point halfway between the centres, we must have $c = \frac{1}{2}a$; and compare the intensity of this field with that at the centre of either coil.

To solve this problem we quote the result established in example 1, using appropriate values of z, and superpose the fields.

For the field at the mid-point between the coils $z = c, -c$,

$$H = \frac{Ia^2}{2(a^2+c^2)^{3/2}} + \frac{Ia^2}{2(a^2+c^2)^{3/2}} = \frac{Ia^2}{(a^2+c^2)^{3/2}}. \tag{1}$$

The field at the centre of each coil is obtained by putting $z = 0, 2c$ in the appropriate expressions and we obtain

$$H = \frac{Ia^2}{2a^3} + \frac{Ia^2}{2(a^2+4c^2)^{3/2}} = \frac{I}{2a}\left\{1+\left(1+\frac{4c^2}{a^2}\right)^{-3/2}\right\}.$$

To investigate the uniformity of the field we find the value of H at a point distant $c+\zeta$ from one coil and $c-\zeta$ from the other, i.e.

$$H = \frac{Ia^2}{2}\left\{\frac{1}{[a^2+(c+\zeta)^2]^{3/2}} + \frac{1}{[a^2+(c-\zeta)^2]^{3/2}}\right\}.$$

Now

$$\frac{1}{[a^2+(c+\zeta)^2]^{3/2}} = \frac{1}{(a^2+c^2)^{3/2}}\left\{1+\frac{\zeta(2c+\zeta)}{a^2+c^2}\right\}^{-3/2}$$

$$= \frac{1}{(a^2+c^2)^{3/2}}\left\{1-\frac{3c\zeta}{a^2+c^2}-\frac{3\zeta^2}{2(a^2+c^2)}+\frac{3}{2}\cdot\frac{5}{2}\cdot\frac{1}{2!}\,\frac{4c^2\zeta^2}{(a^2+c^2)^2}+O(\zeta^3)\right\}$$

$$= \frac{1}{(a^2+c^2)^{3/2}}\left\{1-\frac{3c\zeta}{a^2+c^2}-\frac{3\zeta^2}{2(a^2+c^2)^2}(a^2+c^2-5c^2)+O(\zeta^3)\right\}$$

$$= \frac{1}{(a^2+c^2)^{3/2}}\left\{1-\frac{3c\zeta}{a^2+c^2}-\frac{3\zeta^2(a^2-4c^2)}{2(a^2+c^2)^2}+O(\zeta^3)\right\}.$$

Similarly

$$\frac{1}{[a^2+(c-\zeta)^2]^{3/2}} = \frac{1}{(a^2+c^2)^{3/2}}\left\{1+\frac{3c\zeta}{a^2+c^2}-\frac{3\zeta^2(a^2-4c^2)}{2(a^2+c^2)^2}+O(\zeta^3)\right\}.$$

$$\therefore\ H = \frac{Ia^2}{(a^2+c^2)^{3/2}}\left\{1-\frac{3\zeta^2(a^2-4c^2)}{2(a^2+c^2)^2}+O(\zeta^3)\right\}.$$

If we choose $c = a/2$, the term in ζ^2 disappears and the field varies with ζ only in the term $O(\zeta^3)$. Thus when $c = a/2$ the field at the mid-point is as nearly uniform as possible.

(This discussion has considered variations only along the axis. At points near the mid-point but off the axis considerations of symmetry show that H_x, H_y must vanish when $\zeta = 0$, but these components may have a second order term when evaluated off the axis near to the mid-point. The result above shows that the variations *on* the axis are of the hird o rder.)

Example 3. A uniform wire is bent into the form of a square of side $2a$, and a current I flows round it. Prove that the magnetic field strength at a point on the perpendicular to the plane of the square through its centre and distant d from the plane is

$$\frac{2Ia^2}{\pi(a^2+d^2)\sqrt{(2a^2+d^2)}}.$$

We divide the integral $\oint(\mathbf{ds}\times\mathbf{r})/r^3$ into four contributions, one from each side of the square, Fig. 6.6.

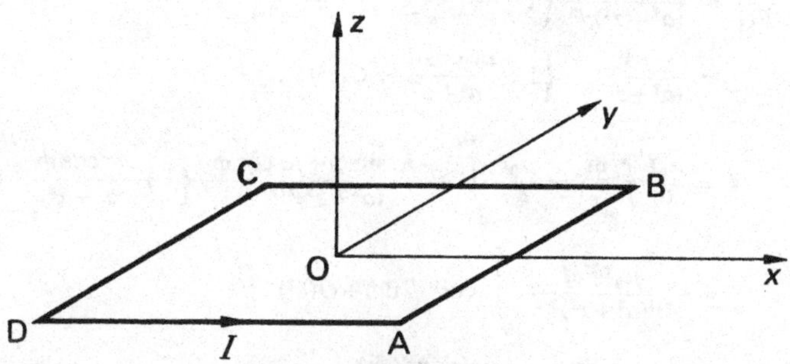

FIG. 6.6

On AB: $\mathbf{ds} = \mathbf{j}\,dy$,

$$\mathbf{r} = -\mathbf{i}a - \mathbf{j}y + \mathbf{k}d; \quad r^2 = (a^2+d^2+y^2).$$

$$\therefore \int_A^B \frac{\mathbf{ds}\times\mathbf{r}}{r^3} = \int_{-a}^{a} \frac{(\mathbf{i}d+\mathbf{k}a)\,dy}{(a^2+d^2+y^2)^{3/2}} = \frac{2a(\mathbf{i}d+\mathbf{k}a)}{(a^2+d^2)\sqrt{(2a^2+d^2)}},$$

using the substitution $y = \sqrt{(a^2+d^2)}\tan\theta$.

Similarly on BC: $\mathbf{ds} = \mathbf{i}\,dx$, $\mathbf{r} = -\mathbf{i}x - \mathbf{j}a + \mathbf{k}d$.

$$\therefore \int_B^C \frac{\mathbf{ds}\times\mathbf{r}}{r^3} = -\int_{+a}^{-a} \frac{(\mathbf{j}d+\mathbf{k}a)\,dx}{(a^2+d^2+x^2)^{3/2}} = +\frac{2a(\mathbf{j}d+\mathbf{k}a)}{(a^2+d^2)\sqrt{(2a^2+d^2)}}.$$

On CD: $\mathbf{ds} = \mathbf{j}\,dy$, $\mathbf{r} = \mathbf{i}a - \mathbf{j}y + \mathbf{k}d$.

$$\therefore \int_C^D \frac{\mathbf{ds}\times\mathbf{r}}{r^3} = -\int_{+a}^{-a} \frac{(-\mathbf{i}d+\mathbf{k}a)\,dy}{(a^2+d^2+y^2)^{3/2}} = +\frac{2a(-\mathbf{i}d+\mathbf{k}a)}{(a^2+d^2)\sqrt{(2a^2+d^2)}}.$$

On DA: $\mathbf{ds} = \mathbf{i}\,dx$, $\mathbf{r} = -\mathbf{i}x + \mathbf{j}a + \mathbf{k}d$.

$$\therefore \int_D^A \frac{\mathbf{ds}\times\mathbf{r}}{r^3} = \int_{-a}^{a} \frac{(-\mathbf{j}d+\mathbf{k}a)\,dx}{(a^2+d^2+x^2)^{3/2}} = \frac{2a(-\mathbf{j}d+\mathbf{k}a)}{(a^2+d^2)\sqrt{(2a^2+d^2)}}.$$

Hence the field $\mathbf{H}$ is given by

$$\mathbf{H} = \frac{I}{4\pi}\oint \frac{\mathbf{ds}\times\mathbf{r}}{r^3} = \frac{I}{4\pi(a^2+d^2)\sqrt{(2a^2+d^2)}}(8a^2\mathbf{k}) = \frac{2Ia^2\mathbf{k}}{\pi(a^2+d^2)\sqrt{(2a^2+d^2)}}.$$

Example 4. A steady current I flows along a circular wire of radius a and centre O. Show that the vector potential at a field point whose position vector relative to O is $z\mathbf{k}+\tilde{\boldsymbol{\omega}}$, where $\mathbf{k}$ is unit vector along the axis of the wire and $\tilde{\boldsymbol{\omega}}$ is a small vector perpendicular to it, is approximately given by

$$\frac{\mu_0 Ia^2\mathbf{k}\times\tilde{\boldsymbol{\omega}}}{4(z^2+a^2)^{3/2}}.$$

We use a coordinate system similar to that of example 1 with $\tilde{\boldsymbol{\omega}}$ parallel to the x-axis, so that

$$\tilde{\boldsymbol{\omega}} = \mathbf{i}\varepsilon, \quad \mathbf{ds} = (-\mathbf{i}a\sin\phi + \mathbf{j}a\cos\phi)\,d\phi,$$

$$PQ^2 = (a\cos\phi-\varepsilon)^2 + a^2\sin^2\phi + z^2 = a^2+z^2-2a\varepsilon\cos\phi+\varepsilon^2.$$

$$\therefore \frac{1}{PQ} = \frac{1}{(a^2+z^2)^{1/2}}\left\{1-\frac{2a\varepsilon\cos\phi}{a^2+z^2}+\frac{\varepsilon^2}{a^2+z^2}\right\}^{-1/2}$$

$$= \frac{1}{(a^2+z^2)^{1/2}}\left\{1+\frac{a\varepsilon\cos\phi}{a^2+z^2}+O(\varepsilon^2)\right\}.$$

But

$$\mathbf{A} = \frac{\mu_0 I}{4\pi}\oint\frac{\mathbf{ds}}{r} = \frac{\mu_0 I}{4\pi}\int_0^{2\pi}\frac{-\mathbf{i}a\sin\phi+\mathbf{j}a\cos\phi}{(a^2+z^2)^{1/2}}\left\{1+\frac{a\varepsilon\cos\phi}{a^2+z^2}+O(\varepsilon^2)\right\}d\phi$$

$$= \frac{\mu_0 Ia^2\varepsilon\mathbf{j}}{4\pi(a^2+z^2)^{3/2}}\int_0^{2\pi}\cos^2\phi\,d\phi + O(\varepsilon^2)$$

$$= \frac{\mu_0 Ia^2\mathbf{j}\varepsilon}{4(a^2+z^2)^{3/2}} = \frac{\mu_0 Ia^2(\mathbf{k}\times\tilde{\boldsymbol{\omega}})}{4(a^2+z^2)^{3/2}}.$$

Note. This result implies that $\boldsymbol{A} = \boldsymbol{0}$ everywhere along the axis of the coil, and near the axis is directed in a plane parallel to the coil.

In this example, where we find $\boldsymbol{A}$ by integration of components, we *must* use cartesian coordinates. If we use any other system for the resolution of vectors, we would be adding, on integration, components resolved in different directions.

We now generalize eqns. (6.16), (6.17) to cases where the currents are not restricted to be linear currents but may be distributed as surface or volume currents. We consider a volume distribution $\boldsymbol{J}$ which must satisfy the (conservation of charge) condition div $\boldsymbol{J} = 0$. The flow pattern of $\boldsymbol{J}$ can be regarded as a collection of narrow tubes of flow, each one of which we treat as a linear circuit to which we apply (6.16), (6.17). The current strength I of a typical filament of cross-section α is $|\boldsymbol{J}|\alpha$ at any point and has the direction of $\boldsymbol{J}$. Hence we generalize (6.16), (6.17) to

$$\boldsymbol{B} = \mu_0\boldsymbol{H} = \sum_{\text{filaments}} \frac{\mu_0|\boldsymbol{J}|}{4\pi} \oint \frac{\alpha\, \mathbf{ds}\times\boldsymbol{r}}{r^3}, \quad \boldsymbol{A} = \sum_{\text{filaments}} \frac{\mu_0|\boldsymbol{J}|}{4\pi} \oint \frac{\alpha\, \mathbf{ds}}{r}.$$

Now $\alpha\,|\mathbf{ds}|$ is the volume of the element $\mathbf{ds}$ of the filament and we may replace $|\boldsymbol{J}|\alpha\, \mathbf{ds}$ at any point by $\boldsymbol{J}\, d\tau$ and so, with the summation over all filaments included in the integration, we obtain

$$\boldsymbol{B} = \mu_0\boldsymbol{H} = \frac{\mu_0}{4\pi}\iiint \frac{\boldsymbol{J}\times\boldsymbol{r}}{r^3}\, d\tau, \quad \boldsymbol{A} = \frac{\mu_0}{4\pi}\iiint \frac{\boldsymbol{J}\, d\tau}{r}. \qquad (6.18a, 19a)$$

If we consider all three kinds of distribution, viz. volume currents $\boldsymbol{J}$, surface currents $\boldsymbol{K}$, and linear currents I, then we have

$$\boldsymbol{B} = \mu_0\boldsymbol{H} = \frac{\mu_0}{4\pi}\iiint \frac{\boldsymbol{J}\times\boldsymbol{r}}{r^3}\, d\tau + \frac{\mu_0}{4\pi}\iint \frac{\boldsymbol{K}\times\boldsymbol{r}}{r^3}\, dS + \sum \frac{\mu_0 I}{4\pi}\oint \frac{\mathbf{ds}\times\boldsymbol{r}}{r^3}, \qquad (6.18)$$

$$\boldsymbol{A} = \frac{\mu_0}{4\pi}\iiint \frac{\boldsymbol{J}\, d\tau}{r} + \frac{\mu_0}{4\pi}\iint \frac{\boldsymbol{K}\, dS}{r} + \sum \frac{\mu_0 I}{4\pi}\oint \frac{\mathbf{ds}}{r}. \qquad (6.19)$$

Provided that r never vanishes, i.e. the field point does not coincide with any current, the integrals in (6.19) are functions of the coordinates of P which can be differentiated and

$$\boldsymbol{B} = \textbf{curl}\, \boldsymbol{A}. \qquad (6.20)$$

Moreover, the integrals in (6.18), even when P does coincide with an interior point of the current distribution $\boldsymbol{J}$, are convergent integrals and eqn. (6.20) is still satisfied for internal points of a volume distribution of current. Hence we have definite values for $\boldsymbol{B}$, $\boldsymbol{H}$ and $\boldsymbol{A}$ at all points except on singularities consisting of linear or surface currents. These formulae also remove the indeterminacy in $\boldsymbol{A}$ mentioned earlier.

From the form of eqn. (6.19) we see that the vector $\boldsymbol{A}$ must satisfy the condition

$$\nabla^2 \boldsymbol{A} = -\mu_0 \boldsymbol{J}. \tag{6.21}$$

Note that where $\boldsymbol{J} = 0$, $\quad \nabla^2 \boldsymbol{A} = \boldsymbol{0}.$

$$\therefore \ \textbf{grad}\,\text{div}\,\boldsymbol{A} - \textbf{curl}\,\textbf{curl}\,\boldsymbol{A} = -\mu_0 \boldsymbol{J}.$$

But, since $\quad \text{div}\,\boldsymbol{J} = 0, \quad \text{div}\,\boldsymbol{A} = 0$

(see example 1, below).

$$\therefore \ \textbf{curl}\,\textbf{curl}\,\boldsymbol{A} = \textbf{curl}\,\boldsymbol{B} = \mu_0 \boldsymbol{J},$$

i.e.

$$\textbf{curl}\,\boldsymbol{H} = \boldsymbol{J}. \tag{6.22}$$

This relation is the magnetic counterpart of div $\boldsymbol{D} = \varrho$ and gives us the general relation between the "capillaries" or $\boldsymbol{H}$ and the currents. We can write this in an integral form using Stokes's theorem, thus

$$\oint \boldsymbol{H} \cdot d\boldsymbol{s} = \iint \boldsymbol{J} \cdot d\boldsymbol{S} = I, \tag{6.23}$$

where I is the total current strength flowing, in the positive sense, through the contour around which the line integral is taken.

Example 1. We prove that div $\boldsymbol{A} = 0$.

Consider $\text{div}_P \iiint (\boldsymbol{J}\, d\tau)/r$. The current $\boldsymbol{J}$ is a function only of the coordinates of $d\tau$, whereas r is a function of the coordinates of P and of $d\tau$. Also, because of the form of r,

$$\textbf{grad}_P\left(\frac{1}{r}\right) = -\textbf{grad}_S\left(\frac{1}{r}\right)$$

where $\textbf{grad}_S$ stands for differentiation w.r. to the coordinates of $d\tau$. Therefore,

$$\begin{aligned}
\text{div}_P \iiint \frac{\boldsymbol{J}\, d\tau}{r} &= \iiint \text{div}_P \left(\frac{\boldsymbol{J}}{r}\right) d\tau \\
&= \iiint \boldsymbol{J} \cdot \textbf{grad}_P \left(\frac{1}{r}\right) d\tau = -\iiint \boldsymbol{J} \cdot \textbf{grad}_S \left(\frac{1}{r}\right) d\tau \\
&= \iiint \left\{\frac{1}{r}\,\text{div}_S\,\boldsymbol{J} - \text{div}_S \left(\frac{\boldsymbol{J}}{r}\right)\right\} d\tau.
\end{aligned}$$

But $\text{div}_S\,\boldsymbol{J} = 0$ and we obtain

$$\text{div}_P \iiint \frac{\boldsymbol{J}\, d\tau}{r} = -\oiint \frac{\boldsymbol{J} \cdot \hat{\boldsymbol{\nu}}}{r}\, dS,$$

where $\hat{\boldsymbol{\nu}}$ is the outward normal to the boundary surface of the region where $\boldsymbol{J} \neq \boldsymbol{0}$. But the boundary conditions satisfied by $\boldsymbol{J}$ include $\boldsymbol{J} \cdot \hat{\boldsymbol{\nu}} = 0$. (See p. 154.)

$$\therefore \ \text{div}_P \iiint \frac{\boldsymbol{J}\, d\tau}{r} = 0.$$

We regard surface currents and linear currents as limiting cases of regions with $\boldsymbol{J} \neq \boldsymbol{0}$. In the case of surface currents $\boldsymbol{K}$ the region has been "compressed" into a surface, the condition $\boldsymbol{K} \cdot \hat{\boldsymbol{\nu}} = 0$ being satisfied always. Similar considerations imply that the result is zero for linear currents also, and so div $\boldsymbol{A} = 0$ everywhere.

Example 2. A long (infinite) conductor has a circular section of radius a and carries a current of total strength I which is uniformly distributed over the area of section.

This is a current distribution with complete cylindrical symmetry and is uniform in the direction of the current, Fig. 6.7. To find the magnetic field $\boldsymbol{H}$ we use eqn. (6.23) here in much the same way as we used Gauss's theorem in electrostatics for distributions with axial and spherical symmetry.

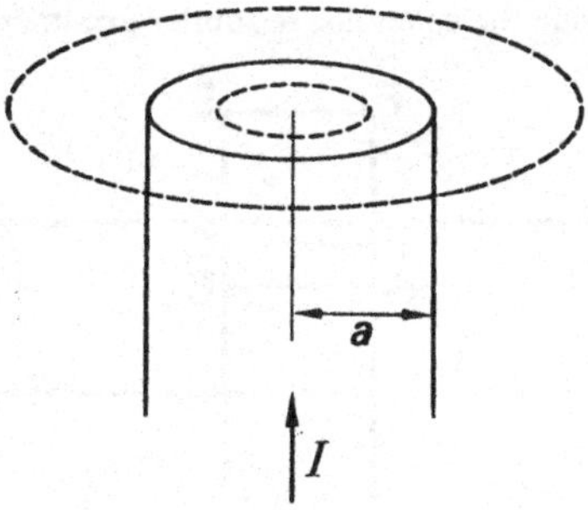

FIG. 6.7

From the symmetry of the field we conclude that $\boldsymbol{H}$ is directed everywhere along a circle centred on the axis of the conductor, and the magnitude H depends only on the radius r of this circle. We use the circle (which may typically be either of the dotted circles in Fig. 6.7) as a contour to which we apply (6.23) thus:

$$\oint \boldsymbol{H}\cdot\mathbf{ds} = 2\pi r H.$$

Also

$$\iint \boldsymbol{J}\cdot\mathbf{dS}\begin{cases} = Ir^2/a^2 & \text{for} \quad r \leqslant a, \\ = I & \text{for} \quad r \geqslant a, \end{cases}$$

and hence we conclude that, in cylindrical polars,

$$\boldsymbol{H}\begin{cases} = \{0 \quad Ir/(2\pi a^2) \quad 0\} & \text{when} \quad r \leqslant a, \\ = \{0 \quad I/(2\pi r) \quad 0\} & \text{when} \quad r \geqslant a. \end{cases}$$

In a similar manner we can deduce that there is no magnetic field inside a long cylindrical shell which carries a current flowing parallel to its generators and uniformly distributed on the shell. For points outside, the field is identical with that above when $r \geqslant a$.

Example 3. We find the field inside a long solenoid carrying a current K per unit length.

The vector $\boldsymbol{B} = \mu_0 \boldsymbol{H}$ everywhere so that div $\boldsymbol{H} = 0$ everywhere inside and outside the solenoid. Because of the cylindrical symmetry, the magnitudes of the components of H depend only on the distance r of the field point from the axis. First we consider a circuit Γ consisting of a circle of radius r with centre on the axis. This circuit encloses no net current and so $\oint_\Gamma \boldsymbol{H}\cdot\mathbf{ds} = 2\pi r H_\phi = 0$. Therefore there is no azimuthal field either inside or outside the solenoid.

Second, the relation div $\boldsymbol{H} = 0$ implies that $\mathrm{d}(rH_r)/\mathrm{d}r = 0$. The only solution of this equation which remains finite everywhere is $H_r = 0$.

Third, we choose a rectangle *ABCD* (see Fig. 6.8) to which we apply eqn. (6.23). This circuit encircles no current, and so $\oint \boldsymbol{H}\cdot d\boldsymbol{s} = 0$. Because of the absence of a radial component

$$\oint \boldsymbol{H}\cdot d\boldsymbol{s} = (H_A - H_O)AB = 0,$$

where H_A, H_O are the magnitudes of $\boldsymbol{H}$ at *A* and *C*. This result implies that $H_A = H_O$ and that the field is uniform inside the solenoid. A similar argument, using a rectangle whose

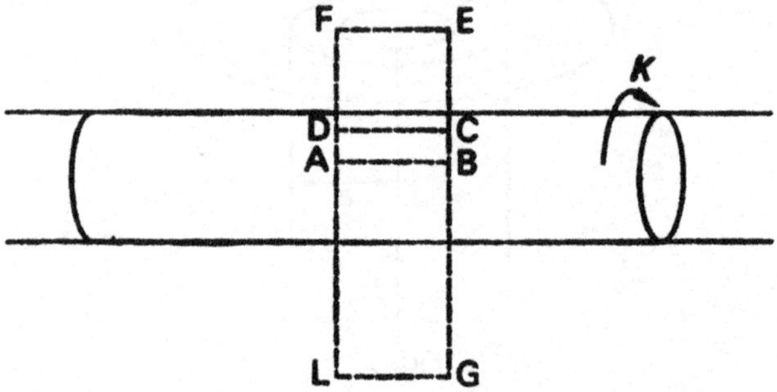

FIG. 6.8

sides are entirely outside the solenoid, shows that the field is uniform outside; since the field is zero at infinity it must be zero everywhere outside the solenoid. When we use a contour such as *EFLG* which intersects the solenoid twice we still have the result $\oint \boldsymbol{H}\cdot d\boldsymbol{s} = 0$ since this contour encircles no net current.

Now we consider a contour such as *ABEF*, one side *AB* lying within the contour, the other *EF* lying outside. In this case

$$\oint \boldsymbol{H}\cdot d\boldsymbol{s} = H_A \cdot AB.$$

For this contour

$$\iint \boldsymbol{J}\cdot d\boldsymbol{S} = K\cdot AB.$$

$$\therefore \quad H_A = K$$

or

$$\boldsymbol{H} = \boldsymbol{k}K,$$

where $\boldsymbol{k}$ is a unit vector along the axis of the solenoid. This equation gives the value of the uniform field inside the solenoid. (This result is an example of the discontinuity in the tangential component of $\boldsymbol{H}$ arising from a surface current, see § 6.4.)

Example 4. We calculate the vector potential arising from a uniform distribution of current in a long circular cylinder of radius *a*.

We cannot use formula (6.16) because the integral arising from the conductor of infinite length does not converge. (We met this difficulty in connection with the similar infinite charge distribution. Then we ignored an infinite constant.) We give two methods.

(a) We use cylindrical polar coordinates (r, ϕ, z) with $\boldsymbol{A} = \{0\,0\,A\}$, where *A* depends only on *r*. Find $\boldsymbol{B} = \text{curl}\,\boldsymbol{A}$ and equate it to the expression for $\boldsymbol{B}$ already obtained in example 2 above for $B = \mu_0 H$ thus:

$$\text{curl}\, A = \left\{0 \quad -\frac{dA}{dr} \quad 0\right\}.$$

$$\therefore \quad \frac{dA}{dr} = -\frac{\mu_0 Ir}{2\pi a^2} \qquad \text{for} \quad r \leqslant a.$$

$$\therefore \quad A = -\frac{\mu_0 Ir^2}{4\pi a^2} + C_1 \qquad \text{for} \quad r \leqslant a.$$

Also

$$\frac{dA}{dr} = -\frac{\mu_0 I}{2\pi r} \qquad \text{for} \quad r \geqslant a.$$

$$\therefore \quad A = -\frac{\mu_0 I}{2\pi} \ln r + C_2 \qquad \text{for} \quad r \geqslant a.$$

The constant C_2 is either ignored or given some convenient value. Then the fact that A is continuous at $r = a$ determines C_1.

(b) Alternatively, we use the fact that $\nabla^2 A = -\mu_0 J$ and write $A = Ak$ since the integrals of (6.16) indicate that A must be directed parallel to the current. Hence

$$\nabla^2 A = -\frac{\mu_0 I}{4\pi a^2} \qquad \text{for} \quad r \leqslant a.$$

In terms of cylindrical coordinates

$$\frac{1}{r}\frac{d}{dr}\left(r\frac{dA}{dr}\right) = -\frac{\mu_0 I}{4\pi a^2}, \qquad \frac{dA}{dr} = -\frac{\mu_0 Ir}{2\pi a^2} - \frac{C_3}{r}.$$

$$\therefore \quad B = \left\{0 \quad \frac{\mu_0 Ir}{2\pi a^2} + \frac{C_3}{r} \quad 0\right\} = \mu_0 H.$$

Since $\oint H \cdot ds = Ir^2/a^2$ we see that $C_3 = 0$.

$$\therefore \quad \frac{dA}{dr} = -\frac{\mu_0 Ir}{2\pi a^2}$$

and the remainder follows as above.

Exercises 6.3

1. A current I flows in a coil formed by winding n turns of insulated wire per unit length onto a circular cylinder of radius a and length $2a$. Show that the magnetic field H at the centre of the cylinder is of magnitude $nI/\sqrt{2}$.

2. If a plane circuit is a circle of radius a and carries a steady current I and P is a point in the plane of the wire at a distance $c\,(< a)$ from the centre of the circuit, prove that the magnitude of B at P is

$$\frac{\mu_0 I}{\pi(a^2 - c^2)} \int_0^{\pi/2} (a^2 - c^2 \sin^2\theta)^{1/2}\, d\theta.$$

3. Unit current flows in a loop of wire which has the shape of a regular polygon with n sides inscribed in a circle of radius a. Show that the field B at a point at distance z from

the plane of the loop, and on the axis of symmetry of the loop, is directed along the axis and has the magnitude

$$\frac{\mu_0 nbc}{2\pi(z^2+a^2)^{1/2}(z^2+b^2)},$$

where $b = a\cos(\pi/n)$, $c = a\sin(\pi/n)$.

4. Three long straight parallel wires intersect a plane normal to the direction of the wires in the vertices of an equilateral triangle ABC of side a. Each of these wires carries a steady current I, and a fourth parallel wire passing through the centroid G of the triangle carries the return current $-3I$. Find the magnetic field H due to these currents at a general point P of a plane containing the central conductor and intersecting the plane at A.

5. A current I flows in a circuit in the shape of an ellipse of semi-axes a, b, the length of the circuit being L. Show that the magnitude of the field H at the centre of the circuit is $IL/(4\pi ab)$.

6. A plane circuit C is defined in plane polar coordinates as

$$r = a,\quad 0 \leqslant \theta \leqslant \pi;\qquad r\cos\theta = -a,\quad \pi \leqslant \theta \leqslant 3\pi/2;\qquad r\cos\theta = a,\quad 3\pi/2 \leqslant \theta \leqslant 2\pi.$$

Show that the vector H at the origin, when current I flows round C in the sense θ increasing, is

$$I(\pi+2)\boldsymbol{k}/(4\pi a).$$

7. A current I flows in a U-shaped circuit consisting of two parallel, semi-infinite wires connected at their ends by a perpendicular straight piece of length $2b$, the specific resistance being the same throughout the circuit. Show that the magnetic field B produced at a point in the plane of the circuit and within it, midway between the long wires and at a distance a from the short piece, is of magnitude

$$\frac{\mu_0 I}{2\pi}\left\{\frac{1}{b}+\sqrt{\left(\frac{1}{a^2}+\frac{1}{b^2}\right)}\right\}$$

and is perpendicular to the plane of the circuit.

8. Two coaxial circular coils of radii a and b and of m and n turns, respectively, are placed with their centres at a distance $\frac{1}{2}(a+b)$ apart. They carry the same current I in the same sense. The height of each coil and the thickness of the wire may be neglected in comparison with the radii. Find the magnetic field at a point on the axis between the two coils and distant $\frac{1}{2}a$ from the centre of the coil of radius a.

Given that $m/a^2 = n/b^2$, show that for points on the axis distant $\frac{1}{2}a+c$ from the centre of the coil of radius a, where $|c|$ is small compared with a and b, the field is uniform if the third powers of c/a and c/b are neglected.

6.4 Boundary conditions

We have so far derived two equations satisfied by the field vectors, viz.

$$\operatorname{div} \boldsymbol{B} = 0, \qquad \textbf{curl}\, \boldsymbol{H} = \boldsymbol{J},$$

and, for a vacuum, the relation $\boldsymbol{B} = \mu_0 \boldsymbol{H}$ has been assumed. Also we have obtained expressions for $\boldsymbol{B}$, $\boldsymbol{H}$ in terms of the current distribution. Now we investigate the relations which hold between the vectors at discontinuities and appeal to results already established for similar integral expressions in connection with electrostatics.

The discontinuities which may occur are:

(i) a boundary of $\boldsymbol{J}$ where $\boldsymbol{J}\cdot\hat{\boldsymbol{\nu}}$ is zero and $\boldsymbol{J}\times\hat{\boldsymbol{\nu}}$ is discontinuous (cf. a boundary of a volume distribution of charge ϱ in electrostatics),

(ii) a surface carrying a surface current $\boldsymbol{K}$ (cf. a surface charge σ in electrostatics),

(iii) a linear current.

(The difficulty which arose in electrostatics of having an insulator to which a space charge density ϱ could be attached does not occur in connection with volume currents $\boldsymbol{J}$. It is possible to have such currents inside gases and liquids, where, presumably, it is possible to measure magnetic fields. Measurements inside solid conductors, however, present serious difficulties.) We do not define field vectors "inside" singularities such as surface or line currents and the latter do not give rise to discontinuities, or jumps, in the values of field quantities any more than do point charges in electrostatics. Consequently we do not consider case (iii) above.

(a) *Discontinuities in the vector potential*

The three component integrals of $\boldsymbol{A} = (\mu_0/4\pi)\iiint (\boldsymbol{J}/r)\,d\tau$ are similar to the integrals $\iiint (\varrho/r)\,d\tau$ in electrostatics and so give rise to no discontinuities in $\boldsymbol{A}$. Hence, $\boldsymbol{A}$ is continuous as the field point crosses a boundary of $\boldsymbol{J}$.

The contribution to $\boldsymbol{A}$ from a surface current, i.e.

$$(\mu_0/4\pi)\iint (\boldsymbol{K}/r)\,dS,$$

consists of components each of which is similar in form to that for the potential V in electrostatics. Since V is continuous across a simple layer of charge, we conclude that $\boldsymbol{A}$ is continuous across a surface current. Therefore

$$\boldsymbol{A}_+ - \boldsymbol{A}_- = \boldsymbol{0}. \tag{6.24}$$

The integrals for the components of $\boldsymbol{A}$ are identical in form to the integrals (2.15) for the electrostatic potential. These integrals, in that case, gave rise to a discontinuity

$$\left(\frac{\partial V}{\partial \nu}\right)_+ - \left(\frac{\partial V}{\partial \nu}\right)_- = -\frac{\sigma}{\varepsilon_0},$$

[eqn. (2.33)]. Hence, we can deduce that a component of A is subject to a discontinuity

$$\left(\frac{\partial A_i}{\partial \nu}\right)_+ - \left(\frac{\partial A_i}{\partial \nu}\right)_- = -\mu_0 K_i.$$

By combining the three components of this equation we obtain

$$\left(\frac{\partial \boldsymbol{A}}{\partial \nu}\right)_+ - \left(\frac{\partial \boldsymbol{A}}{\partial \nu}\right)_- = -\mu_0 \boldsymbol{K}. \tag{6.25}$$

The two equations (6.24), (6.25) are equivalent to the discontinuities obtained below, by other reasoning, in the field vectors.

(b) *Discontinuities in the field vectors*

These are derived from the field equations as in the electrostatic case. From div $\boldsymbol{B} = 0$, by applying the divergence theorem to a suitable "pill box" surface, we deduce that

$$\hat{\boldsymbol{\nu}}\cdot(\boldsymbol{B}_+ - \boldsymbol{B}_-) = 0. \tag{6.26}$$

From the relation **curl** $\boldsymbol{H} = \boldsymbol{J}$, and by treating $\boldsymbol{K}$ as the limit of a volume current "compressed" onto a surface, we deduce that

$$\hat{\boldsymbol{\nu}}\times(\boldsymbol{H}_+ - \boldsymbol{H}_-) = \boldsymbol{K} \tag{6.27}$$

thus. We choose a contour $ABCD$ whose normal $\hat{\boldsymbol{t}}$ lies in the surface, Fig. 6.9. The sides BC, DA of the contour are of length dl and the sides AB, CD

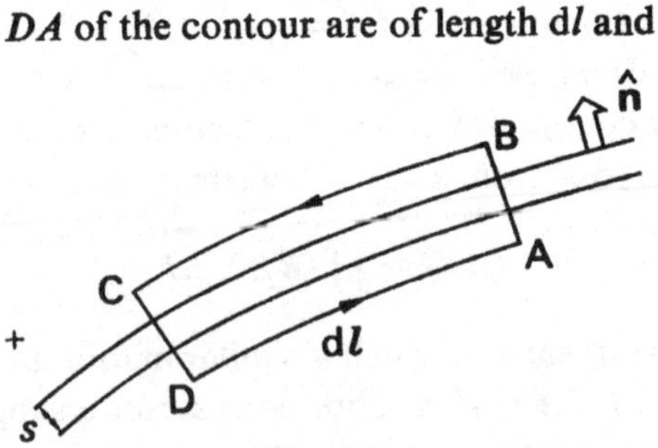

FIG. 6.9

are of the second order in dl. The current is confined in a layer of thickness s which tends to zero in such a way that

$$\lim_{s\to 0} s\boldsymbol{J} = \boldsymbol{K}.$$

The vector $\overrightarrow{DA}$ is dl $(\hat{\boldsymbol{\nu}}\times\hat{\boldsymbol{t}})$, $\overrightarrow{BC} = -\text{d}l(\hat{\boldsymbol{\nu}}\times\boldsymbol{t})$.

Hence $$\oint \boldsymbol{H}\cdot\mathbf{ds} = \text{d}l(\hat{\boldsymbol{\nu}}\times\boldsymbol{t})\cdot(\boldsymbol{H}_- - \boldsymbol{H}_+).$$

From the integral form of **curl** $\boldsymbol{H} = \boldsymbol{J}$ we obtain

$$\oint \boldsymbol{H}\cdot\mathbf{ds} = \iint \boldsymbol{J}\cdot\mathbf{dS} = \lim_{s\to 0} s\, \mathrm{d}l\boldsymbol{J}\cdot\hat{\boldsymbol{t}}.$$

In the limit we obtain

$$\lim \hat{\boldsymbol{t}}\cdot[\hat{\boldsymbol{\nu}}\times(\boldsymbol{H}_+ - \boldsymbol{H}_-)] = \lim s\boldsymbol{J}\cdot\hat{\boldsymbol{t}}.$$

Since $\hat{\boldsymbol{t}}$ is arbitrary, this reduces to

$$\hat{\boldsymbol{\nu}}\times(\boldsymbol{H}_+ - \boldsymbol{H}_-) = \boldsymbol{K}.$$

Example 1. A superconductor is defined as a body with the property that when placed in a magnetic field currents develop immediately on its surface so that there is no magnetic field at points within the superconductor. A superconducting sphere of radius a is introduced in an otherwise uniform magnetic field $\boldsymbol{H}_0$. Show that the density of the surface currents is

$$\tfrac{3}{2}\hat{\boldsymbol{n}}\times\boldsymbol{H}_0,$$

where $\hat{\boldsymbol{n}}$ is a unit vector along the outward normal at the point considered.

We suppose that the original uniform field everywhere is $\boldsymbol{B}_0 = \mu_0\boldsymbol{H}_0$. Then the surface current $\boldsymbol{K}$ generated on the surface of the sphere must neutralize this uniform field inside the sphere.

Since **curl** $(\boldsymbol{e}\times\boldsymbol{r}) = 2\boldsymbol{e}$ where $\boldsymbol{e}$ is a constant vector, we choose a vector potential $\boldsymbol{A}$ for the field generated by the surface current such that **curl** $\boldsymbol{A} = -\boldsymbol{B}_0$ for $|\boldsymbol{r}| \leqslant a$, i.e.

$$\boldsymbol{A} = -\tfrac{1}{2}(\boldsymbol{B}_0\times\boldsymbol{r}).$$

For $|\boldsymbol{r}| \geqslant a$ we choose a form for $\boldsymbol{A}$ which tends to zero as $|\boldsymbol{r}| \to \infty$ and which is continuous with the first expression across the surface of the sphere. We *try* a form

$$\boldsymbol{A} = \lambda(\boldsymbol{B}_0\times\boldsymbol{r})/r^3,$$

where λ is a coefficient to be determined.

At an arbitrary point on the surface of the sphere $\boldsymbol{r} = a\hat{\boldsymbol{n}}$ and so

$$-\frac{1}{2}(\boldsymbol{B}_0\times\hat{\boldsymbol{n}})a = \frac{\lambda}{a^3}(\boldsymbol{B}_0\times\hat{\boldsymbol{n}})a,$$

i.e. $\lambda = -a^3/2$, so that

$$\boldsymbol{A} = \begin{cases} -\tfrac{1}{2}(\boldsymbol{B}_0\times\boldsymbol{r}) & \text{for } |\boldsymbol{r}| \leqslant a, \\ -\dfrac{a^3}{2}\dfrac{(\boldsymbol{B}_0\times\boldsymbol{r})}{r^3} & \text{for } |\boldsymbol{r}| \geqslant a. \end{cases}$$

Since

$$\mathbf{curl}\,\frac{(\boldsymbol{B}_0\times\boldsymbol{r})}{r^3} = -\frac{\boldsymbol{B}_0}{r^3} + \frac{3(\boldsymbol{B}_0\cdot\boldsymbol{r})\boldsymbol{r}}{r^5},$$

the resultant field outside the sphere is

$$\boldsymbol{B}_0 - \frac{a^3}{2}\left\{-\frac{\boldsymbol{B}_0}{r^3} + \frac{3(\boldsymbol{B}_0\cdot\boldsymbol{r})\boldsymbol{r}}{r^5}\right\}$$

and on the surface of the sphere this field is

$$\boldsymbol{B}_0 + \frac{1}{2}\boldsymbol{B}_0 - \frac{3a^3}{2}\frac{(\boldsymbol{B}_0\cdot\hat{\boldsymbol{n}})\hat{\boldsymbol{n}}}{a^3} = \frac{3}{2}\{\boldsymbol{B}_0 - (\boldsymbol{B}_0\cdot\hat{\boldsymbol{n}})\hat{\boldsymbol{n}}\}.$$

The corresponding expression for $\boldsymbol{H}$ is, on dividing by μ_0,

$$\tfrac{3}{2}\{\boldsymbol{H}_0-(\boldsymbol{H}_0\cdot\hat{\boldsymbol{n}})\hat{\boldsymbol{n}}\}.$$

From the discontinuity relation for $\boldsymbol{H}$, since the field inside the sphere is zero, we obtain

$$\boldsymbol{K} = \hat{\boldsymbol{n}}\times(\boldsymbol{H}_+-\boldsymbol{H}_-) = \tfrac{3}{2}\hat{\boldsymbol{n}}\times\{\boldsymbol{H}_0-(\boldsymbol{H}_0\cdot\hat{\boldsymbol{n}})\hat{\boldsymbol{n}}\} = \tfrac{3}{2}\hat{\boldsymbol{n}}\times\boldsymbol{H}_0.$$

Example 2. A sphere of radius a, which carries a surface charge of uniform density σ, rigidly attached, is rotated about an axis through its centre with constant angular velocity $\boldsymbol{\omega}$. Show that the magnetic field at an external point is that due to a dipole at the centre, and find its moment.

At a point P, where the unit normal is $\hat{\boldsymbol{n}}$, the velocity of the charge is $a(\boldsymbol{\omega}\times\hat{\boldsymbol{n}})$ since the position vector of P is $a\hat{\boldsymbol{n}}$, Fig. 6.10. The charge density σ with this velocity constitutes a surface current,

$$\boldsymbol{K} = a\sigma(\boldsymbol{\omega}\times\hat{\boldsymbol{n}}). \quad (1)$$

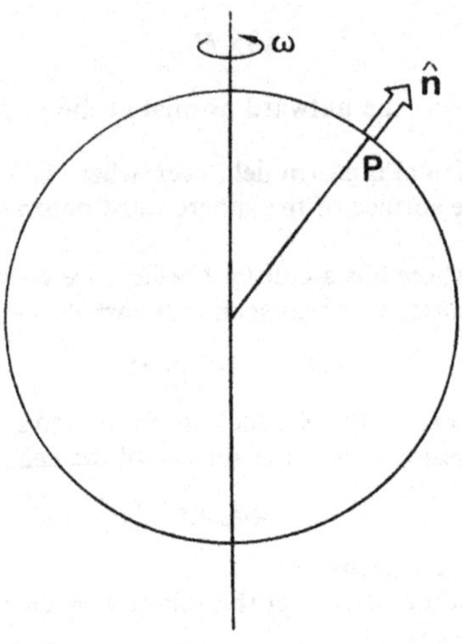

FIG. 6.10

We have to find a vector potential which satisfies $\nabla^2\boldsymbol{A} = \boldsymbol{0}$ both inside and outside the sphere, and which corresponds to a surface current given by $\boldsymbol{K}$. If we denote by $\boldsymbol{A}_-$, $\boldsymbol{A}_+$ the vector potential inside and outside the sphere, the following functions satisfy the continuity conditions and $\boldsymbol{A}_+ \to \boldsymbol{0}$ as $r \to \infty$,

$$\boldsymbol{A}_- = \boldsymbol{e}\times\boldsymbol{r}, \quad \boldsymbol{A}_+ = \frac{a^3(\boldsymbol{e}\times\boldsymbol{r})}{r^3}, \quad (2)$$

where $\boldsymbol{e}$ is a constant, at present unknown, vector. The corresponding values for $\boldsymbol{B}$ are

$$\boldsymbol{B}_- = \text{curl}\,\boldsymbol{A}_- = 2\boldsymbol{e}, \quad \boldsymbol{B}_+ = \text{curl}\,\boldsymbol{A}_+ = a^3\left\{-\frac{\boldsymbol{e}}{r^3}+\frac{3(\boldsymbol{e}\cdot\boldsymbol{r})\boldsymbol{r}}{r^5}\right\}.$$

At the surface of the sphere

$$\boldsymbol{B}_+-\boldsymbol{B}_- = -\boldsymbol{e}+3(\boldsymbol{e}\cdot\hat{\boldsymbol{n}})\hat{\boldsymbol{n}}-2\boldsymbol{e} = 3\{(\boldsymbol{e}\cdot\hat{\boldsymbol{n}})\hat{\boldsymbol{n}}-\boldsymbol{e}\}.$$

$$\therefore\ \hat{\boldsymbol{n}}\cdot(\boldsymbol{B}_+-\boldsymbol{B}_-) = 0.$$

Also $$\hat{\boldsymbol{n}}\times(\boldsymbol{H}_+-\boldsymbol{H}_-) = (3/\mu_0\ \hat{\boldsymbol{n}}\times\{(\boldsymbol{e}\cdot\hat{\boldsymbol{n}})\hat{\boldsymbol{n}}-\boldsymbol{e}\} = -(3/\mu_0)\hat{\boldsymbol{n}}\times\boldsymbol{e}.$$

The boundary condition on $\boldsymbol{H}$ implies that

$$a\sigma(\boldsymbol{\omega}\times\hat{\boldsymbol{n}}) = -(3/\mu_0)(\hat{\boldsymbol{n}}\times\boldsymbol{e}).$$

$$\therefore \quad \boldsymbol{e} = \tfrac{1}{3}\mu_0 a\sigma\boldsymbol{\omega}.$$

From (2) we see that, for external points, the sphere behaves like a magnetic particle of moment $(4\pi/\mu_0)\boldsymbol{e}a^3 = \frac{4}{3}\pi a^4\sigma\boldsymbol{\omega}$.

6.5 The magnetostatic scalar potential

From eqn. (6.22) it follows that in regions where $\boldsymbol{J} = \boldsymbol{0}$, **curl** $\boldsymbol{H} = \boldsymbol{0}$ and there exists a scalar function Ω, such that

$$\boldsymbol{H} = -\,\mathbf{grad}\,\Omega. \tag{6.28}$$

This function Ω, which is arbitrary to the extent of an additive constant, is the magnetostatic scalar potential. Two points must be emphasized; first, Ω does *not* exist in a region where $\boldsymbol{J} \neq \boldsymbol{0}$; second, since currents must flow in closed loops, the space in which **curl** $\boldsymbol{H} = \boldsymbol{0}$ must be multiply connected if there are any linear currents present, and so Ω must be a cyclic potential.

In obtaining the force on a finite circuit from that on a small circuit we replaced the finite circuit by a spanning surface of magnetic particles of uniform strength. We can also use this same equivalent magnetic shell to determine the magnetostatic scalar potential Ω by finding the potential produced by a surface of dipoles. In fact, this result has already been obtained for electric dipoles in § 2.11, p. 61. We verify the result by some manipulations on the expression (6.14) for $\boldsymbol{H}$.

We take the scalar product of $\boldsymbol{H}$ with an arbitrary constant vector $\boldsymbol{e}$ and obtain

$$\frac{4\pi}{I}(\boldsymbol{e}\cdot\boldsymbol{H}) = \oint \frac{\boldsymbol{e}\cdot(\hat{\boldsymbol{t}}\times\boldsymbol{r})}{r^3}\,\mathrm{d}s = \oint \frac{\boldsymbol{r}\times\boldsymbol{e}}{r^3}\cdot\hat{\boldsymbol{t}}\,\mathrm{d}s.$$

$$= \iint \hat{\boldsymbol{n}}\cdot\mathbf{curl}_S\left(\frac{\boldsymbol{r}\times\boldsymbol{e}}{r^3}\right)\mathrm{d}S,$$

where $\hat{\boldsymbol{n}}$ is the unit normal vector to the surface spanning the circuit and the suffix S denotes that differentiations are taken w.r. to the origin of $\boldsymbol{r}$, the position of the element of integration. But

$$\mathbf{curl}_S\left(\frac{\boldsymbol{r}\times\boldsymbol{e}}{r^3}\right) = +\,\mathbf{curl}_P\left(\frac{\boldsymbol{r}\times\boldsymbol{e}}{r^3}\right) = -\frac{3(\boldsymbol{r}\times\boldsymbol{e})\times\boldsymbol{r}}{r^5} + \frac{2\boldsymbol{e}}{r^3}.$$

$$\therefore \quad \frac{4\pi}{I}(\boldsymbol{e}\cdot\boldsymbol{H}) = \iint \hat{\boldsymbol{n}}\cdot\left\{\frac{3(\boldsymbol{e}\cdot\boldsymbol{r})\boldsymbol{r} - 3\boldsymbol{e}r^2}{r^5} + \frac{2\boldsymbol{e}}{r^3}\right\} \mathrm{d}S$$

$$= \iint \hat{\boldsymbol{n}}\cdot\left\{-\frac{\boldsymbol{e}}{r^3} + \frac{3(\boldsymbol{e}\cdot\boldsymbol{r})\boldsymbol{r}}{r^5}\right\} \mathrm{d}S$$

$$= \iint \left\{\frac{3(\boldsymbol{e}\cdot\boldsymbol{r})(\hat{\boldsymbol{n}}\cdot\boldsymbol{r})}{r^5} - \frac{\hat{\boldsymbol{n}}\cdot\boldsymbol{e}}{r^3}\right\} \mathrm{d}S$$

$$= \iint \boldsymbol{e}\cdot\left\{\frac{3(\hat{\boldsymbol{n}}\cdot\boldsymbol{r})\boldsymbol{r}}{r^5} - \frac{\hat{\boldsymbol{n}}}{r^3}\right\} \mathrm{d}S.$$

Now $\hat{\boldsymbol{n}}$ is a constant vector as far as differentiations w.r. to the field end of $\boldsymbol{r}$ are concerned and

$$\mathbf{grad}_P\left(\frac{\hat{\boldsymbol{n}}\cdot\boldsymbol{r}}{r^3}\right) = \frac{\hat{\boldsymbol{n}}}{r^3} - \frac{3(\hat{\boldsymbol{n}}\cdot\boldsymbol{r})\boldsymbol{r}}{r^5}.$$

$$\therefore \quad \frac{4\pi}{I}(\boldsymbol{e}\cdot\boldsymbol{H}) = -\boldsymbol{e}\cdot\mathbf{grad}_P\left[\iint \frac{\hat{\boldsymbol{n}}\cdot\boldsymbol{r}}{r^3}\,\mathrm{d}S\right].$$

Since $\boldsymbol{e}$ is arbitrary, we deduce

$$\boldsymbol{H} = -\mathbf{grad}_P\left[\frac{I}{4\pi}\iint \frac{\hat{\boldsymbol{n}}\cdot\boldsymbol{r}}{r^3}\,\mathrm{d}S\right],$$

i.e. $$\boldsymbol{H} = -\mathbf{grad}\,\Omega, \quad \Omega = \frac{I}{4\pi}\iint \frac{\hat{\boldsymbol{n}}\cdot\boldsymbol{r}}{r^3}\,\mathrm{d}S. \tag{6.29}$$

The form of the integrand in (6.29) shows on comparison with (2.41) that Ω is the scalar potential of a surface of dipoles of uniform strength I per unit area. The integral in (6.29) can be evaluated as in eqn. (2.41) and is equal to the solid angle ω subtended at the field point by the circuit. Therefore

$$\Omega = I\omega/(4\pi).$$

When the change in Ω is traced out as P follows a closed contour we see that, since ω changes either by zero or a multiple of 4π when the contour does not or does interlink with the circuit I, the cyclic constant for Ω is I. If there is more than one circuit, the space has a higher connection than 2, and the cyclic constants in the potential are the current strengths in the various circuits. In these cases we see that

$$\oint \boldsymbol{H}\cdot\mathbf{ds} = \sum_r I_r, \tag{6.30}$$

where the sum extends over those circuits which are linked with the contour of integration. The result (6.30) is a special case of eqn. (6.23).

The potential Ω is calculated as though the dipole surface (magnetic shell) consisted of dipoles comprising positive and negative "magnetic charges" —magnetic poles—which produce the field H according to the inverse square law (with $\varepsilon_0 = 1$). This scalar potential is often used to calculate fields in current free regions, after replacing linear circuits by the equivalent magnetic shells, because a scalar potential is easier to handle than a vector potential with three components.

Since $\boldsymbol{B} = \mu_0 \boldsymbol{H}$, the field equation div $\boldsymbol{B} = 0$ implies that

$$\nabla^2 \Omega = 0,$$

and all the usual methods for solving Laplace's equation may be used. The boundary conditions applicable to magnetostatic problems are less simple than those for electrostatics (see Chapter 8 on potential theory). It must be remembered that $\boldsymbol{H}$, not $\boldsymbol{B}$, is the gradient of Ω (unlike the electrostatic case where $\boldsymbol{E}$, not $\boldsymbol{D}$, is the gradient of V) and that the permeability μ_0 does not occur in the expression for Ω.

Example. We consider again the example of a uniform distribution of charge carried on the surface of a rotating sphere, which has angular velocity ω.

We consider a zone on the surface which subtends an angle $d\theta$ at the centre, Fig. 6.11. This zone has a width $a\, d\theta$ on the surface and it has a speed $a\omega \sin\theta$. Hence the motion of the charge on this zone constitutes a current of strength $\omega\sigma a^2 \sin\theta\, d\theta$ circulating in this zone.

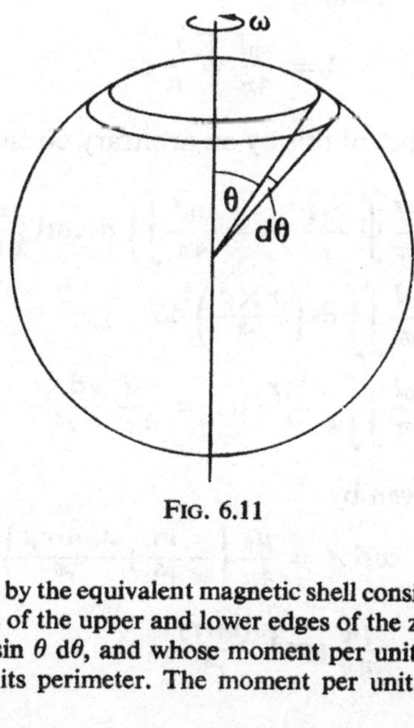

FIG. 6.11

We replace this current by the equivalent magnetic shell consisting of magnetization enclosed between the planes of the upper and lower edges of the zone. This is a disc of radius $a \sin\theta$ and thickness $a \sin\theta\, d\theta$, and whose moment per unit area is $\omega\sigma a^2 \sin\theta\, d\theta$, the current strength around its perimeter. The moment per unit *volume* of the "material" of the disc is thus

$$\frac{\omega\sigma a^2 \sin\theta \, d\theta}{a \sin\theta \, d\theta} = \omega\sigma a$$

and is directed parallel to the axis of rotation. Thus the whole distribution of moving charge must produce a field outside the sphere equivalent to that from a uniformly magnetized sphere, whose magnetization is $\omega\sigma a$. This is equivalent to a magnetic dipole at the centre of magnetic moment

$$m = \tfrac{4}{3}\pi a^3 \omega\sigma a = \tfrac{4}{3}\pi a^4 \sigma\omega$$

directed along the axis of rotation.

6.6 Small magnets

Our use of the couple, and force, on a small circuit as the starting point for our investigation of the magnetic field implies that the behaviour of a small circuit, or magnetic particle, in a magnetic field is exactly analogous to that of an electric dipole in an electrostatic field, see § 2.10. The moment of the circuit of area α with unit normal $\hat{\boldsymbol{e}}$ and carrying a current I is $\boldsymbol{m} = I\alpha\hat{\boldsymbol{e}}$. The forces and couples acting on magnetic particles are those corresponding to a "work function" $-\boldsymbol{m}\cdot\boldsymbol{B}$, $\boldsymbol{m}$ being a constant vector in the displacements used.

We now show that the field produced by a magnetic particle also has the form of that produced by an electric dipole. The vector potential $\boldsymbol{A}$ is given by (6.19), with $\mathbf{ds} = \hat{\boldsymbol{t}}\, ds$. In fact,

$$\boldsymbol{A} = \frac{\mu_0 I}{4\pi} \oint \frac{\hat{\boldsymbol{t}}}{r}\, ds.$$

We take the scalar product of this by an arbitrary constant vector $\boldsymbol{c}$ thus

$$\boldsymbol{c}\cdot\boldsymbol{A} = \frac{\mu_0 I}{4\pi} \oint \frac{\boldsymbol{c}\cdot\hat{\boldsymbol{t}}}{r}\, ds = \frac{\mu_0 I}{4\pi} \iint \hat{\boldsymbol{n}}\cdot\mathbf{curl}\left(\frac{\boldsymbol{c}}{r}\right) dS$$

$$= \frac{\mu_0 I}{4\pi} \iint \hat{\boldsymbol{n}}\cdot\left(\frac{\boldsymbol{r}\times\boldsymbol{c}}{r^3}\right) dS.$$

$$\therefore\ \boldsymbol{A} = \frac{\mu_0 I}{4\pi} \iint \frac{\hat{\boldsymbol{n}}\times\boldsymbol{r}}{r^3}\, dS = \frac{\mu_0 I}{4\pi}\,\frac{\alpha\hat{\boldsymbol{n}}\times\boldsymbol{r}}{r^3} = \frac{\mu_0}{4\pi}\,\frac{\boldsymbol{m}\times\boldsymbol{r}}{r^3}. \tag{6.31}$$

The field produced is given by

$$\boldsymbol{B} = \mathbf{curl}\,\boldsymbol{A} = \frac{\mu_0}{4\pi}\left\{-\frac{\boldsymbol{m}}{r^3} + \frac{3(\boldsymbol{m}\cdot\boldsymbol{r})\boldsymbol{r}}{r^5}\right\}$$

$$= -\frac{\mu_0}{4\pi}\,\mathbf{grad}\,\frac{(\boldsymbol{m}\cdot\boldsymbol{r})}{r^3}. \tag{6.32}$$

Equation (6.32) corresponds to the expression

$$\Omega = \frac{1}{4\pi}\frac{\boldsymbol{m}\cdot\boldsymbol{r}}{r^3} \tag{6.33}$$

for the scalar potential of the field produced by a magnetic dipole at the origin.

The work function for a particle $\boldsymbol{m}_2$ in the field of another particle $\boldsymbol{m}_1$ is

$$W = -\boldsymbol{B}\cdot\boldsymbol{m}_2 = \frac{\mu_0}{4\pi}\left\{\frac{\boldsymbol{m}_1\cdot\boldsymbol{m}_2}{r^3} - \frac{3(\boldsymbol{m}_1\cdot\boldsymbol{r})(\boldsymbol{m}_2\cdot\boldsymbol{r})}{r^5}\right\}. \tag{6.34}$$

The decrease in W consequent upon some displacement of either of the particles is equal to the work done by the forces acting on that particle. Assuming that changes of position, etc., do not alter $\boldsymbol{m}_1$ or $\boldsymbol{m}_2$ (i.e. ignoring the effects of electromagnetic induction), we investigate the stability of a given position of equilibrium by determination of whether or not W has a *minimum* in that position. The use of this method is illustrated in the examples which follow.

The magnetic field prevailing over the surface of the earth closely resembles the field which would arise from a magnetic dipole at the centre of the earth with its positive (N.-seeking) pole directed towards the S. end of the axis of rotation. The above formulae are used in approximate calculations based on this model.

Example 1. Find the couple $\boldsymbol{\Gamma}_2$ experienced by one magnet of moment $\boldsymbol{m}_2$ in the field of a second of moment $\boldsymbol{m}_1$.

To determine this couple we keep $\boldsymbol{m}_1$ fixed and give $\boldsymbol{m}_2$ a small rotation $\boldsymbol{\delta\theta}$. In such a rotation the change in $\boldsymbol{m}_2$ is $\boldsymbol{\delta m}_2 = \boldsymbol{\delta\theta}\times\boldsymbol{m}_2$. Now, if $\boldsymbol{r}$ is the position vector of $\boldsymbol{m}_2$ referred to $\boldsymbol{m}_1$

$$\frac{4\pi}{\mu_0}W = \frac{\boldsymbol{m}_1\cdot\boldsymbol{m}_2}{r^3} - \frac{3(\boldsymbol{m}_1\cdot\boldsymbol{r})(\boldsymbol{m}_2\cdot\boldsymbol{r})}{r^5},$$

$$\frac{4\pi}{\mu_0}\delta W = \frac{\boldsymbol{m}_1\cdot\boldsymbol{\delta\theta}\times\boldsymbol{m}_2}{r^3} - \frac{3(\boldsymbol{m}_1\cdot\boldsymbol{r})\,[\boldsymbol{\delta\theta}\times\boldsymbol{m}_2\cdot\boldsymbol{r}]}{r^5}$$

$$= \boldsymbol{\delta\theta}\cdot\left\{\frac{\boldsymbol{m}_2\times\boldsymbol{m}_1}{r^3} - \frac{3(\boldsymbol{m}_1\cdot\boldsymbol{r})(\boldsymbol{m}_2\times\boldsymbol{r})}{r^5}\right\}.$$

But $\delta W = -\boldsymbol{\Gamma}_2\cdot\boldsymbol{\delta\theta}$ and so

$$\boldsymbol{\Gamma}_2 = \mu_0\frac{\boldsymbol{m}_1\times\boldsymbol{m}_2}{4\pi r^3} + \frac{3\mu_0(\boldsymbol{m}_1\cdot\boldsymbol{r})(\boldsymbol{m}_2\times\boldsymbol{r})}{4\pi r^5} \tag{1}$$

$$= \frac{\mu_0\boldsymbol{m}_2}{4\pi}\times\left\{-\frac{\boldsymbol{m}_1}{r^3} + \frac{3(\boldsymbol{m}_1\cdot\boldsymbol{r})\boldsymbol{r}}{r^5}\right\}.$$

This confirms the expression $\boldsymbol{m}\times\boldsymbol{B}$ for the couple experienced by a dipole in a field. If we rotate $\boldsymbol{m}_1$ keeping $\boldsymbol{m}_2$ fixed we obtain the couple $\boldsymbol{\Gamma}_1$ acting on $\boldsymbol{m}_1$ as

$$\boldsymbol{\Gamma}_1 = -\frac{\mu_0\boldsymbol{m}_1\times\boldsymbol{m}_2}{4\pi r^3} + \frac{3\mu_0(\boldsymbol{m}_1\times\boldsymbol{r})(\boldsymbol{m}_2\cdot\boldsymbol{r})}{4\pi r^5}. \tag{2}$$

However, $\boldsymbol{\Gamma}_1 \neq -\boldsymbol{\Gamma}_2$ as we might be led to expect by the law of action and reaction.

The explanation of the discrepancy lies in the fact that a force also acts on each dipole. This force $\boldsymbol{F}$ and the couple $\boldsymbol{\Gamma}_1$ constitute a force system (wrench) reduced to a base at the position of $\boldsymbol{m}_1$. The force of $-\boldsymbol{F}$ and the couple $\boldsymbol{\Gamma}_2$ is a system reduced to a base at the position of $\boldsymbol{m}_2$. Taken together the force and couple on $\boldsymbol{m}_1$ *are* equal and opposite to the system on $\boldsymbol{m}_2$. We leave the reader to verify this result in detail.

Example 2. Two small magnets each of moment m are fixed in position at the vertices B and C of an equilateral triangle ABC of side a, with their axes along the internal bisectors of ABC and ACB and pointing inwards. A third equal magnet is pivoted at A so that it can oscillate in the plane of the triangle. Show that it is in a position of stable equilibrium when its direction is outwards along the bisector of BAC, and find the period of small oscillations about this position.

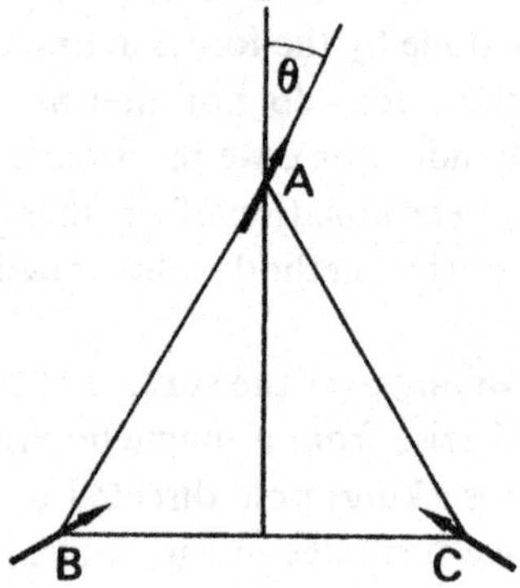

FIG. 6.12

The field produced at A is, Fig. 6.12,

$$\frac{\mu_0}{4\pi}\left\{-\frac{\boldsymbol{m}_B}{a^3}+\frac{3(\boldsymbol{m}_B\cdot\overrightarrow{BA})\overrightarrow{BA}}{a^5}-\frac{\boldsymbol{m}_C}{a^3}+\frac{3(\boldsymbol{m}_C\cdot\overrightarrow{CA})\overrightarrow{CA}}{a^5}\right\}.$$

The work function for the magnet of moment $\boldsymbol{m}$ situated at A is

$$W=-\boldsymbol{m}\cdot\boldsymbol{B}=\frac{\mu_0}{4\pi}\left\{\frac{\boldsymbol{m}_B\cdot\boldsymbol{m}}{a^3}+\frac{\boldsymbol{m}_C\cdot\boldsymbol{m}}{a^3}-\frac{3(\boldsymbol{m}_B\cdot\overrightarrow{BA})(\boldsymbol{m}\cdot\overrightarrow{BA})+3(\boldsymbol{m}_C\cdot\overrightarrow{CA})(\boldsymbol{m}\cdot\overrightarrow{CA})}{a^5}\right\}$$

$$=\frac{\mu_0}{4\pi}\left\{\frac{m^2}{a^3}\cos\left(\theta-\frac{\pi}{3}\right)+\frac{m^2}{a^3}\cos\left(\theta+\frac{\pi}{3}\right)\right\}$$

$$-\frac{\mu_0}{4\pi}\,\frac{3m^2}{a^3}\left\{\cos\frac{\pi}{6}\cos\left(\theta-\frac{\pi}{6}\right)+\cos\frac{\pi}{6}\cos\left(\theta+\frac{\pi}{6}\right)\right\}=-\frac{7\mu_0 m^2\cos\theta}{8\pi a^3}.$$

Hence the work function has a minimum value where $\theta=0$, i.e. when the magnet at A points outwards.

When the magnet makes a small angle χ with the bisector of the angle at A the equation of motion is

$$I\ddot{\chi}=-\frac{7\mu_0 m^2}{8\pi a^3}\chi,$$

where I is the moment of inertia of the magnet about the pivot. This is an oscillation of period

$$2\pi\sqrt{\left(\frac{8\pi a^3 I}{7m^2\mu_0}\right)} = \frac{4\pi a}{m}\left(\frac{2\pi I a}{7\mu_0}\right)^{1/2}.$$

Example 3. **A magnetic doublet of strength M is held in a uniform magnetic field of magnitude H with its axis parallel to the field. Show that, according to the direction in which the doublet points, the resultant magnetic field will vanish either on a certain circle or at two points.**

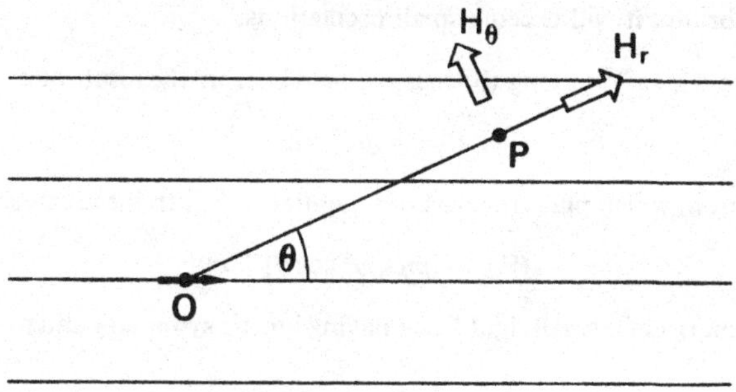

FIG. 6.13

At a typical point P (Fig. 6.13) the resultant field is

$$\boldsymbol{H}-\frac{\boldsymbol{M}}{r^3}+\frac{3(\boldsymbol{M}\cdot\boldsymbol{r})\boldsymbol{r}}{r^5}$$

and this field resolves into radial and transverse components

$$H_r = H\cos\theta-\frac{M\cos\theta}{r^3}+\frac{3M\cos\theta}{r^3} = \left(H+\frac{2M}{r^3}\right)\cos\theta,$$

$$H_\theta = H\sin\theta-\frac{M\sin\theta}{r^3} = \left(H-\frac{M}{r^3}\right)\sin\theta.$$

For zero resultant field both H_r and H_θ must vanish, i.e.

$$\left(H+\frac{2M}{r^3}\right)\cos\theta = 0, \quad \left(H-\frac{M}{r^3}\right)\sin\theta = 0.$$

Hence we have two possibilities:

(a) $$H+\frac{2M}{r^3} = 0 \quad \text{and} \quad \sin\theta = 0.$$

When $M > 0$, the first of these equations has no real solution (for r).
When $M < 0$, i.e. the magnet points opposite to the field,

$$r = (-2M/H)^{1/3} \quad \text{and} \quad \theta = 0 \quad \text{or } \pi.$$

The field vanishes at two points on the axis separated by a distance

$$d = 2(-2M/H)^{1/3}.$$

(b) $$H-\frac{M}{r^3} = 0, \quad \cos\theta = 0.$$

When $M < 0$ the first equation gives no real values for r.

When $M > 0$, i.e. M is parallel to the field,

$$r = r_1 = (M/H)^{1/3} \quad \text{and} \quad \theta = \pi/2.$$

This defines a circle of radius $(M/H)^{1/3}$ whose axis is along the magnet.

Example 4. A current j flows in a circular loop of wire of radius a, and a small magnet of moment m is pivoted freely at its centre, which is on the axis of the loop at a fixed distance r from points of the wire. Find the intensity at the position of the magnet of the magnetic field due to the current, and show that, if the magnet is slightly disturbed from its position of stable equilibrium, it will execute small oscillations.

At a distance $(r^2-a^2)^{1/2}$ along the axis the field is, from the result of example 1, p. 207,

$$\boldsymbol{B} = \mu_0 ja^2\boldsymbol{k}/(2r^3).$$

The work function, which plays the part of potential energy in these cases, is

$$W = -(m\mu_0 ja^2 \cos\theta)/(2r^3).$$

The kinetic energy of a small rigid body having kinetic symmetry and principal moments of inertia $\{I\ I\ 0\}$ is

$$T = \tfrac{1}{2}I(\dot{\theta}^2 + \dot{\psi}^2 \sin^2\theta),$$

where ψ is the azimuthal angle of spherical polar coordinates, based at the magnet, θ, ψ fixing the direction of its axis.

The Lagrangian function for the motion of the system is

$$L = \tfrac{1}{2}I(\dot{\theta}^2 + \dot{\psi}^2 \sin^2\theta) + (\mu_0 mja^2 \cos\theta)/(2r^3)$$

from which the equations of motion are

$$\frac{d}{dt}(\dot{\psi}\sin^2\theta) = 0, \qquad I\ddot{\theta} - I\dot{\psi}^2 \sin\theta\cos\theta + \frac{\mu_0 mja^3 \sin\theta}{2r^3} = 0.$$

Since displacements from the equilibrium position are small we neglect $\dot{\psi}^2 \sin\theta$ and have the approximate equation of motion

$$I\ddot{\theta} + \mu_0 mja^2\theta/(2r^3) = 0,$$

which corresponds to an oscillation of period

$$2\pi\sqrt{\left(\frac{2Ir^3}{\mu_0 mja^2}\right)}.$$

Exercises 6.6

1. A thin straight magnetized needle AB of length $2l$ is held at right angles to the plane of the magnetic meridian. On AB produced and at a distance a from the mid-point of AB a small magnet of moment M is free to turn in a horizontal plane. When in stable equilibrium the small magnet makes an angle θ with the needle. Show that the period of

the small oscillations of the small magnet about its position of stable equilibrium is $2\pi\sqrt{\{(I\sin\theta)/(MB)\}}$, where I is the moment of inertia of the small magnet about its pivot and B is the horizontal component of the earth's field.

2. Two small magnets, each of moment $\mathbf{m}$, are pivoted at points A and B, where $AB = a$, so that they are free to turn in a plane. An infinite straight wire normal to this plane cuts AB at its middle point and carries a current I. Show that there is a position of equilibrium in which both magnets are perpendicular to AB and pointing in senses corresponding to a positive rotation about the current, and that this position is stable provided that $m < 4Ia^2$.

3. Two magnetic particles A, A' of moments $\mathbf{m}, \mathbf{m}'$ are free to rotate in the same plane about their centres, which are fixed at a distance a apart. Show that the position given by $\theta = 0 = \theta'$, where θ, θ' are the angles between $\overrightarrow{AA'}$ and $\mathbf{m}, \mathbf{m}'$, is one of stable equilibrium.

If $|\mathbf{m}| = 5p$, $|\mathbf{m}'| = 8p$ and if a uniform field H is applied in the direction of $\overrightarrow{A'A}$ prove that this position remains stable provided $H < 6p/a^3$.

4. Oxy are rectangular axes in a plane and a magnetic doublet of moment P is fixed along Ox at O whilst a second doublet of moment P is fixed along Oy at O. The centre of a small magnet is pivoted at a point A of the plane Oxy and the magnet is free to turn about A in this plane. The magnetic moment of the magnet is M and it performs small oscillations in the plane Oxy in the field of the two doublets at O. If $OA = a$, and the angle $AOx = \theta$ and the moment of inertia of the magnet about A is MP/a^2g, prove that the oscillations are isochronous with those of a simple pendulum of length

$$a/(5+6\cos\theta\sin\theta)^{1/2}.$$

5. Two small magnets of equal moment M are pivoted at A and B and can turn freely in a plane through AB. There is a uniform magnetic field H in the direction BA. If in a position of equilibrium the axes of the magnets make angles $\pm 60°$ with AB, prove that $M = 2Ha^3$, where $AB = a$. Show further that this position is unstable for arbitrary small angular displacements of the magnets.

6.7 Inductance

We have seen that the "work function" from which the forces and couples acting on a linear circuit can be derived involve the magnetic flux N through the circuit. When the field arises from a number (m) of linear circuits the vector potential of the field is given by

$$\mathbf{A} = \sum_{p=1}^{m} \frac{\mu_0 I_p}{4\pi} \oint_{C_p} \frac{\mathbf{ds}_p}{r}.$$

The flux through a circuit q is

$$N_q = \oint_{C_q} \mathbf{A}\cdot\mathbf{ds}_q = \sum_p L_{pq} I_p \tag{6.35}$$

since $\mathbf{A}$ is a linear combination of the I_p. The coefficients L_{pq} are called coefficients of inductance; L_{pp} is the coefficient of self-inductance of circuit

p and $L_{pq}(p \neq q)$ is the coefficient of mutual inductance of circuits p and q. This definition of L_{pq} can be stated thus:

The mutual inductance is the flux through circuit p when unit current flows in circuit q, all other currents being zero. Since

$$N_q = \oint_{C_q} \left\{ \frac{\mu_0 I_p}{4\pi} \oint_{C_p} \frac{\mathbf{ds}_p}{r} \right\} \cdot \mathbf{ds}_q$$

$$= \frac{\mu_0}{4\pi} \sum_{p,q} \oint_{C_p} \oint_{C_q} I_p \frac{\mathbf{ds}_p \cdot \mathbf{ds}_q}{r},$$

$$L_{pq} = \frac{\mu_0}{4\pi} \oint_{C_p} \oint_{C_q} \frac{\mathbf{ds}_p \cdot \mathbf{ds}_q}{r}. \tag{6.36}$$

Formula (6.36) shows that

$$L_{pq} = L_{qp}. \tag{6.37}$$

Example 1. We calculate the self-inductance of a long straight solenoid of length l and area of cross-section α, carrying n turns in all.

When the current is I the field inside this solenoid is [see example 3, p. 213] $B = \mu_0 nI/l$ since effectively the surface current is nI/l. The flux through one turn of the solenoid is therefore $B\alpha = \mu_0 nI\alpha/l$, and the total flux through the whole circuit is

$$N = LI = \mu_0 n^2 I\alpha/l.$$

$$\therefore\ L = \mu_0 n^2 \alpha/l.$$

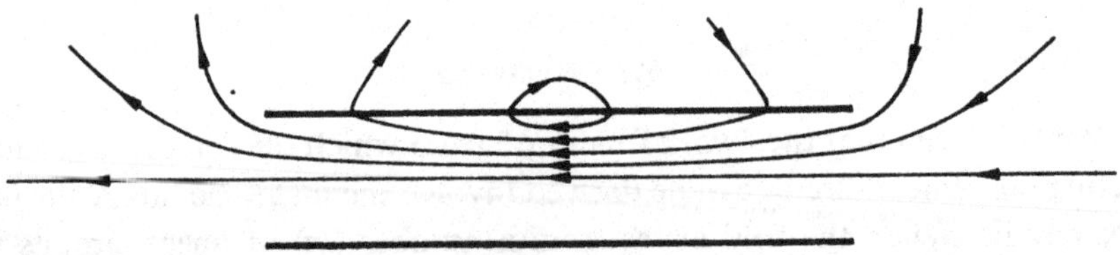

FIG. 6.14

This expression is approximate because we have used an expression for B which assumes an infinite value for l. In fact, B is not uniform along the length l and some of the lines "leak" through the sides, since the field is not exactly zero outside the solenoid (Fig. 6.14).

Example 2. We calculate a similar expression for a toroidal solenoid with n turns, made by the revolution of a circle of radius b about an axis in its plane so that its centre follows a circle of radius a ($> b$), Fig. 6.15.

By the use of Ampère's circuital relation $\oint \mathbf{H} \cdot \mathbf{ds} = I$ we find that at a distance x from the axis Oz the field H is

$$H = nI/(2\pi x).$$

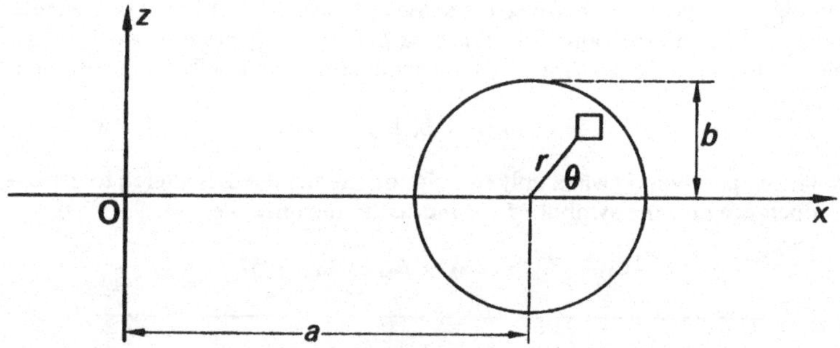

FIG. 6.15

Hence the flux through a single circle is

$$F = \iint \frac{\mu_0 nI}{2\pi x}\,dx\,dz = \frac{\mu_0 nI}{2\pi}\iint \frac{dx\,dz}{x}.$$

To evaluate the integral we write $x = a+r\cos\theta$, $z = r\sin\theta$ and find

$$F = \frac{\mu_0 nI}{2\pi}\iint \frac{r\,dr\,d\theta}{a+r\cos\theta} = \frac{\mu_0 nI}{2\pi}\int_0^b\left\{2\int_0^\pi \frac{d\theta}{a+r\cos\theta}\right\} r\,dr.$$

But
$$\int_0^\pi \frac{d\theta}{a+r\cos\theta} = \int_0^\infty \frac{2\,dt}{(a+r)+(a-r)t^2} = \frac{\pi}{\sqrt{(a^2-r^2)}},$$

giving

$$F = \mu_0 nI\int_0^b \frac{r\,dr}{\sqrt{(a^2-r^2)}} = \mu_0 nI\{a-\sqrt{(a^2-b^2)}\}.$$

This flux threads all n turns and so

$$N = LI = \mu_0 n^2 I\{a-\sqrt{(a^2-b^2)}\},$$

i.e.
$$L = \mu_0 n^2\{a-\sqrt{(a^2-b^2)}\}.$$

Example 3. Two currents (1) and (2) are wound, with their turns closely and uniformly interleaved on the same solenoid of length l and (small) area of cross-section α, and they have n_1 and n_2 turns respectively.

From example 1 the self-inductances are

$$L_{11} = \mu_0 n_1^2\alpha/l, \qquad L_{22} = \mu_0 n_2^2\alpha/l.$$

When a current I_1 flows in the first circuit the flux through each turn of the second circuit is $\mu_0 n_1 I_1\alpha/l$, so that the total flux through the second circuit is

$$N_2 = L_{12}I_1 = \pm\mu_0 n_1 n_2 I_1\alpha/l.$$

$$\therefore\ L_{12} = \pm\mu_0 n_1 n_2\alpha/l.$$

(The ambiguous sign arises because the senses of positive circulation of currents in the two circuits may not be the same.) In this case $L_{11}L_{22} = L_{12}^2$ because we have assumed no leakage of flux whatsoever—this is an approximation. Generally there is some leakage and

$$L_{12}^2 < L_{11}L_{22}.$$

The notation, particularly when only two circuits are involved, is altered to use one suffix for self-inductance and the symbol M for mutual inductance, i.e.

$$L_{11} \to L_1, \quad L_{22} \to L_2, \quad L_{12} \to M.$$

We have developed our theory using a linear current circuit, in which the current *density* vector has infinite magnitude, as the basic element corresponding to the point charge, in which the charge *density* has an infinite value. We found, in connection with electrostatics, that the energy required to "compress" a finite charge into a point was infinite, but that forces between bodies in the field could be obtained by ignoring these "infinite self-energies". Because we treated the charges as rigid the forces holding the charges together contributed nothing to the work performed in any of the displacements.

Similarly in magnetostatics the work function, which we use for finding the interactions between rigid circuits, has an infinite value when we calculate the coefficients of self-inductance. The integral (6.36) used to give L_{pp} is

$$L_{pp} = \frac{\mu_0}{4\pi} \oint\oint \frac{\mathbf{ds}_p \cdot \mathbf{ds}'_p}{r}. \tag{6.36a}$$

The integrand has a singularity, $r = 0$, when the elements of integration $\mathbf{ds}_p$ and $\mathbf{ds}'_p$ coincide. Consequently, the integral is not used to calculate self-inductances; rather we use the method of the examples above.

When the field arises from a number of rigid linear circuits the coefficient of self-inductance L_{pp} depends only on the shape of circuit p and does not alter when the circuits are displaced. The coefficients of mutual inductance depend upon the parameters which specify the relative dispositions in space of the various circuits. We obtain now the work function for a collection of such circuits.

Consider two rigid circuits carrying currents I_1 and I_2, Fig. 6.16. The field produced by circuit 1 is $\boldsymbol{B}_1$ and the resultant force on I_2 is

$$\boldsymbol{F} = I_2 \oint \mathbf{ds}_2 \times \boldsymbol{B}_1 = I_2 \oint (\hat{\boldsymbol{t}}_2 \times \boldsymbol{B}_1)\, ds_2,$$

where

$$\boldsymbol{B}_1 = \frac{\mu_0 I_1}{4\pi} \oint \frac{\mathbf{ds}_1 \times \boldsymbol{r}_{12}}{r_{12}^3} = \frac{\mu_0 I_1}{4\pi} \oint \frac{(\hat{\boldsymbol{t}}_1 \times \boldsymbol{r}_{12})}{r_{12}^3}\, ds_1.$$

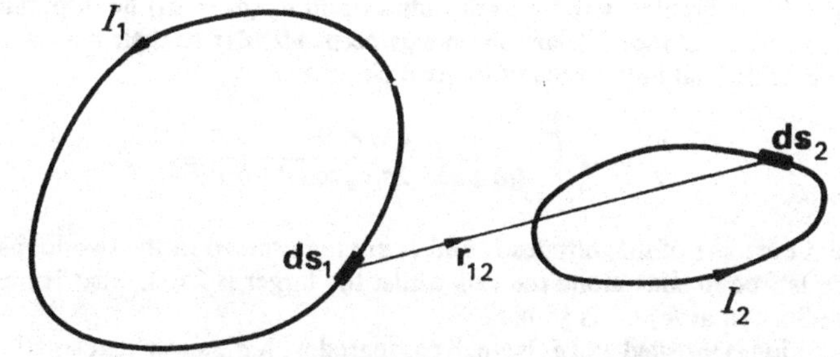

FIG. 6.16

We give the second circuit a small displacement $\boldsymbol{e}$. In this displacement the work done by $\boldsymbol{F}$ is

$$\boldsymbol{F}\cdot\boldsymbol{e} = \frac{\mu_0 I_1 I_2}{4\pi}\oint\oint \frac{\boldsymbol{e}\cdot\{\hat{\boldsymbol{t}}_2\times(\hat{\boldsymbol{t}}_1\times\boldsymbol{r}_{12})\}}{r_{12}^3}\,ds_1\,ds_2$$

$$= \frac{\mu_0 I_1 I_2}{4\pi}\oint\oint \frac{(\boldsymbol{e}\cdot\hat{\boldsymbol{t}}_1)(\hat{\boldsymbol{t}}_2\cdot\boldsymbol{r}_{12})-(\hat{\boldsymbol{t}}_1\cdot\hat{\boldsymbol{t}}_2)(\boldsymbol{e}\cdot\boldsymbol{r}_{12})}{r_{12}^3}\,ds_1\,ds_2.$$

But

$$\oint \frac{\hat{\boldsymbol{t}}_2\cdot\boldsymbol{r}_{12}}{r_{12}^3}\,ds_2 = -\oint \hat{\boldsymbol{t}}_2\cdot\mathbf{grad}_2\left(\frac{1}{r_{12}}\right)ds_2 = 0.$$

$$\therefore\ \boldsymbol{F}\cdot\boldsymbol{e} = -\frac{\mu_0 I_1 I_2}{4\pi}\oint\oint \frac{(\hat{\boldsymbol{t}}_1\cdot\hat{\boldsymbol{t}}_2)(\boldsymbol{e}\cdot\boldsymbol{r}_{12})}{r_{12}^3}\,ds_1\,ds_2$$

$$= \frac{\mu_0 I_1 I_2}{4\pi}\oint\oint (\hat{\boldsymbol{t}}_1\cdot\hat{\boldsymbol{t}}_2)\left\{\boldsymbol{e}\cdot\mathbf{grad}_2\left(\frac{1}{r_{12}}\right)\right\}ds_1\,ds_2.$$

But $\boldsymbol{e}\cdot\mathbf{grad}_2\,(1/r_{12})$ is the change in $(1/r_{12})$ consequent upon the displacement $\boldsymbol{e}$.

$$\therefore\ \boldsymbol{F}\cdot\boldsymbol{e} = \Delta\frac{\mu_0 I_1 I_2}{4\pi}\oint\oint \frac{(\hat{\boldsymbol{t}}_1\cdot\hat{\boldsymbol{t}}_2)}{r_{12}}\,ds_1\,ds_2$$

$$= \Delta(L_{12}I_1I_2) = (\Delta L_{12})\,I_1I_2$$

since the currents are kept constant. Hence the work function for the force between the two circuits is $-L_{12}I_1I_2$. By a generalization of this result the work function for a collection of circuits is

$$W' = -\sum_{p>q} L_{pq}I_pI_q. \tag{6.38a}$$

Since the expressions $L_{pp}I_p^2$ do not depend upon the relative dispositions of the circuits we may also use the expression

$$W = -\tfrac{1}{2}\sum_{p,\,q} L_{pq}I_pI_q, \tag{6.38}$$

which is more commonly used than (6.38a).

Example 1. Two circular wire loops of radii a_1 and a_2 ($a_2 < a_1$) lie in parallel planes; their centres lie at a distance b apart on an axis perpendicular to their planes. Show that the coefficient of mutual inductance of the two circuits is

$$\frac{\mu_0}{2}a_1a_2\int_0^{2\pi}\frac{\cos\theta\,d\theta}{(a_1^2+a_2^2-2a_1a_2\cos\theta+b^2)^{1/2}}.$$

If constant currents of magnitudes I_1 and I_2 are maintained in the two loops and if the smaller loop is free to slide along the axis whilst the larger is fixed, what is the condition that the equilibrium at $b = 0$ is stable?

If the condition is satisfied and a_2 is small compared with a_1, show that small oscillations about $b = 0$ have frequency $\{3\mu_0a_2^2I_1I_2/(8\pi m a_1^3)\}^{1/2}$, where m is the mass of the moving loop.

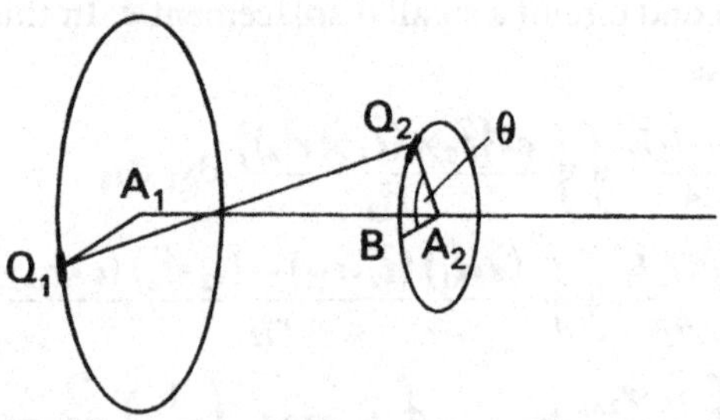

FIG. 6.17

We consider two typical elements $\mathbf{ds}_1$ and $\mathbf{ds}_2$ on the respective circuits, Fig. 6.17. Initially we choose $\mathbf{ds}_1$, at Q_1, and denote the position of $\mathbf{ds}_2$ at Q_2 by the angle θ where A_2B is parallel to A_1Q_1.

The distance Q_1Q_2 is r, where

$$r^2 = (a_1-a_2\cos\theta)^2+(a_2\sin\theta)^2+b^2 = a_1^2+a_2^2-2a_1a_2\cos\theta+b^2.$$

We use the integral form (6.36) to calculate the mutual inductance and perform the integration w.r. to s_2 first.

$$\mathbf{ds}_1\cdot\oint\frac{\mathbf{ds}_2}{r} = ds_1\int_0^{2\pi}\frac{a_2\cos\theta\,d\theta}{r}$$

since $|\mathbf{ds}_2| = a_2\,d\theta$, and the angle between $\mathbf{ds}_1$ and $\mathbf{ds}_2$ is θ.

$$\therefore\ \mathbf{ds}_1\cdot\oint\frac{\mathbf{ds}_2}{r} = a_2\,ds_1\int_0^{2\pi}\frac{\cos\theta\,d\theta}{(a_1^2+a_2^2-2a_1a_2\cos\theta+b^2)^{1/2}}.$$

Since the integral does not depend upon the position of $\mathbf{ds}_1$ the second integration, w.r. to s_1, simply reduces to multiplication by $2\pi a_1$.

$$\therefore\ \oint\oint\frac{\mathbf{ds}_1\cdot\mathbf{ds}_2}{r} = 2\pi a_1a_2\int_0^{2\pi}\frac{\cos\theta\,d\theta}{(a_1^2+a_2^2-2a_1a_2\cos\theta+b^2)^{1/2}}.$$

$$\therefore\ L_{12} = \frac{\mu_0}{4\pi}\oint\oint\frac{\mathbf{ds}_1\cdot\mathbf{ds}_2}{r} = \frac{\mu_0a_1a_2}{2}\int_0^{2\pi}\frac{\cos\theta\,d\theta}{(a_1^2+a_2^2-2a_1a_2\cos\theta+b^2)^{1/2}}.$$

The work function for the force between the circuits is

$$W = -L_{12}I_1I_2 = -\frac{\mu_0 a_1 a_2 I_1 I_2}{2}\int_0^{2\pi} \frac{\cos\theta\, d\theta}{(a_1^2+a_2^2-2a_1a_2\cos\theta+b^2)^{1/2}}.$$

$$\therefore\ \frac{dW}{db} = \frac{\mu_0 a_1 a_2 I_1 I_2}{2}\int_0^{2\pi} \frac{b\cos\theta\, d\theta}{(a_1^2+a_2^2-2a_1a_2\cos\theta+b^2)^{3/2}},$$

$$\frac{d^2W}{db^2} = \frac{\mu_0 a_1 a_2 I_1 I_2}{2}(Z_1-3b^2Z_2)$$

where

$$Z_1 = \int_0^{2\pi} \frac{\cos\theta\, d\theta}{(a_1^2+a_2^2-2a_1a_2\cos\theta+b^2)^{3/2}},\quad Z_2 = \int_0^{2\pi} \frac{\cos\theta\, d\theta}{(a_1^2+a_2^2-2a_1a_2\cos\theta+b^2)^{5/2}}.$$

The condition for stability, viz. $d^2W/db^2 > 0$, then reduces to the condition that, with $b = 0$,

$$\frac{\mu_0 a_1 a_2 I_1 I_2}{2}\int_0^{2\pi} \frac{\cos\theta\, d\theta}{(a_1^2+a_2^2-2a_1a_2\cos\theta)^{3/2}} > 0.$$

The denominator of the integral is positive for all values of θ, so we divide the range of integration into two parts for which $\cos\theta > 0$ and for which $\cos\theta < 0$. Now

$$(Z_1)_{b=0} = \int_{-\pi/2}^{\pi/2} \frac{\cos\theta\, d\theta}{(a_1^2+a_2^2-2a_1a_2\cos\theta)^{3/2}} + \int_{\pi/2}^{3\pi/2} \frac{\cos\theta\, d\theta}{(a_1^2+a_2^2-2a_1a_2\cos\theta)^{3/2}},$$

where we have replaced the original range of integration $(0, 2\pi)$ by the range $(-\pi/2, 3\pi/2)$. In the second integral we change the variable of integration to $\phi = \theta-\pi$ and obtain

$$(Z_1)_{b=0} = \int_{-\pi/2}^{\pi/2} \frac{\cos\theta\, d\theta}{(a_1^2+a_2^2-2a_1a_2\cos\theta)^{3/2}} - \int_{-\pi/2}^{\pi/2} \frac{\cos\phi\, d\phi}{(a_1^2+a_2^2+2a_1a_2\cos\phi)^{3/2}}.$$

Since $\cos\theta$, $\cos\phi$ are both positive in the ranges of integration the second integral must be smaller than the first because of the positive sign in the denominator. Hence $(Z_1)_{b=0} > 0$ and so the equilibrium with $b = 0$, which makes $dW/db = 0$, is stable if $\frac{1}{2}\mu_0 a_1 a_2 I_1 I_2 > 0$. The condition is therefore that I_1, I_2 circulate in the same sense.

The equation of motion, regarding b as the variable, is

$$m\ddot{b} = -dW/db = -(\mu_0 a_1 a_2 I_1 I_2/2)\, b(Z_1)_{b=0}.$$

When a_2 is small compared with a_1,

$$(a_1^2+a_2^2-2a_1a_2\cos\theta)^{-3/2} = \frac{1}{a_1^3}\left\{1-2\left(\frac{a_2}{a_1}\right)\cos\theta+O\left(\frac{a_2}{a_1}\right)^2\right\}^{-3/2}$$

$$= \frac{1}{a_1^3}\left\{1+3\left(\frac{a_2}{a_1}\right)\cos\theta+O\left(\frac{a_2^2}{a_1^2}\right)\right\}.$$

$$\therefore\ (Z_1)_{b=0} = \frac{1}{a_1^3}\int_0^{2\pi}\left\{1+3\left(\frac{a_2}{a_1}\right)\cos\theta+O\left(\frac{a_2^2}{a_1^2}\right)\right\}\cos\theta\, d\theta$$

$$= \frac{1}{a_1^3}\left\{3\left(\frac{a_2}{a_1}\right)\pi+O\left(\frac{a_2^2}{a_1^2}\right)\right\}.$$

The equation of motion, correct to the first order in b, is

$$\ddot{b} = -\frac{3\pi\mu_0 a_2^2 I_1 I_2}{2ma_1^3}\, b.$$

This corresponds to oscillations about $b = 0$ of frequency

$$\frac{a_2}{2\pi}\left(\frac{3\pi\mu_0 I_1 I_2}{2ma_1^3}\right)^{1/2}.$$

Example 2. A long straight current intersects at right angles a diameter of a circular current, and the plane of the circle makes an acute angle α with the plane through this diameter and the straight current. Show that the coefficient of mutual induction is

$$\mu_0\{c \sec\alpha - (c^2\sec^2\alpha - a^2)^{1/2}\} \quad \text{or} \quad \mu_0 c(\sec\alpha - \tan\alpha)$$

according as the straight current passes within or without the circle, a being the radius of the circle and c the distance of the straight current from its centre.

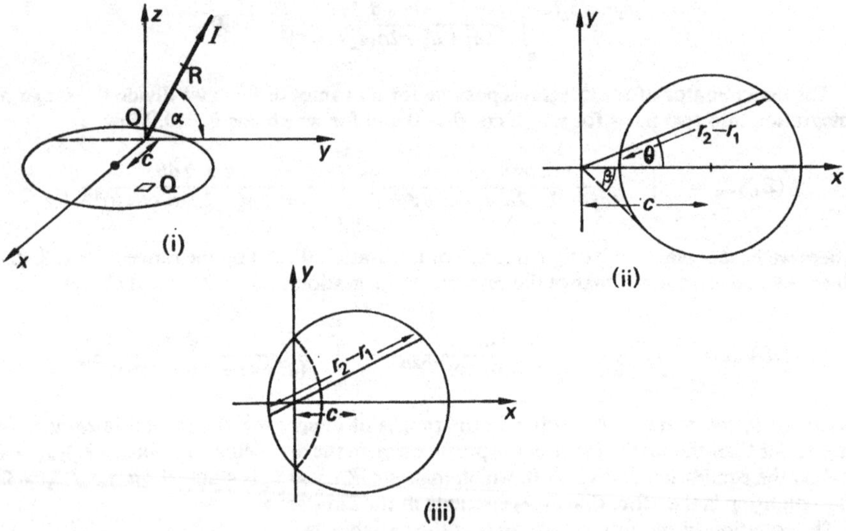

FIG. 6.18

We choose the axes of coordinates Fig. 6.18 (i) so the straight conductor passes through the origin, its direction being given by the unit vector,

$$\hat{\mathbf{e}} = \mathbf{j}\cos\alpha + \mathbf{k}\sin\alpha.$$

First we find the field produced at an arbitrary point $Q\{x\ y\ 0\}$ in the plane of the circle. The vector $\overrightarrow{RQ}$ joining the point R, with position $l\hat{\mathbf{e}}$ on the conductor, to Q is

$$\mathbf{i}x + \mathbf{j}y - l(\mathbf{j}\cos\alpha + \mathbf{k}\sin\alpha) = \mathbf{i}x + \mathbf{j}(y - l\cos\alpha) - \mathbf{k}l\sin\alpha.$$

This is perpendicular to the conductor if

$$(y - l\cos\alpha)\cos\alpha - l\sin^2\alpha = 0, \quad \text{i.e.} \quad l = y\cos\alpha.$$

Hence the vector joining some point R on the conductor to Q such that $\overrightarrow{RQ}$ is perpendicular to $\hat{e}$ is

$$\boldsymbol{i}x+\boldsymbol{j}y\sin^2\alpha-\boldsymbol{k}y\sin\alpha\cos\alpha.$$

Therefore the field at Q produced by the current I is

$$\boldsymbol{B}=\frac{\mu_0 I}{2\pi}\,\frac{(\boldsymbol{j}\cos\alpha+\boldsymbol{k}\sin\alpha)\times(\boldsymbol{i}x+\boldsymbol{j}y\sin^2\alpha-\boldsymbol{k}y\sin\alpha\cos\alpha)}{RQ^2}$$

$$=\frac{\mu_0 I}{2\pi}\,\frac{(-\boldsymbol{i}y\sin\alpha+\boldsymbol{j}x\sin\alpha-\boldsymbol{k}x\cos\alpha)}{x^2+y^2\sin^2\alpha}.$$

The flux of $\boldsymbol{B}$ through any circuit in $z=0$ is $N=\iint Bz\,dx\,dy$, where the integral is taken over the area enclosed by the circuit.

$$\therefore\quad M=-\frac{\mu_0}{2\pi}\iint\frac{x\cos\alpha\,dx\,dy}{x^2+y^2\sin^2\alpha}.$$

We evaluate the integral in terms of polar coordinates thus:

$$M=-\frac{\mu_0}{2\pi}\cos\alpha\iint\frac{r\cos\theta\,r\,dr\,d\theta}{r^2(\cos^2\theta+\sin^2\alpha\sin^2\theta)}$$

$$=-\frac{\mu_0}{2\pi}\cos\alpha\int\frac{\cos\theta\,d\theta}{\cos^2\theta+\sin^2\alpha\sin^2\theta}\int dr.$$

In the evaluation of the integral we distinguish between the cases for which (1) $c>a$ Fig. 6.18 (ii), and (2) $c<a$, Fig. 6.18 (iii).

(1) $c>a$; the limits of r are (r_1, r_2) and of θ are $(-\beta, \beta)$, where r_1, r_2 are the (positive) roots of

$$r^2-2cr\cos\theta+c^2=a^2,$$

i.e.

$$r_1+r_2=2c\cos\theta,\qquad r_1r_2=c^2-a^2.$$

$$\therefore\quad (r_2-r_1)^2=4(c^2\cos^2\theta-c^2+a^2)=4(a^2-c^2\sin^2\theta).$$

Also the angle β is given by $\sin\beta=a/c$.

$$\therefore\quad M=-\frac{\mu_0\cos\alpha}{2\pi}\int_{-\beta}^{\beta}\frac{\cos\theta\,d\theta}{\cos^2\theta+\sin^2\alpha\sin^2\theta}\int_{r_1}^{r_2}dr.$$

$$=-\frac{2\mu_0\cos\alpha}{\pi}\int_0^{\beta}\frac{(a^2-c^2\sin^2\theta)^{1/2}\cos\theta\,d\theta}{\cos^2\theta+\sin^2\alpha\sin^2\theta}.$$

By means of the substitution $c\sin\theta=a\sin\phi$, this leads to

$$M=-\mu_0\{c\sec\alpha-(c^2\sec^2\alpha-a^2)^{1/2}\}.$$

(2) $c<a$. In this case, Fig. 6.18 (iii), the flux over the segment of the circle for which $x<0$ cancels out the flux for the symmetrical positive segment, so we may take the following limits of integration. Those for r are (r_1, r_2), those for θ are $(-\pi/2, \pi/2)$. In this case r_2 and $-r_1$ $(r_1>0)$ are the roots of

$$r^2-2cr\cos\theta+c^2=a^2,\quad\text{i.e.}\quad r^2-2cr\cos\theta-(a^2-c^2)=0$$

so that

$$r_2-r_1=2c\cos\theta,\qquad r_1r_2=a^2-c^2.$$

Hence

$$M = -\frac{\mu_0 \cos\alpha}{2\pi} 2\int_0^{\pi/2} \frac{2c\cos^2\theta\, d\theta}{\sin^2\alpha + \cos^2\alpha\cos^2\theta}.$$

On evaluation this gives

$$M = -\mu_0 c(\sec\alpha - \tan\alpha).$$

Exercises 6.7

1. Show that for two parallel wires

$$-a/2 \leqslant x \leqslant a/2, \quad y = 0; \qquad -a/2 \leqslant x \leqslant a/2, \quad y = h$$

in the plane $z = 0$,

$$\iint \frac{\mathbf{ds}_1 \cdot \mathbf{ds}_2}{r} = 2a \sinh^{-1}(a/h) + 2h - 2(a^2+h^2)^{1/2},$$

where r denotes the distance between elements $\mathbf{ds}_1$, $\mathbf{ds}_2$ of the two wires.

Hence, or otherwise, determine the mutual inductance of two square circuits, each of side a, in parallel planes distance a apart and forming the boundaries of opposite faces of a cube.

2. A loop of wire carrying a current I_1 lies in the plane $z = 0$, and a long straight wire carrying a current I_2 lies along the line $x = f$, $z = 0$. The wires are *in vacuo* and do not intersect. Show that the mechanical action on the loop is statically equivalent to a force F in the line $y = h$, $z = 0$, where F and h are given by

$$F = \frac{\mu_0 I_1 I_2}{2\pi}\int \frac{dy}{f-x}, \qquad h\int \frac{dy}{f-x} = \int \frac{y}{f-x}\, dy,$$

and the integrals are taken round the loop.

When the loop is the square circuit with sides $x+y = \pm a$, $x-y = \pm a (a < f)$ and a/f is small, show that the approximate value of F is $\mu_0 I_1 I_2 a^2/(\pi f^2)$.

3. A current I flows along a circular wire of radius a, and a current I' flows along a straight wire which is coplanar with the circular wire and at a distance b from its centre ($b > a$). Show that the resultant reaction on the circular wire is a force

$$\mu_0 I I' \left\{ \frac{b}{(b^2-a^2)^{1/2}} - 1 \right\}$$

along the diameter normal to the straight wire.

4. A long straight conductor in which a current I is flowing passes through a circular conductor through which a current J is flowing. The straight conductor is perpendicular to the plane of the circle and intersects that plane at a small distance ξ from its centre. Prove that the torque on the circular conductor is $\frac{1}{2}\mu_0 IJ\xi$.

5. A current I flows in a wire in the form of an ellipse of axes $2a$, $2b$ ($a > b$) and a current I' flows in an infinite straight wire parallel to the minor axis of the elliptic wire, in the same plane with it, and at a distance c ($> a$) from its centre. The currents in the two wires are in the same sense at the points nearest together.

Show that the mutual attraction between the wires is

$$\frac{\mu_0 II'b}{a}\left\{\left(1-\frac{a^2}{c^2}\right)^{-1/2}-1\right\}.$$

6. Two currents I and J flow in the same sense around two circuits which form the edges of two opposite square faces of a rectangular block of edges (a, a, b). Show that their mutual action is a repulsive force of magnitude

$$\frac{2\mu_0 IJ}{\pi}\left\{\frac{b(b^2+2a^2)^{1/2}}{b^2+a^2}+1-\frac{a^2+2b^2}{b(a^2+b^2)^{1/2}}\right\}.$$

7. Two circular wires of radii a, b have a common centre, and are free to turn on an insulating axis which is a diameter of both. Show that when the wires carry currents I, I' a couple of magnitude

$$\frac{\pi b^2}{2a}\left(1-\frac{9}{16}\frac{b^2}{a^2}\right)\mu_0 II'$$

is required to hold them with their planes at right angles, it being assumed that b/a is so small that its fifth power may be neglected.

8. A current I flows in a rectangular circuit whose sides are of length $2a$, $2b$, and the circuit is free to rotate about an axis through its centre parallel to the sides of length $2a$. Another current I' flows in a long straight wire parallel to the axis and at a distance d from it. Prove that the couple required to keep the plane of the rectangle inclined at an angle ϕ to the plane through its centre and the straight current is

$$\frac{2\mu_0 II'abd(b^2+d^2)\sin\phi}{\pi(b^4+d^4-2b^2d^2\cos 2\phi)}.$$

Miscellaneous Exercises VI

1. Two circular loops of wire of radius a have their planes perpendicular to Oz and their centres at $(0, 0, \pm b)$. Equal currents I flow through the wires in opposite senses, that in the wire with positive z being in the positive direction relative to Oz. Show that, for small z, the magnetic flux density at the point $\{0\ 0\ z\}$ is $\{0\ 0\ \lambda z\}$, where

$$\lambda = \frac{3\mu_0 Ia^2b}{(a^2+b^2)^{5/2}}.$$

Show also that the magnetic flux density at any point (x, y, z) near the origin is

$$\{-\tfrac{1}{2}\lambda x\ \ -\tfrac{1}{2}\lambda y\ \ \lambda z\}.$$

2. Two circular loops, each of radius a and carrying a current J, are placed in parallel planes with the line joining their centres perpendicular to the planes and the currents are in the same sense. The centres are at a distance $2a$ apart, and a small magnet of moment M is held midway between them with its axis perpendicular to the line joining them. Show that the magnitude of the couple necessary to hold the magnet in this position is

$$\frac{\sqrt{2}\mu_0 JM}{4\pi ca}.$$

3. A non-magnetic sphere of centre O and radius a carries a charge q fixed on its surface and uniformly distributed over it, and rotates with constant angular velocity ω about an axis OA. Show that the magnetic scalar potential at a point P inside the sphere is

$$-\frac{q\omega}{6\pi}\frac{R}{a}\cos\theta,$$

where $OP = R$ and the angle $AOP = \theta$, and find its value outside the sphere.

4. Prove that at a point $\boldsymbol{r}$ referred to the centre O of a spherical distribution of magnetic poles of uniform density σ the magnetic field strength is $\sigma\boldsymbol{r}/3$, the point being inside the distribution. Hence show that if a second such distribution, overlapping with the first, of density $-\sigma$ has its centre at a point $\boldsymbol{x}$ referred to O, the intensity at any point inside both distributions is $\boldsymbol{x}\sigma/3$.

Hence, or otherwise, show that the magnetic field strength inside a sphere permanently magnetized to uniform intensity $\boldsymbol{I}$ is $-\boldsymbol{I}/3$, while if a spherical cavity is made anywhere inside the sphere, the intensity is zero in the cavity.

5. A circular cylinder with surface $r = a$ in cylindrical coordinates (r, θ, z) carries a uniform volume current density j flowing parallel to its length in the positive z-direction. Show that at points outside the cylinder the vector potential $\boldsymbol{A}$ is $\{0\;0\; -\frac{1}{2}\mu_0 a^2 j \ln r\}$. Give the corresponding expression for the value of $\boldsymbol{A}$ within the cylinder.

Determine the magnetic flux density $\boldsymbol{B}$ inside and outside the cylinder.

6. A non-conducting plane disc of radius a carries an electric charge of uniform density σ on one side and rotates about its axis with angular velocity ω. Show that the magnetic scalar potential at a point on the axis distant z from the centre of the disc is

$$\tfrac{1}{2}\sigma\omega\{z^2+\tfrac{1}{2}a^2-z\sqrt{(z^2+a^2)}\}.$$

7. Prove that the potential energy of two small coplanar magnets of moment $\boldsymbol{m}$ and $\boldsymbol{m}'$ with their centres at P and P' is

$$-\frac{\mu_0 mm'}{4\pi r^3}(2\cos\theta\cos\theta'-\sin\theta\sin\theta')$$

where $\overrightarrow{PP'} = \boldsymbol{r}$ and θ, θ' are the angles between $\boldsymbol{r}$ and $\boldsymbol{m}$, $\boldsymbol{m}'$ respectively.

The centres of the magnets are fixed and the magnet of moment $\boldsymbol{m}$ is held fixed by a couple of moment $\boldsymbol{G}$. The other magnet, which is free to rotate about its centre in the plane of the magnets, is in equilibrium. Prove that

$$G = \frac{3\mu_0 mm'}{8\pi r^3}\frac{\sin 2\theta}{(3\cos^2\theta+1)^{1/2}}.$$

8. Two magnetic dipoles, each with moment of magnitude $M > 0$, are free to rotate about their respective centres O_1, O_2 in a plane through O_1O_2 and θ, θ' are the respective angles the axes of the dipoles make with O_1O_2. If there is a uniform external magnetic field of magnitude B in the direction O_1O_2, prove that the work which must be done to move the dipoles to any position from the position $\theta = \theta' = 0$ is

$$MB(2-\cos\theta-\cos\theta')+\frac{\mu_0 M^2}{4\pi a^3}\{2+\cos(\theta-\theta')-3\cos\theta\cos\theta'\}$$

where a is the length O_1O_2. Hence prove that the position $\theta = \theta' = 0$ is stable provided that $Ba^3 > -M$.

9. A thin, spherical shell of radius a having a uniform surface charge of density σ attached to it is rotated about an axis through its centre with angular velocity ω. Show that the magnetic field produced is a uniform field of strength $B = \frac{2}{3}\mu_0 a\sigma\omega$ inside the shell, whereas the field outside the shell is identical with that of a dipole of moment $2\pi a^3 B/\mu_0$ at the centre.

10. Show that the magnitude of the vector potential inside a solid circular cylinder of radius a due to a current of volume density $(0, 0, jr \cos \theta)$, where j is a constant, is $\pi jr(a^2 - \frac{1}{2}r^2) \cos \theta$.

11. A current I flows in the circular wire $x^2+y^2 = a^2$, $z = 0$. Show that the magnetic intensity at (x, y, z) has components

$$3Ixz/(4a^3), \quad 3Iyz/(4a^3), \quad I\{2a^2+\tfrac{3}{2}(x^2+y^2-2z^2)\}/(4a^3)$$

correct to the second order in x, y, and z.

$ABCD$ is a small square circuit of side $2b$ which can turn freely about EF, where E and F are the mid-points of AB and CD respectively. EF lies along the y-axis, and its mid-point is at the origin. If the current in $ABCD$ is I' prove that a couple, of approximate magnitude

$$\frac{\mu_0 II'}{2a^3} b^2(4a^2-3b^2+15b^2 \cos^2 \theta) \sin \theta,$$

is necessary to maintain the plane $ABCD$ at an angle θ to the plane of the circular wire.

12. An infinite straight cable consists of a homogeneous conducting core in the form of a circular cylinder of radius a and a homogeneous conducting sheath in the form of a hollow circular cylinder fitting closely over the core and insulated from it. The internal and external radii of the sheath are a and b, and the relative permeabilities of the core and sheath are both unity. A steady current of strength I flows along the core and returns along the sheath. Show that the vector potential at a point distant r from the axis of the cable is given by

$$-\frac{\mu_0 Ir^2}{4\pi a^2}, \qquad -\frac{\mu_0 Ib^2}{4\pi(b^2-a^2)}\left(1-\frac{r^2}{b^2}+\ln \frac{r^2}{a^2}\right), \qquad -\frac{\mu_0 Ib^2}{4\pi(b^2-a^2)} \ln \frac{b^2}{a^2},$$

according as $r < a$, $a < r < b$, $b < r$ respectively.

13. Show that the magnetic vector potential for an infinite straight wire carrying a current I may be taken as $-(\mu_0 \boldsymbol{k} I \ln r)/(2\pi)$, where $\boldsymbol{k}$ is a unit vector in the direction of the wire, and r is the perpendicular distance from the wire.

A current J flows in an infinite solid circular cylinder whose cross-section is of radius a, the current being uniformly distributed over the cross-section. Show that the magnitude of the magnetic field at a distance r from the axis is

$$\mu_0 Jr/(2\pi a^2), \quad 0 \leqslant r \leqslant a; \qquad \mu_0 J/(2\pi r), \quad r \geqslant a.$$

14. The magnetic field $\boldsymbol{H}$ arising from a finite steady current distribution $\boldsymbol{j}$ is given by

$$\text{curl}\, \boldsymbol{H} = \boldsymbol{j}, \qquad \text{div}\, \boldsymbol{H} = 0, \qquad \boldsymbol{H} \text{ is } O(r^{-2}) \text{ for large } r.$$

Show that $\boldsymbol{H}$ is thereby uniquely determined, and that the above conditions are satisfied completely by

$$4\pi \boldsymbol{H}(\boldsymbol{r}) = \text{curl} \int \frac{\boldsymbol{j}(\boldsymbol{r}')}{|\boldsymbol{r}-\boldsymbol{r}'|}\, d\tau'.$$

If a steady current J flows in a thin closed loop of wire C, show that outside the wire both the above determinations of $\boldsymbol{H}$ lead to the same magnetic field as each of the following:

(i) $$\boldsymbol{H}(\boldsymbol{r}) = \frac{J}{4\pi} \int_C \frac{d\boldsymbol{s}' \times (\boldsymbol{r}-\boldsymbol{r}')}{|\boldsymbol{r}-\boldsymbol{r}'|^3};$$

(ii) the equivalent magnetic shell.

15. A spiral wire of n turns on a circular cylinder of radius a is given by

$$\mathbf{r} = a\cos\theta\,\mathbf{i}+a\sin\theta\,\mathbf{j}+a\theta\tan\alpha\,\mathbf{k}$$

for $-n\pi \leqslant \theta \leqslant n\pi$ (n is an integer), and a current I flows through the wire. Neglecting the contributions of the rest of the closed circuit, prove that the magnetic flux density at the origin is in the axial plane perpendicular to that containing the ends of the wire, and show that the resolute along the axis of the cylinder is

$$(\mu_0 nI)/\{2a(1+n^2\pi^2\tan^2\alpha)^{1/2}\}.$$

CHAPTER 7

MAGNETIZATION

7.1 Permanent magnets

The historical development of our knowledge of magnetism began with investigations into the behaviour of permanent magnets (lodestone). However, the historical development of a subject is not necessarily the most suitable one for a logical theoretical development. In the preceding chapter (§ 6.6) we discussed the properties of, and field produced by, a "magnetic particle", which there was considered to be the limiting case of a small circuit carrying a current around its perimeter; the limit is taken as the current tends to infinity and the area tends to zero. A permanently magnetized body is one in which each (infinitesimal) volume element $d\tau$ can be regarded as a magnetic particle of moment $\boldsymbol{M}\,d\tau$. The vector $\boldsymbol{M}$ is the magnetization of the body; it may, in general, vary from point to point of the body.

A magnetized body always aligns itself, if free to rotate, so that it "points in the direction of the earth's field". This is the property displayed by a compass needle, and the end or face of the body which seeks the north is called the North (or north-seeking) pole. Until quite recently the origin of a magnetic field was taken to be the distribution of magnetic poles near the ends of a magnetized body; N. poles were concentrated at one end and S. poles —equal and opposite in strength—were concentrated at the other. However, we consider magnetic effects to have their origin in current distributions, and a distribution of current on the surface of a uniformly magnetized body (i.e. a solenoid) can produce a field, outside the body, identical to that of the body itself. Such a distribution of current can be used instead of a distribution of poles, and is more in accord with our previous outlook. Furthermore, magnetic poles of one kind only, N. or S., are never found in normal circumstances; they always occur in pairs, one kind at each end of the body.

In § 6.6 we obtained the following results for the vector potential and field of a magnetic particle:

$$A = \frac{\mu_0}{4\pi}\frac{m\times r}{r^3}, \quad B = \mu_0 H = \frac{\mu_0}{4\pi}\left\{-\frac{m}{r^3}+\frac{3(m\cdot r)r}{r^5}\right\}.$$

We generalize these results by means of integration and obtain the following expressions for the field arising from a distribution M:

$$A = \frac{\mu_0}{4\pi}\iiint \frac{M\times r}{r^3}\,d\tau = \frac{\mu_0}{4\pi}\iiint M\times \mathbf{grad}_S\left(\frac{1}{r}\right) d\tau, \tag{7.1}$$

$$B = \mu_0 H = \frac{\mu_0}{4\pi}\iiint\left\{-\frac{M}{r^3}+\frac{3(m\cdot r)r}{r^5}\right\} d\tau. \tag{7.2}$$

The suffix S, as usual, denotes differentiations w.r. to the coordinates of the origin of r, i.e. the coordinates of the volume element.

For field points outside the magnetized volume r never vanishes and A is a continuous function of the coordinates of the field point and may be differentiated. On performing the differentiation we see that $B = \mathbf{curl}\, A$ for such points.

For field points inside the magnetized volume both (7.1), (7.2) have singularities in the region of integration, the integrands being $O(1/r^2)$, $O(1/r^3)$ respectively. This implies that A is defined by (7.1) for internal points but B is a semi-convergent integral whose value depends upon the shape of the cavity used to investigate its convergence (see § 2.12).

From (7.1) we can write

$$A = \frac{\mu_0}{4\pi}\iiint\left\{\frac{1}{r}\,\mathbf{curl}\, M - \mathbf{curl}\left(\frac{M}{r}\right)\right\} d\tau,$$

i.e.
$$A = \frac{\mu_0}{4\pi}\iiint \frac{\mathbf{curl}\, M}{r}\,d\tau + \frac{\mu_0}{4\pi}\iint \frac{M\times \hat{\nu}}{r}\,dS, \tag{7.3}$$

where the differentiations are taken w.r. to the coordinates of the element of integration and the surface integral is taken over the boundary surface of M. Equation (7.3) shows A as the vector potential of a current distribution

$$J' = \mathbf{curl}\, M, \quad K' = M\times \hat{\nu}. \tag{7.4}$$

These two currents produce the same vector potential as the distribution M, and are the magnetic equivalents of the Poisson charge [see eqn. (2.48)].

When we investigate the convergence of the integral for B by drawing a cavity in the volume, so that the field point is external to M, we must take account of the surface currents on the walls of the cavity given by (7.4). We follow the same general procedure as for electric polarization P and use cylindrical cavities with axes parallel to $\boldsymbol{M}$. For such cavities $\boldsymbol{M}\times\hat{\boldsymbol{\nu}}$ is a current flowing around the curved surface in a direction perpendicular to the generators; in fact, this current behaves effectively as the current on a solenoid.

To define $\boldsymbol{B}$ we choose the shape of cavity which reduces the local contribution from the walls of the cavity to negligible proportions (cf. the definition of $\boldsymbol{E}$). To achieve this we use a disc-shaped cavity in which the curved surface is negligible compared with the plane faces. In this case $\boldsymbol{B}$ arises from the equivalent Poisson currents—as we may call them—(7.4) on the distant boundary surfaces alone, i.e.

$$\boldsymbol{B} = \frac{\mu_0}{4\pi}\iiint \frac{(\mathbf{curl}_S\,\boldsymbol{M})\times\boldsymbol{r}}{r^3}\,d\tau + \frac{\mu_0}{4\pi}\iint \frac{(\boldsymbol{M}\times\hat{\boldsymbol{\nu}})\times\boldsymbol{r}}{r^3}\,dS, \tag{7.5}$$

cf. eqn. (2.49), and $\boldsymbol{B} = \mathbf{curl}\,\boldsymbol{A}$, where $\boldsymbol{A}$ is given by eqn. (7.3).

To define $\boldsymbol{H}$ we choose the shape of cavity which makes the local contribution from the walls of the cavity as great as possible. For this to be so the cavity must have end faces whose areas are negligible compared with the area of the curved surface; the cavity thus resembles a long solenoid with a current $\boldsymbol{M}\times\hat{\boldsymbol{\nu}}$ circulating on its walls. The contribution to $\boldsymbol{H}$ from this current is $(-\hat{\boldsymbol{\nu}})\times(\boldsymbol{M}\times\hat{\boldsymbol{\nu}})$. (The field $\boldsymbol{H}$ inside a long solenoid is $\hat{\boldsymbol{n}}\times\boldsymbol{K}$, where $\hat{\boldsymbol{n}}$ is the *outward* normal of the solenoid and $\boldsymbol{K}$ is the surface current and so $\hat{\boldsymbol{\nu}}$ is drawn *into* the cavity.) This contribution from the walls is thus

$$-\hat{\boldsymbol{\nu}}\times(\boldsymbol{M}\times\hat{\boldsymbol{\nu}}) = \hat{\boldsymbol{\nu}}(\boldsymbol{M}\cdot\hat{\boldsymbol{\nu}})-\boldsymbol{M} = -\boldsymbol{M}$$

since $\boldsymbol{M}$ is perpendicular to $\hat{\boldsymbol{\nu}}$. Therefore we obtain for $\boldsymbol{H}$ the expression, on adding the contribution from boundary currents $\boldsymbol{K}'$, etc.,

$$\boldsymbol{H} = \frac{1}{4\pi}\iiint \frac{(\mathbf{curl}_S\,\boldsymbol{M})\times\boldsymbol{r}}{r^3}\,d\tau + \frac{1}{4\pi}\iint \frac{(\boldsymbol{M}\times\hat{\boldsymbol{\nu}})\times\boldsymbol{r}}{r^3}\,dS - \boldsymbol{M} \tag{7.6}$$

[cf. eqn. (2.50)].

From eqns. (7.5), (7.6) we see that

$$\boldsymbol{H} = \boldsymbol{B}/\mu_0 - \boldsymbol{M}, \tag{7.7}$$

which is the magnetic counterpart of $\boldsymbol{D} = \varepsilon_0\boldsymbol{E}+\boldsymbol{P}$. For points outside the magnetization where $\boldsymbol{M} = \boldsymbol{0}$, eqn. (7.7) reduces to eqn. (6.15). In fact, (7.7) may be regarded as the generalization of (6.15) to regions of magnetization. Further, the relation $\boldsymbol{B} = \mathbf{curl}\,\boldsymbol{A}$ applies to internal as well as external points. From eqn. (7.3) for $\boldsymbol{A}$ we see that

$$\nabla^2\boldsymbol{A} = -\mu_0\,\mathbf{curl}\,\boldsymbol{M}.$$

$$\therefore\quad \mathbf{curl}\,\mathbf{curl}\,\boldsymbol{A} = \mathbf{grad}\,\mathrm{div}\,\boldsymbol{A}-\nabla^2\boldsymbol{A} = \mu_0\,\mathbf{curl}\,\boldsymbol{M}$$

since div $\boldsymbol{A} = 0$ (see p. 212). But, from eqn. (7.7),

$$\mathbf{curl}\ \boldsymbol{H} = (1/\mu_0)\ \mathbf{curl}\ \mathbf{curl}\ \boldsymbol{A} = \mathbf{curl}\ \boldsymbol{M} = \boldsymbol{0}$$

and so we can find a scalar potential Ω such that $\boldsymbol{H} = -\mathbf{grad}\ \Omega$ for the field produced by a distribution $\boldsymbol{M}$. The relation between $\boldsymbol{M}$ and Ω is identical to that between $\boldsymbol{P}$ and V, with $\varepsilon_0 = 1$, for electric dipoles. The use of Ω is based on the view that magnetization arises from a distribution of magnetic poles which interact according to the inverse square law, and a magnetic particle is regarded as a "magnetic dipole" in a sense similar to an electric dipole. However, it is the field vector $\boldsymbol{H}$ which is given by Ω and no factor corresponding to ε_0 enters into the expressions.

Example. We give an alternative discussion of the relation between $\boldsymbol{B}$ and $\boldsymbol{H}$.

The volume T is occupied by a distribution $\boldsymbol{M}$ of magnetization. We define a "potential" $\boldsymbol{C}$ by

$$\boldsymbol{C} = \frac{\mu_0}{4\pi}\iiint \frac{\boldsymbol{M}}{r}\,d\tau. \tag{1}$$

From the form (1), for $\boldsymbol{C}$ we can deduce that

$$\nabla^2\boldsymbol{C} = \begin{cases} -\mu_0\boldsymbol{M} & \text{for field points inside } T, \\ \boldsymbol{0} & \text{for field points outside } T. \end{cases}$$

Also

$$\nabla^2\boldsymbol{C} = -\mathbf{curl}\ \mathbf{curl}\ \boldsymbol{C} + \mathbf{grad}\ \text{div}\ \boldsymbol{C}. \tag{2}$$

Now at a field point P,

$$\mathbf{curl}\ \boldsymbol{C} = \frac{\mu_0}{4\pi}\iiint \mathbf{curl}_P\left(\frac{\boldsymbol{M}}{r}\right)d\tau = -\frac{\mu_0}{4\pi}\iiint \boldsymbol{M}\times\mathbf{grad}_P\left(\frac{1}{r}\right)d\tau$$

since $\boldsymbol{M}$ is independent of the coordinates of P. But $\mathbf{grad}_P\,(1/r) = -\mathbf{grad}_S\,(1/r)$.

$$\therefore\quad \mathbf{curl}\ \boldsymbol{C} = \frac{\mu_0}{4\pi}\iiint \boldsymbol{M}\times\mathbf{grad}_S\left(\frac{1}{r}\right)d\tau = \frac{\mu_0}{4\pi}\iiint\left\{\frac{\mathbf{curl}_S\ \boldsymbol{M}}{r} - \mathbf{curl}_S\left(\frac{\boldsymbol{M}}{r}\right)\right\}d\tau$$

$$= \frac{\mu_0}{4\pi}\iiint \frac{\mathbf{curl}_S\ \boldsymbol{M}}{r}\,d\tau + \frac{\mu_0}{4\pi}\iint \frac{\boldsymbol{M}\times\hat{\boldsymbol{\nu}}}{r}\,dS = \boldsymbol{A}, \tag{3}$$

where $\boldsymbol{A}$ is given by (7.3) and is the vector potential.
Also by a similar analysis

$$\text{div}\ \boldsymbol{C} = \frac{\mu_0}{4\pi}\iiint \text{div}_P\left(\frac{\boldsymbol{M}}{r}\right)d\tau = \frac{\mu_0}{4\pi}\iiint \boldsymbol{M}\cdot\mathbf{grad}_P\left(\frac{1}{r}\right)d\tau$$

$$= -\frac{\mu_0}{4\pi}\iiint \frac{\boldsymbol{M}\cdot\boldsymbol{r}}{r^3}\,d\tau. \tag{4}$$

The scalar potential Ω is related to a distribution of dipoles equivalent to $\boldsymbol{M}$ so we can write

$$\Omega = \frac{1}{4\pi}\iiint \frac{\boldsymbol{M}\cdot\boldsymbol{r}}{r^3}\,d\tau.$$

$$\therefore\quad \text{div}\ \boldsymbol{C} = -\mu_0\Omega. \tag{5}$$

Hence, combining (1)–(5) we obtain

$$-\mu_0 \boldsymbol{M} = -\textbf{curl}\, \boldsymbol{A} - \mu_0\, \textbf{grad}\, \Omega,$$

i.e.

$$\boldsymbol{B} = \mu_0(\boldsymbol{H}+\boldsymbol{M})$$

for field points inside T, and

$$\boldsymbol{0} = -\textbf{curl}\, \boldsymbol{A} - \mu_0\, \textbf{grad}\, \Omega, \quad \text{i.e.} \quad \boldsymbol{B} = \mu_0 \boldsymbol{H}$$

for field points outside T.

At a boundary of $\boldsymbol{M}$ or, more generally, on a surface separating regions of different magnetization $\boldsymbol{M}_+, \boldsymbol{M}_-$ there is an equivalent Poisson surface current

$$\boldsymbol{K}' = \hat{\boldsymbol{\nu}}\times(\boldsymbol{M}_+ - \boldsymbol{M}_-),$$

where $\hat{\boldsymbol{\nu}}$ is directed from the negative region into the positive region. The relations of discontinuity therefore give

$$\hat{\boldsymbol{\nu}}\cdot(\boldsymbol{B}_+ - \boldsymbol{B}_-) = 0 \tag{7.8}$$

and

$$\hat{\boldsymbol{\nu}}\times(\boldsymbol{B}_+ - \boldsymbol{B}_-) = \mu_0 \hat{\boldsymbol{\nu}}\times(\boldsymbol{M}_+ - \boldsymbol{M}_-),$$

i.e.

$$\hat{\boldsymbol{\nu}}\times(\boldsymbol{H}_+ - \boldsymbol{H}_-) = \boldsymbol{0}. \tag{7.9}$$

If, in addition to $\boldsymbol{M}$, there are volume and surface distributions $\boldsymbol{J}$, $\boldsymbol{K}$ of "true" current, then, by the principle of superposition, the field vectors must satisfy the field equations

$$\text{div}\, \boldsymbol{B} = 0, \quad \textbf{curl}\, \boldsymbol{H} = \boldsymbol{J},$$
$$\boldsymbol{H} = \boldsymbol{B}/\mu_0 - \boldsymbol{M}$$

and the corresponding surface relations

$$\hat{\boldsymbol{\nu}}\cdot(\boldsymbol{B}_+ - \boldsymbol{B}_-) = 0, \quad \hat{\boldsymbol{\nu}}\times(\boldsymbol{H}_+ - \boldsymbol{H}_-) = \boldsymbol{K}.$$

We do not consider a surface distribution of magnetic dipoles—corresponding to "double layers" in electrostatics—for such distributions correspond, in the case of uniform distributions, to linear currents around the perimeter or, in the non-uniform case, to a surface current $\boldsymbol{K}$. Both of these situations are covered by the field equations, etc., obtained so far.

Example 1. A sphere of radius a and centre O is uniformly magnetized, the intensity of magnetization being $\boldsymbol{M}$. Find the magnetic vector potential at a field point P and prove that the magnetic scalar potential at P is

$$\Omega = \tfrac{1}{3}\boldsymbol{M}\cdot\boldsymbol{r} \quad \text{if} \quad |\boldsymbol{r}| < a,$$

$$\Omega = \frac{a^3}{3r^3}\boldsymbol{M}\cdot\boldsymbol{r} \quad \text{if} \quad |\boldsymbol{r}| > a,$$

where $\boldsymbol{r} = \overrightarrow{OP}$.

The Poisson surface current for a uniformly magnetized body is $\mathbf{K}' = \mathbf{M}\times\hat{\mathbf{n}}$. This is of exactly the same form as that of the current on the surface of a sphere of rotating uniform electrostatic surface charge, which was discussed in example 2, p. 220, where $\mathbf{K} = a\sigma(\boldsymbol{\omega}\times\hat{\mathbf{n}})$. Hence the field produced by a uniformly magnetized sphere can be obtained from the results of that example with $\mathbf{M}$ put equal to $a\sigma\boldsymbol{\omega}$.

Hence the field inside the sphere is given by

$$\mathbf{B}_- = \mathbf{curl}\ \mathbf{A}_- = \tfrac{2}{3}\mu_0\mathbf{M},$$
$$\mathbf{H}_- = \mathbf{B}_-/\mu_0 - \mathbf{M} = -\tfrac{1}{3}\mathbf{M}.$$

Also the field outside the sphere is that of a dipole of moment $\frac{1}{3}\mu_0 a^3\mathbf{M}$, i.e.

$$\mu_0\mathbf{H}_+ = \mathbf{B}_+ = \mathbf{curl}\left(\frac{1}{3}\mu_0 a^3\frac{\mathbf{M}\times\mathbf{r}}{r^3}\right) = -\mathbf{grad}\left(\frac{1}{3}\mu_0 a^3\frac{\mathbf{M}\cdot\mathbf{r}}{r^3}\right).$$

We conclude that

$$\mathbf{H}_+ = -\mathbf{grad}\ \Omega_+, \quad \text{where} \quad \Omega_+ = \frac{a^3}{3}\frac{\mathbf{M}\cdot\mathbf{r}}{r^3} \qquad r \geqslant a,$$

$$\mathbf{H}_- = -\mathbf{grad}\ \Omega_-, \quad \text{where} \quad \Omega_- = \frac{1}{3}\mathbf{M}\cdot\mathbf{r} \qquad r \leqslant a.$$

However, these latter results can be obtained directly by the use of the scalar potential.

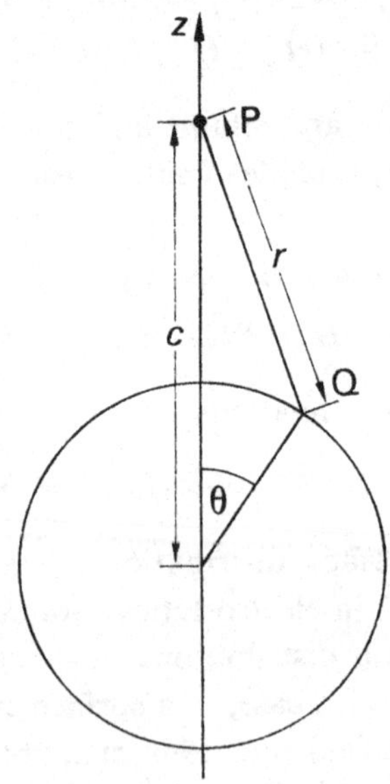

FIG. 7.1

We consider P at a distance c along the z-axis and denote the position of Q by means of spherical polar angles, Fig. 7.1. The Poisson distribution of magnetic poles equivalent to the given magnetization has a surface density $\mathbf{M}\cdot\hat{\mathbf{n}}$ and the magnetostatic potential arising from this distribution is

$$\Omega = \iint \frac{\mathbf{M}\cdot\hat{\mathbf{n}}}{4\pi r^3}\,dS,$$

where the integral is taken over the surface of the sphere and

$$\frac{M}{4\pi}\cdot\iint\frac{\hat{n}}{r}\,dS = \frac{M\cdot C}{4\pi}, \quad C = \iint\frac{\hat{n}}{r}\,dS = \iint\frac{dS}{r}.$$

Here again we must use cartesian resolution to sum these vectors evaluated at different positions on the sphere. Any other resolution would give components resolved in different directions, e.g. radial. In terms of Fig. 7.1 we write

$$\hat{n}\,dS = \{\sin\theta\cos\phi \quad \sin\theta\sin\phi \quad \cos\theta\}a^2\sin\theta\,d\theta\,d\phi,$$

$$r^2 = a^2+c^2-2ac\sin\theta,$$

$$C = \iint\frac{(i\sin\theta\cos\phi+j\sin\theta\sin\phi+k\cos\theta)a^2\sin\theta\,d\theta\,d\phi}{r}.$$

The ϕ integration eliminates the x and y components, leaving

$$C = 2\pi k\int_0^\pi\frac{a^2\sin\theta\cos\theta\,d\theta}{r}.$$

We change the variable of integration from θ to r thus:

$$r\,dr = ac\sin\theta\,d\theta, \quad \cos\theta = (a^2+c^2-r^2)/(2ac),$$

$$C = 2\pi k\int\frac{a^2(a^2+c^2-r^2)}{2acr}\,\frac{r\,dr}{ac} = \frac{\pi k}{c^2}\int(a^2+c^2-r^2)\,dr.$$

The limits of integration of r vary according to whether P is outside or inside the sphere, i.e. whether $c > a$ or $c < a$.

$$\therefore\ C_+ = \frac{\pi k}{c^2}\int_{c-a}^{c+a}(a^2+c^2-r^2)\,dr \qquad (\text{for } c > a)$$

$$= \frac{\pi k}{c^2}\left\{2a(a^2+c^2)-\frac{1}{3}(c+a)^3+\frac{1}{3}(c-a)^3\right\} = \frac{4}{3}\,\frac{\pi a^3}{c^2}k.$$

When P is inside the sphere the limits of integration are $a-c$, $a+c$. Therefore

$$C_- = \frac{\pi k}{c^2}\int_{a-c}^{a+c}(a^2+c^2-r^2)\,dr = \frac{4}{3}\pi ck.$$

The position vector of P is $kc = r$.

$$\therefore\ \Omega_+ = \frac{M\cdot C_+}{4\pi} = \frac{1}{3}\,\frac{a^3}{r^3}M\cdot r, \quad \Omega_- = \frac{M\cdot C_-}{4\pi} = \frac{1}{3}M\cdot r.$$

Example 2. A cavity in the form of a cube is removed from a spherical mass of matter with uniform intensity of magnetization M. Show that the magnetic intensity at the centre P of the cube vanishes, whatever the position of P or the orientation of the cube.

We may use the coordinate directions as the normals to the faces of the cube. Part of the field inside the cavity is due to the equivalent Poisson current $M\times\hat{n}$ on the outer surface of the sphere. Previous examples show that this gives a contribution

$$B = \tfrac{2}{3}\mu_0 M, \quad H = -\tfrac{1}{3}M.$$

Because this field is uniform inside the sphere we can use the centre of the cube as the origin. Considering first the faces of the cube perpendicular to the x-axis, the surface currents are

$$\text{on} \quad x = a, \quad \boldsymbol{M}\times(-\boldsymbol{i}); \qquad \text{on} \quad x = -a, \quad \boldsymbol{M}\times\boldsymbol{i}.$$

The vector from the element to the origin is

$$\boldsymbol{r} = -\boldsymbol{i}a+\boldsymbol{j}y+\boldsymbol{k}z \quad \text{for} \quad x = a, \qquad \boldsymbol{r} = +\boldsymbol{i}a+\boldsymbol{j}y'+\boldsymbol{k}z' \quad \text{for} \quad x = -a.$$

Hence the contribution to the field from these faces is

$$\frac{\mu_0}{4\pi}\iint \frac{[\boldsymbol{M}\times(-\boldsymbol{i})]\times[-\boldsymbol{i}a+\boldsymbol{j}y+\boldsymbol{k}z]}{(a^2+y^2+z^2)^{3/2}}\,dy\,dz$$

$$+\frac{\mu_0}{4\pi}\iint \frac{(\boldsymbol{M}\times\boldsymbol{i})\times[\boldsymbol{i}a+\boldsymbol{j}y'+\boldsymbol{k}z']}{(a^2+y'^2+z'^2)^{3/2}}\,dy'\,dz',$$

where the integrals are taken over the squares $y = \pm a$, $z = \pm a$, and $y' = \pm a$, $z' = \pm a$. Because the denominator must be taken positively and depends only on y^2, z^2 those integrals having y, z in the numerator vanish by the symmetry of the area of integration. Hence we reduce the expression to

$$\frac{\mu_0}{2\pi}\iint \frac{(\boldsymbol{M}\times\boldsymbol{i})\times\boldsymbol{i}a\,dy\,dz}{(a^2+y^2+z^2)^{3/2}} = -\frac{\mu_0(\boldsymbol{j}M_2+\boldsymbol{k}M_3)a}{2\pi}\iint \frac{dy\,dz}{(a^2+y^2+z^2)^{3/2}}.$$

On insertion of the limits the last integral is

$$I = 4\int_0^a dy \int_0^a \frac{dz}{(a^2+y^2+z^2)^{3/2}}$$

into which we substitute $z = \sqrt{(a^2+y^2)}\tan u$, obtaining

$$I = 4\int_0^a \frac{a\,dy}{(a^2+y^2)\sqrt{(2a^2+y^2)}}.$$

This is evaluated by the substitution $y = \sqrt{2}a\tan v$ thus:

$$I = \frac{4}{a}\int_0^{\tan^{-1}(1/\sqrt{2})} \frac{\cos v\,dv}{1+\sin^2 v}$$

$$= \frac{4}{a}\Big[\tan^{-1}(\sin v)\Big]_0^{\tan^{-1}(1/\sqrt{2})} = \frac{4}{a}\tan^{-1}\left(\frac{1}{\sqrt{3}}\right) = \frac{2\pi}{3a}.$$

Hence the contribution to the field from these faces is

$$-\frac{\mu_0(\boldsymbol{j}M_2+\boldsymbol{k}M_3)a}{2\pi}\,\frac{2\pi}{3a} = -\frac{1}{3}\mu_0(\boldsymbol{j}M_2+\boldsymbol{k}M_3).$$

Combining with this the corresponding results for the two other pairs of faces we obtain for the field contributed by the currents on the faces of the cube

$$-\frac{\mu_0}{3}2(\boldsymbol{i}M_1+\boldsymbol{j}M_2+\boldsymbol{k}M_3) = -\frac{2}{3}\mu_0\boldsymbol{M}.$$

Hence the final resultant is zero.

(This problem can, of course, be done by using the vector $\boldsymbol{H}$ and the Poisson distribution of poles. To evaluate the contribution from the faces of the cube leads to the same integral, and there is effectively little difference between the methods.)

Example 3. A given system of magnets lies entirely in one plane. C is a closed curve lying in this plane and enclosing no magnetic poles. When a small compass needle, free to rotate in the plane of C, is carried once round the curve C, it makes one complete rotation. Show that C encloses at least one point of equilibrium.

Since the field arises from a number of magnets, and there are no currents, the field must satisfy the conditions

$$\text{curl}\, \boldsymbol{B} = \mu_0\, \text{curl}\, \boldsymbol{H} = \boldsymbol{0}, \qquad \text{div}\, \boldsymbol{B} = 0$$

everywhere except where there are magnets.

Since all the magnets lie in one plane (which we choose to be Oxy) and we are concerned only with the field points in this plane we deduce that, at a point $(x, y, 0)$,

$$\boldsymbol{B} = \{B_x\, B_y\, 0\},$$

$$\frac{\partial B_x}{\partial z} = 0, \quad \frac{\partial B_y}{\partial z} = 0, \quad \frac{\partial B_y}{\partial x} = \frac{\partial B_x}{\partial y}, \quad \frac{\partial B_x}{\partial x} = -\frac{\partial B_y}{\partial y}. \tag{1}$$

Suppose we write the complex number B defined by

$$B = B_y + iB_x.$$

The last two of conditions (1) imply that B is an analytic function of the complex variable $x+iy$. Also

$$\arg B = \tan^{-1}(B_x/B_y).$$

The given information implies that when the field point traces out the contour C the change in $\arg B$ is 2π. But there is a theorem of functions of a complex variable which implies that

$$[\arg B]_C = 2\pi(N-P) = 2\pi \tag{2}$$

in our case. In this formula N is the number of zeros and P the number of poles (in the sense of complex variable theory) within the contours. These poles, or singularities, of B occur wherever there is a magnet. From (2)

$$N = P+1$$

and so the minimum possible value for N is one. At a zero $B = 0$, i.e. $B_x = B_y = 0$, and corresponds to a point of equilibrium.

Example 4. Regarding the earth as a uniformly and rigid magnetized sphere of radius a, and denoting the intensity of the magnetic flux density at the equator by B, show that a wire surrounding the earth along the parallel of south latitude λ, and carrying a current I from west to east, would experience a resultant force towards the south pole of the heavens of amount

$$6\pi aIB \sin\lambda \cos^2\lambda.$$

We use a cartesian frame of reference with origin at the centre of the sphere, Ox through the Greenwich meridian, Oy through the 90°W. meridian, and Oz through the S. pole, Fig. 7.2. We use the azimuthal angle ψ, and the angle of southerly latitude λ to denote position on the sphere. We denote the position of the element Q by

$$\boldsymbol{r} = a\{\cos\lambda \sin\psi \quad \cos\lambda\cos\psi \quad \sin\lambda\}.$$

$$\therefore\ \mathbf{ds} = a\{\cos\lambda\cos\psi \quad -\cos\lambda\sin\psi \quad 0\}\, d\psi,$$

where $\mathbf{ds}$ is an element of arc of the circuit at Q.

The field of the earth is that due to a magnet of moment $\{0\ 0\ m\}$, i.e. pointing due S., at the origin. The field on the equator then has magnitude B, and points due N., with $B = \mu_0 m/(4\pi a^3)$.

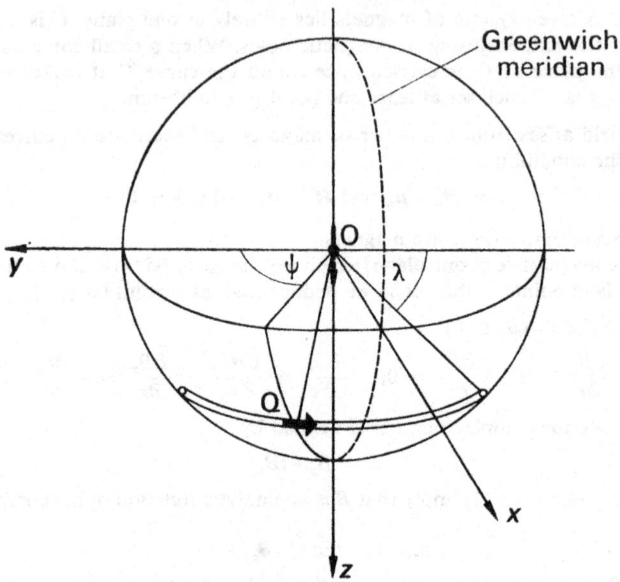

FIG. 7.2

The force on an element of the circuit is $I\,\mathbf{ds}\times\boldsymbol{B}$ where

$$\boldsymbol{B} = \frac{\mu_0}{4\pi}\left\{-\frac{\boldsymbol{m}}{a^3}+\frac{3(\boldsymbol{m}\cdot\boldsymbol{r})\boldsymbol{r}}{a^5}\right\}.$$

But $\boldsymbol{m}\cdot\boldsymbol{r} = ma\sin\lambda$,

$$\therefore \qquad \boldsymbol{B} = \frac{\mu_0 m \sin\lambda}{4\pi a^3}\left\{3\cos\lambda\sin\psi \quad 3\cos\lambda\cos\psi \quad 3\sin\lambda-\frac{1}{\sin\lambda}\right\}.$$

$$\therefore \quad I\,\mathbf{ds}\times\boldsymbol{B} = \frac{\mu_0 Im}{4\pi a^2}\{-(3\sin^2\lambda-1)\cos\lambda\sin\psi$$

$$-(3\sin^2\lambda-1)\cos\lambda\cos\psi \quad 3\cos^2\lambda\sin\lambda\}\,d\psi.$$

On integration w.r. to ψ we obtain

$$I\oint \mathbf{ds}\times\boldsymbol{B} = \frac{\mu_0 Im}{4\pi a^2}\{0 \quad 0 \quad 6\pi\cos^2\lambda\sin\lambda\} = IBa\{0 \quad 0 \quad 6\pi\cos^2\lambda\sin\lambda\},$$

i.e. a southerly force $6\pi IBa\sin\lambda\cos^2\lambda$.

Example 5. Show that if the external field of the earth were due to an eccentric dipole obtained by displacing a central dipole through a *small* distance c in a direction perpendicular to its axis, the geographical latitude λ of the magnetic poles would be approximately given by

$$\cos\lambda = \frac{3c}{a}\sin^2\lambda.$$

The dipole at P produces a field at Q which is directed along CQ when Q is at the magnetic pole. (See Fig. 7.3, which shows the meridian section of the earth containing the

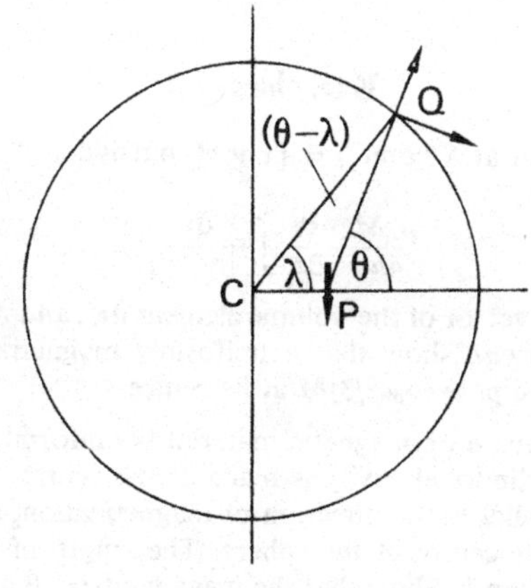

FIG. 7.3

dipole.) The field in the direction QP is

$$H_r = \frac{2M\sin\theta}{r^3}, \qquad (r = PQ)$$

and the field in the perpendicular direction is

$$H_\theta = \frac{M\cos\theta}{r^3}.$$

Hence, for Q to be at the magnetic pole

$$\frac{H_\theta}{H_r} = \frac{M\cos\theta}{r^3}\Big/\left(\frac{2M\sin\theta}{r^3}\right) = \tan(\theta-\lambda)$$

i.e. $$1 = 2\tan\theta\tan(\theta-\lambda)$$

or $$1+3\tan\theta\tan\lambda-2\tan^2\theta = 0. \qquad (1)$$

From the triangle CPQ,

$$\frac{a}{\sin\theta} = \frac{c}{\sin(\theta-\lambda)},$$

i.e. $$\sin\theta\cos\lambda-\cos\theta\sin\lambda = \frac{c}{a}\sin\theta$$

or $$\sin\theta(\cos\lambda-c/a) = \cos\theta\sin\lambda.$$

Substitution into (1) leads to

$$\cos\lambda(\cos\lambda-c/a)^2+3\sin^2\lambda(\cos\lambda-c/a)-2\sin^2\lambda\cos\lambda = 0,$$

i.e. $$\cos^3\lambda+\sin^2\lambda\cos\lambda-3(c/a)\sin^2\lambda-2(c/a)\cos^2\lambda+(c/a)^2\cos\lambda = 0.$$

But (c/a) is small and $\lambda \approx \pi/2$, i.e. $\cos\lambda$ is also small. We therefore neglect the last two terms and retain only the terms, as a first approximation,

$$\cos\lambda-3(c/a)\sin^2\lambda = 0.$$

Exercises 7.1

1. Show that the potential at a point $r = \{x\ y\ z\}$ outside a body of uniform magnetization M is

$$\frac{\mu_0 M}{4\pi} \frac{\partial}{\partial z} \int \frac{d\tau'}{|r-r'|},$$

where r' is the position vector of the volume element $d\tau'$, and M has the direction of the z-axis. Hence or otherwise show that a uniformly magnetized sphere has the same external potential as a dipole $(\mu_0 a^3/3)M$ at its centre.

2. A solid sphere of radius a of magnetic material is uniformly magnetized with intensity M, and a small cylindrical cavity is made at the centre of the sphere, the axis of the cylinder being parallel to the direction of magnetization, and the mid-point of the axis coinciding with the centre of the sphere. The length of the cylinder is l and the radius of the cross-section b. Show that the magnetic field B at the centre of the sphere is of magnitude

$$\mu_0 \left\{ \frac{M}{3} - \frac{Ml}{\sqrt{(l^2+b^2)}} \right\}$$

and is directed along the direction of magnetization.

3. A sphere of radius a and centre O is magnetized such that the magnetization M at a point P of the sphere is

$$M = r \times (k \times r),$$

where r is the vector $\overrightarrow{OP}$ and k is a constant vector. Show that outside the sphere the magnetic field is identical with that of a dipole at O of moment $2\mu_0 a^5 k/15$.

4. A spherical shell of magnetic material, of internal and external radii a and b respectively, is uniformly magnetized to an intensity M. Show that the magnetic field at points in empty space inside and outside the shell is zero and $\frac{1}{3}\mu_0(b^3-a^3)M$ respectively.

5. A spherical specimen of magnetic material has uniform intensity of magnetization M. A cavity in the form of a thin plane disc is cut from its interior, the normal to the plane making an angle θ with the direction of M. Show that the magnetic field inside the cavity has magnitude

$$\frac{\mu_0 I}{3}(1+3\cos^2\theta)^{1/2}$$

and makes an angle ϕ with the direction of M, where

$$\cot\phi = \cot\theta - \tfrac{1}{3}\sec\theta \operatorname{cosec}\theta.$$

6. A point P lies initially within a spherical cavity inside a magnet. The shape of the cavity is then altered to a cubical form, the change being made in such a way that the cube has centre at P. Show that the magnetic field at P is unaffected by the change, provided the intensity of magnetization is uniform and remains unaffected by the change.

7. A permanent magnet in the form of a circular cylinder of length $2l$ and radius r has uniform magnetization M parallel to the axis. There is a spherical cavity with its centre at the middle point of the axis of the cylinder. Show that the field at the centre of the cavity is of magnitude

$$\mu_0\left(\frac{Ml}{\sqrt{(l^2+r^2)}} - \frac{2M}{3}\right)$$

in the direction of magnetization of the magnet.

7.2 Permeable media

This section is concerned with the magnetic counterpart of the dielectric. The basic phenomenon is the increase of the magnetic field at points inside a solenoid when a core of "soft iron" is inserted within the solenoid. This effect is of great importance in all machines or other devices which depend upon the production of strong magnetic fields in certain regions, e.g. electric motors, electromagnets, loudspeakers, magnetic relays or switches.

Briefly stated, the phenomenon is that a material such as "soft iron" becomes strongly magnetized when it is placed in a field produced, say, by a solenoid. The field arising from the magnetized core then reinforces the field of the solenoid. This corresponds to the effect of a dielectric, except that the presence of the latter tends to reduce the electrostatic effect of a charge at a distance.

In our discussion of permanent magnetization in § 7.1 we replaced the distribution $\boldsymbol{M}$ by a current distribution **curl** $\boldsymbol{M}$ inside and $\boldsymbol{M}\times\hat{\boldsymbol{n}}$ on the surface of the magnetized region. We can give a qualitative explanation of the effect in terms of molecular or atomic currents which circulate in the molecules, atoms, or in groups of them, which constitute the material.

If a small plane coil carrying a current I is placed in a magnetic field, it turns under the action of the field until its axis (i.e. the vector moment $I\alpha\hat{\boldsymbol{e}}$) is parallel to the field. In this position the field produced by the coil itself near its centre is in the same direction as the external field. Thus we should expect that many such coils, each of which is free to rotate, would take up similar positions when placed in a magnetic field and so would increase the field strength at points outside the material. (We assume here that the current I in a coil is unaffected by the change in the flux through it when the coil alters its direction. This, in general, is not the case. However, qualitatively the effect is the same so long as I does not vanish or change sign as a result of its rotation in the field.)

It is broadly true that the atoms and molecules of all substances have electrons circulating within them; these circulating electrons can be regarded as constituting small circuits in the atoms and molecules. For certain substances some of these electronic orbits give rise to a magnetic moment which is not zero for an individual atom, molecule, or group of such particles. In the absence of an external field the resultant magnetic moment of a large number of such particles is zero because the individual moments are directed at random. In the presence of a field the "molecular currents" tend to align themselves with the field and reinforce the effect of the field as described above for a single coil. Such an effect is displayed by the so-called paramagnetic and ferromagnetic substances. In the paramagnetic substances the contribution to the field from the material is comparatively small and is roughly proportional to the external field strength. For ferromagnetic substances, because of their particular atomic and crystalline structure, the effect can be very large. We refer later to some of the characteristic features of ferromagnetism, but cannot give here anything like a complete discussion of the subject. For this readers should consult special books and texts.

There is a class of (diamagnetic) substances which behave differently. They possess the property of opposing or reducing the effect of a magnetic field in which they are situated. This occurs because the establishment of a magnetic field (through the working of electromagnetic induction, see Chapter 9) affects the strengths of the currents circulating in the atoms and molecules in such a way as to reduce the resultant magnetic field. These diamagnetic effects are comparatively small; the metal bismuth can be shown to have diamagnetic properties.

On the basis of this discussion, and using the mathematical equivalence of a suitable current distribution to a distribution of magnetization, we build our theory of induced magnetism by assuming the appearance of magnetization or of equivalent Poisson currents in a medium which is placed in a magnetic field. This corresponds to the assumption of the appearance of electric dipoles, or the equivalent Poisson charge, on a dielectric placed in an electrostatic field.

7.3 Induced magnetization

First we consider a long narrow solenoid in which the "interior volume" is filled by a magnetic material, Fig. 7.4. Because the solenoid is long, the field $\boldsymbol{H}$ inside the solenoid is effectively uniform and the field $\boldsymbol{H}$ outside

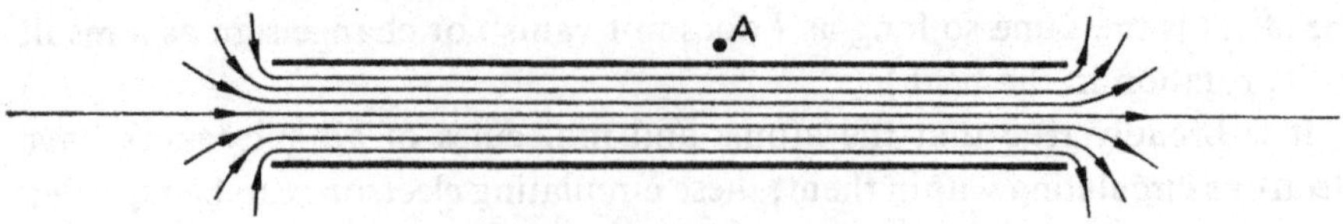

FIG. 7.4

(in the immediate vicinity of the solenoid), e.g. at A, is very small. In the absence of a core the fields, both inside and outside the solenoid, are related by $\boldsymbol{B} = \mu_0 \boldsymbol{H}$; for points well outside, the solenoid behaves as a bar magnet. For internal points the field is given by

$$|\boldsymbol{H}| = H = K,$$

where K is the strength of the surface current flowing around the walls of the solenoid.

When a core is placed in the solenoid, at points well outside the solenoid the field resembles that of a bar magnet, but one of much greater strength than before. We assume, therefore, that the core has become magnetized and so produces this augmented external field. Inside the solenoid we have

$$H = K, \quad B = \mu_0(H+M').$$

Outside, the fields are related by $\boldsymbol{B} = \mu_0\boldsymbol{H}$ and the field is augmented by that of a bar magnet of moment $\tau \boldsymbol{M}'$, where τ is the volume of the core. We assume that, inside the core, all the vectors are uniform and directed parallel to the axis of the solenoid.

Because we have assumed that $\boldsymbol{M}'$ is uniform, the equivalent Poisson current $\boldsymbol{M}'\times\hat{\boldsymbol{n}}$ augments the "true" current $\boldsymbol{K}$ and so increases the magnetic field outside the solenoid. This discussion is the magnetic counterpart of the discussion of a parallel plate condenser given in § 4.1.

We now consider a distribution of "true" current $\boldsymbol{J}, \boldsymbol{K}$ and linear currents $\boldsymbol{I}$, giving rise to a field in a region which may contain magnetizable media. We assume that, as a result of the action of the field on the elementary currents in the material, magnetization $\boldsymbol{M}'$ is induced at each point. We use the formulae established in § 7.1 to include the contributions from $\boldsymbol{M}'$ so that

$$\operatorname{div} \boldsymbol{B} = 0, \quad \mathbf{curl}\, \boldsymbol{H} = \boldsymbol{J}, \tag{7.10, 11}$$

$$\boldsymbol{H} = \boldsymbol{B}/\mu_0 - \boldsymbol{M}'. \tag{7.12}$$

The corresponding surface relations are

$$\hat{\boldsymbol{\nu}}\cdot(\boldsymbol{B}_+ - \boldsymbol{B}_-) = 0, \quad \hat{\boldsymbol{\nu}}\times(\boldsymbol{H}_+ - \boldsymbol{H}_-) = \boldsymbol{K}. \tag{7.13, 14}$$

Equations (7.11), (7.14) indicate that Ampère's circuital relation holds unchanged and may be used to calculate fields with cylindrical symmetry as in examples 2, 3 of pp. 213–214. This is the counterpart of the use of Gauss's theorem in electrostatics.

Since the magnetization $\boldsymbol{M}'$ depends upon the field, the relations (7.10), (7.11), (7.12) are insufficient to determine the field, and, as with dielectrics, there is a relation between two of the vectors $\boldsymbol{B}, \boldsymbol{H}$ and $\boldsymbol{M}'$. This relation may take several forms, the simplest such relation, corresponding to (4.12), being

$$\boldsymbol{H} = \boldsymbol{B}/\mu = \boldsymbol{B}/\mu_0 - \boldsymbol{M}', \quad \boldsymbol{M}' = \boldsymbol{B}(1/\mu_0 - 1/\mu), \tag{7.15}$$

where μ is a constant independent of the field. A medium for which such a relation holds is a *linear, isotropic medium*. A *linear, anisotropic* medium is one for which $\boldsymbol{B}$, $\boldsymbol{H}$ are in different directions, the relation between their components being linear.

The quantity μ is the *permeability* of the medium and corresponds to the permittivity of a dielectric. For paramagnetic substances μ is slightly greater than μ_0, i.e.

$$\mu = \mu_0(1+\varkappa)$$

where $\varkappa > 0$, but is small. For diamagnetic substances $\varkappa < 0$ and is small. If we write $\mu = \mu_0\mu_r$, the (dimensionless) quantity μ_r is called the *relative* permeability and is given in terms of the magnetic susceptibility $\varkappa$ by $\mu_r = 1+\varkappa$.

For the most important class of magnetic materials called *ferromagnetic materials*, which comprises "soft" iron, cobalt, and a large number of alloys, the simple linear relation (7.15) does not represent the relation between $\boldsymbol{B}$, $\boldsymbol{H}$, $\boldsymbol{M}'$ even approximately. This is due to the fact that $\boldsymbol{M}$ depends not only upon the prevailing values of $\boldsymbol{B}$, $\boldsymbol{H}$ but also upon the history of the specimen. We discuss this phenomenon of *hysteresis* in a little more detail in § 7.4.

If current flows in an infinite medium of constant permeability μ (i.e. μ depends neither upon the field strength nor upon position), the equation relating $\boldsymbol{A}$ and the current is

$$\textbf{curl}\left(\frac{1}{\mu}\,\textbf{curl}\,\boldsymbol{A}\right) = \boldsymbol{J}.$$

$$\therefore\quad \nabla^2\boldsymbol{A} = -\mu\boldsymbol{J}.$$

This equation replaces μ_0 in eqn. (6.21) with μ. Hence all expressions for inductance must be multiplied by the factor $\mu/\mu_0\,(=\mu_r)$. In practice the medium of permeability $\mu(\neq \mu_0)$ is never of infinite extent but is confined to those regions where the field is large such as the interior of a solenoid, or the core of a transformer. When the field outside such regions is negligibly small, the effect of using cores of a highly permeable material is to multiply all inductances by this factor μ_r. (Arrangements in which the magnetic field is effectively confined to certain regions of high permeability are called *magnetic circuits*, and the field strengths and magnetic fluxes satisfy laws similar to those for electric currents.)

7.4 Hysteresis

When a specimen of magnetizable material is placed as a core in a solenoid, by varying the current in the solenoid and by measuring the field produced at a suitable external point it is possible to obtain a curve which relates two of the three vectors $\boldsymbol{B}$, $\boldsymbol{H}$ and $\boldsymbol{M}'$.

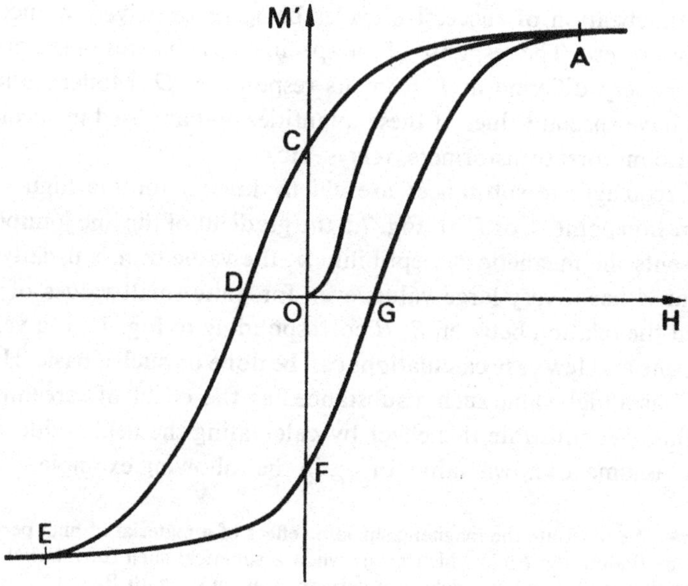

FIG. 7.5

A typical curve relating M', H, shown in Fig. 7.5, is called a hysteresis curve. It shows that $M' = |\boldsymbol{M}'|$ is a double (or multiple valued) function of H. The curve is symmetrical and displays certain important features. The section OA relates M', H as the fields are increased from zero at O, with the specimen unmagnetized, to ***saturation*** at A. At A further increases in the magnetizing current produce no further magnetization of the specimen; roughly speaking, at A all the elementary circulating currents are aligned with their axes parallel to the field. As the magnetizing current is reduced to zero $|\boldsymbol{B}|$ does not fall to zero but takes the value OC representing the ***residual magnetization***. It requires a magnetizing current in the negative direction to remove the magnetization. Further changes of the current produce saturation, in the opposite direction at E, then a residual magnetization at F, and complete the ***hysteresis cycle*** when the current reaches the value corresponding to A once again.

Different materials have hysteresis curves with different characteristics but the same general shape; they may differ in the saturation value, or in the lengths OC, OA. The effect of the history of the specimen on its state at any time is illustrated by considering the "unmagnetized" states represented by O and D. To reach the state at D the magnetizing field is reduced to a negative value from saturation at A; to reduce the system to the unmagnetized state at O it must be taken repeatedly around cycles of magnetization, the maximum magnetization of successive cycles being progressively reduced from saturation to zero. The response of the specimen to changes in the magnetizing field is very different at D from its response at O. Modern alloys are made to have special values of these quantities and are used in special ways for electric motors, transformers, relays, etc.

The ferromagnetic substances are chiefly known for the high value of the saturation point A, or E. In Fig. 7.5 the gradient of the line joining O and A represents the magnetic susceptibility $\varkappa$; the value of $\varkappa$ is usually $\gg 1$ so that μ_r at A has a very large value, even for quite small values of H. It is clear that the relation between B, H corresponding to Fig. 7.5 is a very complicated one and few easy calculations can be done on such a basis. However, when μ_r has a high value such a substance has the effect of screening magnetic fields. We illustrate this effect by calculating the field inside a cavity when we assume a known value for μ_r, in the following example.

Example. We illustrate the magnetic shielding effect of a material of high permeability ($\mu \gg \mu_0$) by finding the fields which occur when a spherical shell (of internal radius a, external radius b) is placed in a field of otherwise uniform strength $\boldsymbol{B}_0$.

Since there are no currents in the system, only distributions of magnetization, we use the method of the scalar potential (see example 1, p. 264) to calculate the field strengths in the three regions. We assume a scalar potential Ω, where

$$\boldsymbol{B} = \mu \boldsymbol{H} = -\mu\, \mathbf{grad}\, \Omega, \qquad \text{inside the material,}$$
$$\boldsymbol{B} = \mu_0 \boldsymbol{H} = -\mu_0\, \mathbf{grad}\, \Omega, \qquad \text{elsewhere,}$$

with Ω given by

$$\Omega = \begin{cases} \Omega_2 = A_2 r \cos\theta, & \text{for} \quad 0 < r < a, \\ \Omega_1 = \left(\dfrac{B_1}{r^2} + C_1 r\right)\cos\theta, & \text{for} \quad a < r < b, \\ \Omega_0 = \left(\dfrac{D_0}{r^2} - \dfrac{B_0 r}{\mu_0}\right)\cos\theta, & \text{for} \quad b < r. \end{cases}$$

This potential function corresponds to fields, all directed parallel to $\boldsymbol{B}_0$ of the magnitudes given below (see Fig. 7.6) at the various positions:

inside the cavity	$-\mu_0 A_2$;
in the material at E	$-\mu(B_1/a^3 + C_1)$;
in the material at F	$-\mu(B_1/b^3 + C_1)$;
outside the material at F	$B_0 - \mu_0 D_0/b^3$.

In the method of solution (see example 1, p. 264) the condition for continuity of Ω at $r = a$ gives

$$A_2 = B_1/a^3 + C_1$$

and continuity at $r = b$ gives

$$B_0 - \mu_0 D_0/b^3 = \mu_0(B_1/b^3 + C_1).$$

This shows that the field strength inside the material is μ/μ_0 times that outside the material.

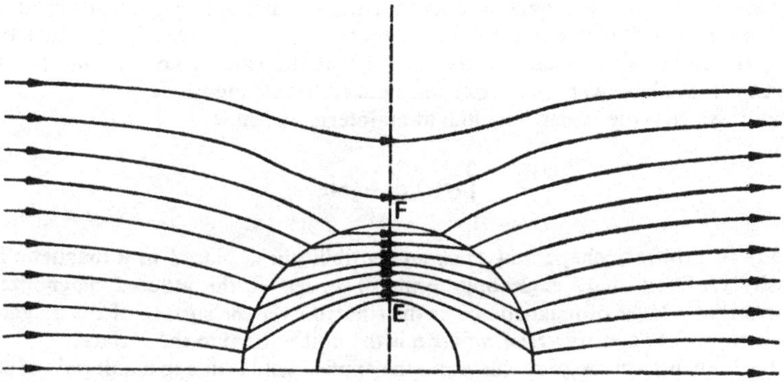

FIG. 7.6

The solution of the boundary conditions (which we omit here) gives

$$A_2 = -\frac{9\mu b^3}{b^3(\mu+2\mu_0)(2\mu+\mu_0)-2a^3(\mu-\mu_0)^2} B_0.$$

$$\therefore \quad -\mu_0 A_2 = \frac{9}{\left(1+\frac{2\mu_0}{\mu}\right)\left(2+\frac{\mu_0}{\mu}\right)-\frac{2a^3}{b^3}\left(1-\frac{\mu_0}{\mu}\right)^2}\cdot\frac{\mu_0 B_0}{\mu}.$$

This shows that the field strength inside the cavity, even when a is a substantial fraction of b (i.e. a fairly thin shell), is a very small fraction of the external field strength B_0, because μ_0/μ is very small for a material of high permeability. The diagram in Fig. 7.6 shows how the field is concentrated into the material of the shell. (The figure shows a section through the centre of the shell, the complete figure being symmetrical about the diameter of the semicircle shown.)

Miscellaneous Exercises VII

1. (i) Prove that the magnetic field of a body magnetized with intensity $\boldsymbol{I}$ at an external point is equal to that of a volume distribution of fictitious magnetic poles of density $-\text{div}\,\boldsymbol{I}$ together with a surface density of fictitious poles $\boldsymbol{I}\cdot\boldsymbol{n}$ on the surface of the body, $\boldsymbol{n}$ being the outward-drawn normal.

(ii) Show that the magnetic scalar potential of a normally magnetized shell of strength ϕ at an external point P is

$$\frac{1}{4\pi}\oiint_S \phi\left(\mathbf{grad}_Q \frac{1}{r}\right)\cdot\mathbf{d}\boldsymbol{S},$$

where Q is a point of the shell, $r = PQ$ and $\mathbf{d}\boldsymbol{S}$ is an oriented element of the surface of the shell S.

(iii) A body is uniformly magnetized with intensity $\boldsymbol{I}$. Show that the potential at a field point outside it is the same as the potential of a normally magnetized shell coinciding with the surface of the body and of strength $\boldsymbol{I}\cdot\boldsymbol{r}$, where $\boldsymbol{r}$ is the position vector of a point of the surface of the body relative to the field point.

2. A sphere of radius a is magnetized so that the intensity of magnetization at a point whose position vector is $\boldsymbol{r}$ referred to the centre is $f(r)\boldsymbol{r}+\boldsymbol{c}$, where $f(r)$ is any function of $r = |\boldsymbol{r}|$ and $\boldsymbol{c}$ is a constant vector. Prove that the field at an external point is the same as if the sphere were uniformly magnetized to the intensity $\boldsymbol{c}$.

Show also that the scalar potential at an internal point is

$$\int_r^a rf(r)\,dr+\tfrac{1}{3}\boldsymbol{r}\cdot\boldsymbol{c}.$$

3. A body of arbitrary shape and *small* susceptibility $\varkappa$ is placed in a magnetic field of intensity $\boldsymbol{H}$. Show that, neglecting terms of order $\varkappa^2$, the induced magnetization is equivalent to a layer of magnetic poles distributed over the surface of the magnetizable body with surface density $\varkappa\boldsymbol{H}\cdot\boldsymbol{n}$, where $\boldsymbol{n}$ is the unit normal to the surface.

If the body is a sphere of radius a and the applied field is due to a unit pole of strength m at a distance c from the centre of the sphere, show that the force exerted on the body is

$$F = (\mu_0\varkappa m^2/16c^2)\,(\sinh 2y-2y),$$

where $\tanh \tfrac{1}{2}y = a/c$.

4. Show that the magnetic vector potential for a uniform magnetic field $\{0\ 0\ B\}$ has components A_r, A_θ, A_ϕ in spherical polar coordinates given by

$$A_r = 0, \quad A_\theta = 0, \quad A_\phi = \tfrac{1}{2}Br\sin\theta.$$

Show also that, for the field due to a small magnet of moment M at the origin pointing along the axis of the coordinates, the vector potential is

$$A_r = 0, \quad A_\theta = 0, \quad A_\phi = (\mu_0/4\pi)Mr^{-2}\sin\theta.$$

A sphere is uniformly magnetized with permanent magnetization of intensity $\boldsymbol{M}$ and is placed *in vacuo* in a uniform magnetic field of flux density $\boldsymbol{B}$. Prove that the resultant couple on the sphere is $\tfrac{4}{3}\pi a^3\boldsymbol{M}\times\boldsymbol{B}$.

5. A sphere has uniform magnetization $\boldsymbol{M}$. Show that, at all points in the interior of the sphere, $\boldsymbol{B} = -2\mu_0\boldsymbol{H} = \tfrac{2}{3}\mu_0\boldsymbol{M}$.

Hence or otherwise show that, when a sphere made of material of constant permeability μ is placed in a uniform external magnetic field $\boldsymbol{B}_0$, the magnetization is

$$\frac{3(\mu-\mu_0)}{\mu_0(\mu+2\mu_0)}\boldsymbol{B}_0.$$

6. When the plane of a magnetic dip-circle is vertical and inclined to the magnetic meridian at an angle α the needle comes to rest at an angle θ to the horizontal. When the plane is turned through a right angle about a vertical axis the needle comes to rest at an angle ϕ to the horizontal. Prove that $\tan\alpha = \tan\theta\cot\phi$ and that the angle of dip δ is given by

$$\cot^2\delta = \cot^2\theta+\cot^2\phi.$$

If T_1, T_2 are the periods of small oscillations of the needle in the two positions and T is the period when the needle oscillates in the magnetic meridian, prove that

$$(\cos^2\theta)/T_1^4+(\cos^2\phi)/T_2^4 = (\cos^2\delta)/T^4.$$

CHAPTER 8

POTENTIAL PROBLEMS

8.1 Fundamental theorems and techniques

In this chapter we illustrate the results of Chapters 1–7 by solving a selection of problems in electrostatics, magnetostatics and steady current flow and also give a brief account of the theory of gravitation. Special techniques are introduced for solving different classes of problem and the illustrative examples contain some important theorems and results which should be regarded as an integral part of the text.

We state the problem of electrostatics thus: Given

(1) the dielectric constant K and the density of space charge ϱ everywhere,

(2) the fixed positions of point charges of known strengths (these charges may be regarded as distributed on small spheres),

(3) the positions of all conductors, for each of which the total charge *or* the potential is known,

then the electrostatic potential V satisfies the following conditions:

(i) $$\operatorname{div}\,(K\,\mathbf{grad}\,V) = -\varrho/\varepsilon_0 \tag{8.1}$$

except at surface (or point) charges,

(ii) $$\oiint_{S_i} K\,\frac{\partial V}{\partial n}\,\mathrm{d}S = -Q_i/\varepsilon_0, \tag{8.2}$$

S_i being a surface enclosing a conductor carrying charge Q_i,

(iii) $$V = \text{constant on each conductor,} \tag{8.3}$$

(iv) if relevant, the standard boundary conditions

$$V = O(1/R), \quad E = O(1/R^2) \quad \text{at infinity,} \tag{8.4}$$

(v) $$V \text{ is continuous at a discontinuity of } K, \tag{8.5}$$

(vi) near a point charge Q,

$$V = \frac{Q}{4\pi K\varepsilon_0 r} + O(1). \tag{8.6}$$

If the field is of infinite extent and is uniform at great distances from any conductors, say of magnitude E_0 directed along Oz, then the standard boundary condition at infinity is replaced by $V = -E_0 z + O(1/R)$.

Our problem is to determine the function V which satisfies the conditions (i)–(vi). Essentially, we guess trial solutions and adjust functions and disposable parameters so that the conditions (8.1)–(8.6) are satisfied. Then we appeal to the uniqueness theorems (p. 74, Vol. 1, etc.) to justify our results. Before proceeding further, the reader should re-examine the examples of Vol. 1, § 3.2, etc., where this basic idea was first introduced. A more detailed justification for cases with spherical or axial symmetry is given in § 8.3. First, however, the reader must become thoroughly familiar with the method, and to serve this end we solve now some further problems. Usually the method involves obtaining separable solutions of Laplace's equation (or an equation of similar type) in an orthogonal coordinate system in which the surface of at least one of the conductors or a surface of discontinuity assumes a particularly simple form.

Example 1. An earthed conducting sphere of radius a is coated with a uniform thickness $b-a$ of dielectric of dielectric constant K. The sphere and dielectric are placed in a uniform electric field $\mathbf{E}$. Show that the change in the field outside the dielectric is the same as that produced by an electric dipole of moment

$$4\pi\varepsilon_0 E b^3 \frac{(2K+1)a^3+(K-1)b^3}{2(K-1)a^3+(K+2)b^3}$$

at the centre of the sphere. Show also that the surface density of charge at a point on the conductor is

$$\frac{9K\varepsilon_0 E \cos\theta}{[K+2+2(K-1)a^3/b^3]},$$

where θ is the angle between the radius to the point and the direction of the field $\mathbf{E}$.

Here we use spherical polar coordinates (r, θ, ψ) with O at the centre of the sphere and the polar axis Oz along the direction of the imposed field $\mathbf{E}$, Fig. 8.1. In the absence of the sphere, the potential of the uniform field can be taken to be $-Ez = -Er\cos\theta$. But, except on the spheres $r = a$, $r = b$, there is no charge and so the potential V satisfies Laplace's equation $\nabla^2 V = 0$. Also the boundary conditions to be satisfied are

$$V = 0 \quad \text{on} \quad r = a, \tag{1}$$

$$V \text{ is continuous at } r = b, \tag{2}$$

$$D_r \text{ is continuous at } r = b, \tag{3}$$

$$V \to -Er\cos\theta \quad \text{at infinity}. \tag{4}$$

In future we use the notation $[V]_{r=b}$ to denote $(V)_+ - (V)_-$ so that boundary conditions (2) and (3) can be written

$$[V]_{r=b} = 0, \quad [D_r]_{r=b} = 0.$$

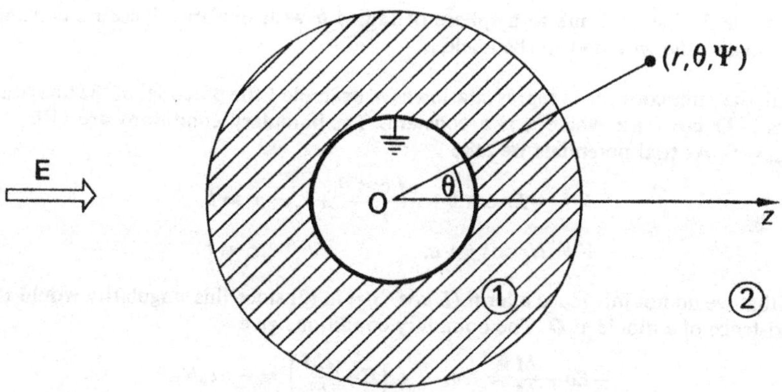

FIG. 8.1

Since boundary conditions (1), (2), (3) are imposed at surfaces r = constant and the potential of the exciting field involves θ through the factor $\cos\theta$ only, we seek a solution of the problem by adding to the exciting potential $-Er\cos\theta$ other (disturbance) potential functions which are obtained as separable solutions of Laplace's equation and involve θ through the factor $\cos\theta$ only. Such solutions are $r\cos\theta$, $(\cos\theta)/r^2$. In fact, we take

$$V = -Er\cos\theta + \frac{A\cos\theta}{r^2} \quad \text{for} \quad b \leqslant r, \tag{5}$$

$$V = Br\cos\theta + \frac{C\cos\theta}{r^2} \quad \text{for} \quad a \leqslant r \leqslant b, \tag{6}$$

$$V = 0 \quad \text{for} \quad r \leqslant a. \tag{7}$$

Here A, B, C are constants to be determined and we introduce the term $(A\cos\theta)/r^2$ only into (5) since $V \to -Er\cos\theta$ at great distances from the sphere. Boundary conditions (1), (2), (3) give respectively

$$Ba + \frac{C}{a^2} = 0, \tag{8}$$

$$Bb + \frac{C}{b^2} = -Eb + \frac{A}{b^2}, \tag{9}$$

$$\varepsilon_0 K\left(-B + \frac{2C}{b^3}\right) = \varepsilon_0\left(E + \frac{2A}{b^3}\right). \tag{10}$$

[Note that the factor $\cos\theta$ has been cancelled throughout.] Solving (8), (9), (10), we find

$$\frac{A}{(2K+1)a^3+(K-1)b^3} = \frac{B}{-3} = \frac{C}{3a^3} = \frac{b^3E}{(2K-1)a^3+(K+2)b^3}.$$

Outside the dielectric the additional field, derived from a potential $(A\cos\theta)/r^2$, is that due to a dipole of moment $4\pi\varepsilon_0 A$ situated at O and directed along Oz. This is equivalent to the stated result (cf. § 2.10). By the uniqueness theorem this superposed field gives the unique solution of the problem.

The induced surface charge σ at the surface of the conductor is given by (cf. § 4.2)

$$\sigma = -K\varepsilon_0\left(\frac{\partial V}{\partial r}\right)_{r=a},$$

i.e.

$$\sigma = \left(\frac{2C}{a^3} - B\right)K\varepsilon_0\cos\theta,$$

which leads to the required result.

Example 2. The field due to a sphere of radius b with uniform dielectric constant K placed in an otherwise undisturbed field $\mathbf{E}$.

With the same coordinate representation as in example 1 the potential of the undisturbed field is $-Er\cos\theta+\alpha$ where α is a constant. The boundary conditions are $[V]_{r=b}=0$, $[D_r]_{r=b}=0$. As trial potentials we take

$$V=-Er\cos\theta+\frac{M\cos\theta}{r^2}+\alpha, \qquad r\geqslant b, \tag{1}$$

$$V=Nr\cos\theta+a, \qquad r\leqslant b. \tag{2}$$

Note that we do not introduce a term $(L\cos\theta)/r^2$ in (2) since this singularity would imply the existence of a dipole at O. The boundary conditions give

$$-Eb+\frac{M}{b^2}=Nb, \qquad \varepsilon_0\left(E+\frac{2M}{b^3}\right)=-K\varepsilon_0 N,$$

whence

$$\frac{M}{(K-1)b^3}=\frac{N}{-3}=\frac{E}{(K+2)}.$$

As before, by the uniqueness theorem, this gives the unique solution. [As a check, the case $K=1$ corresponds to the case of an undisturbed field.]

Notes

1. The disturbance field outside the dielectric is the same as that of a dipole of moment $4\pi\varepsilon_0 M$ situated at O and directed along Oz. Inside the dielectric the field is uniform, of intensity N and directed along Oz.

2. If we let $K\to\infty$, so that $|\mathbf{E}|\to 0$ within the sphere, we obtain the solution of the case in which an uncharged conductor at potential α is placed in a uniform field. The reader should verify this result by direct analysis.

Example 3. A sphere of radius a whose dielectric constant is $K_0(a/r)^2$, where K_0 is a constant and r is the distance from the centre of the sphere, is placed in a uniform electric field of intensity $\mathbf{E}$. Show that, at points outside the sphere, the part of the electric field due to the presence of the dielectric is the same as that of a dipole of moment

$$4\pi\varepsilon_0\frac{2K_0-1}{2K_0+2}Ea^3$$

situated at the centre of the sphere.

Show also that the equation of a line of force inside the sphere is

$$r\sin^2\theta=\text{constant},$$

where θ is the angle made by the radius vector and the direction of $\mathbf{E}$.

We use the same coordinate system as in example 1. Also the undisturbed field can be taken as $-Er\cos\theta$. This time, however, the potential within the sphere satisfies the equation

$$\text{div}\,(K\varepsilon_0\,\mathbf{grad}\,V)=0. \tag{1}$$

Since K is not constant, (1) *does not reduce to Laplace's equation.* However, the form of the exciting potential, together with the fact that the boundary conditions $[V]_{r=a}=0$, $[D_r]_{r=a}=0$ are imposed at the surface $r=a$, suggests that we seek a solution of (1) in the form

$$V=R(r)\cos\theta, \tag{2}$$

where $R(r)$ is a function of r only. Using the formulae for divergence and gradient in spherical polars, eqn. (1) becomes

$$\frac{1}{r^2 \sin\theta}\left\{\frac{\partial}{\partial r}\left(\frac{K_0\varepsilon_0 a^2 r^2 \sin\theta}{r^2}\frac{\partial V}{\partial r}\right)+\frac{\partial}{\partial\theta}\left(\frac{K_0\varepsilon_0 a^2 r\sin\theta}{r^2}\frac{1}{r}\frac{\partial V}{\partial\theta}\right)\right\}=0.$$

Then assumption (2) gives as the equation for R,

$$\frac{d^2R}{dr^2}-\frac{2R}{r^2}=0 \tag{3}$$

after cancelling the factor $\sin\theta\cos\theta$ throughout. The solution of (3) is

$$R = Ar^2+B/r,$$

where A, B are arbitrary constants. Since the potential can have no singularity at O, $B = 0$. We take, therefore, the potential for $r < a$ to be

$$V = Ar^2\cos\theta. \tag{4}$$

Outside the dielectric the potential satisfies Laplace's equation $\nabla^2 V = 0$ and so we try

$$V = -Er\cos\theta+\frac{C\cos\theta}{r^2} \quad \text{for} \quad r > a. \tag{5}$$

The boundary conditions give

$$Aa^2 = -Ea+\frac{C}{a^2}, \qquad -2K_0\varepsilon_0 aA = \varepsilon_0\left(E+\frac{2C}{a^3}\right),$$

whence

$$\frac{A}{-3/a}=\frac{C}{(2K_0-1)a^3}=\frac{E}{2K_0+2}.$$

Again the uniqueness theorem implies that this is the only possible solution. The disturbance field for $r > a$ is equivalent to that of a dipole of moment $4\pi\varepsilon_0 C$ situated at O and directed along Oz.

The differential equations of the lines of force for $r < a$ are

$$\frac{dr}{\dfrac{\partial V}{\partial r}}=\frac{r\,d\theta}{\dfrac{1}{r}\dfrac{\partial V}{\partial\theta}}$$

or, using (4),

$$\frac{dr}{2Ar\cos\theta}=\frac{r^2\,d\theta}{-Ar^2\sin\theta},$$

i.e.

$$dr/r = -2\cot\theta\, d\theta.$$

This integrates to give $r = \lambda\operatorname{cosec}^2\theta$ as required.

Example 4. **Earnshaw's theorem** and associated results.

If the electric field in vacuo is due solely to charged conductors, then the greatest and least values of the potential occur on the conductors.

Since $\nabla^2 V = 0$ inside and outside the conductors this result follows from the mean value theorem for harmonic functions.

Notes (under the conditions of Earnshaw's theorem)

1. If the field extends to infinity and all charges are of the same sign, either the greatest or the least value of V will be at infinity.

2. It follows that if there exists positive charge in the field at least one of the conductors must be positively charged everywhere on its surface; this conductor is the one on which V attains its greatest value. A similar result holds if there is negative charge in the field. This result follows by considering the fall of potential along the lines of force starting on positive charge and ending on negative charge.

3. Since V cannot have a minimum at a point P not on a conductor, the potential energy, QV, of a point charge Q cannot have a minimum at P. Hence a point charge cannot rest in stable equilibrium in the electrostatic field.

4. Since, referred to cartesian coordinates,

$$\mathbf{E} = \{E_x \ \ E_y \ \ E_z\} = -\{\partial V/\partial x \ \ \partial V/\partial y \ \ \partial V/\partial z\},$$

$$\nabla^2 E_x = \nabla^2\left(-\frac{\partial V}{\partial x}\right) = -\frac{\partial}{\partial x}(\nabla^2 V) = 0.$$

It follows by analysis similar to that given above that E_x cannot have a maximum or a minimum at a point within the field. *It follows that* $|\mathbf{E}|$ *attains its greatest value on one of the conductors bounding the field.*

We prove this theorem as follows: Suppose the theorem is false so that $|\mathbf{E}|$ attains its greatest value at an internal point O. Choose the coordinate axes $Oxyz$ so that, at O, $\mathbf{E} = \{E \ 0 \ 0\}$. Then, since E_x cannot have a maximum or minimum at a point within the field, at some point, P say, near O, $|E_x|$ must exceed $|\mathbf{E}|$.

$$\therefore \quad |\mathbf{E}|_P = \sqrt{\{(E_x)_P^2+(E_y)_P^2+(E_z)_P^2\}} > |\mathbf{E}| = |\mathbf{E}_O|,$$

contrary to our supposition that $|\mathbf{E}|$ has a maximum at O. The theorem must therefore be true.

5. *The field* $\mathbf{E}$ *can vanish only at isolated points outside charged conductors and not within a volume of finite extent.* [This implies that the positions of equilibrium for a point charge form a discrete set.] Assume the theorem to be false so that, in a domain D containing O, $\mathbf{E}$ vanishes, and so V is constant, V_0, within D. Outside D, the potential V is not equal to V_0. Suppose that it exceeds V_0 in a domain D' abutting on D. This domain D' must be of finite extent since V is continuous. Then we can draw a surface Σ in $D+D'$ in which V in some finite region exceeds V_0. Nowhere on Σ is V less than V_0. But the mean value theorem for harmonic functions implies that the mean value of V on Σ is V_0. Hence we have a contradiction and the assumption that the theorem is false cannot hold. Hence V cannot be constant in a region of non-vanishing extent between the conductors. It follows that $\mathbf{E} = -\mathbf{grad}\ V$ vanishes only at isolated points.

Exercises 8.1

Exercises 1–5 are essentially theorems concerning electrostatic potential functions *in vacuo*. Hints are given for their solutions in the answers to the exercises.

1. If the electrostatic potential, V, is constant, V_0, over a surface Σ not containing charge, then the potential takes the same constant value everywhere inside Σ.

2. Show that the mean value of the electrostatic potential over the surface of a sphere is equal to the potential at the centre of the sphere due to the charge outside the sphere plus the total charge inside the sphere divided by its radius.

3. If the electrostatic potential is constant throughout a region R unoccupied by charge, then the potential is constant over any region which can be reached from R without crossing charge.

4. If two different distributions of charge have the same potential (not necessarily constant) over a closed surface Σ not containing charge, they have the same potential inside the surface and over all space which can be reached without crossing charge.

5. If two electrostatic distributions of charge have the same potential over any surface Σ enclosing both distributions, then they have equal potentials at all points external to Σ.

6. An uncharged conducting sphere of radius R is introduced into a uniform electric field of intensity E. Verify that the variable part of the potential at any point P outside the sphere is given by

$$E\left(\frac{R^3}{r^2}-r\right)\cos\theta,$$

where r is the distance of P from the centre O of the sphere, and θ is the angle between OP and the original direction of the field.

Find the smallest charge which, when placed on the sphere, will ensure that its surface charge density is nowhere negative.

7. A hollow sphere of material, of dielectric constant 3, whose outer radius is double the inner, is introduced into a uniform field of electric force E. Prove that the field in the hollow is $\frac{27}{34}E$.

8. A sphere of radius a and centre O carries a non-uniform distribution of surface charge Inside the sphere the resulting electric field is of uniform strength E_0 in the direction opposite to OZ. Outside the sphere the field is equivalent to that of a dipole of moment E_0a^3 at O along OZ. Find the surface charge density at a point P of the sphere such that the angle ZOP is θ.

9. A spherical shell of dielectric has inner and outer radii a, λa. The dielectric constant is $(\lambda a/r)^2$, r being the distance from the centre. The shell is placed in a uniform electric field E. Show that in the central cavity the field is

$$\frac{9\lambda^4E}{4\lambda^5+4\lambda^3+2\lambda^2-1}.$$

10. Prove that the necessary and sufficient condition for the family of surfaces

$$f(x, y, z) = \lambda$$

(where λ is a variable parameter) to be equipotentials in a region of space where there is no charge is that

$$\nabla^2 f/(\mathbf{grad}\, f)^2$$

is a function of f. Deduce the relation between the potential and λ.

Show that the surfaces S_λ

$$\frac{x^2}{a^2+\lambda}+\frac{y^2}{b^2+\lambda}+\frac{z^2}{c^2+\lambda} = 1, \quad \text{where} \quad \lambda \geqslant 0,$$

can be equipotentials. Hence, or otherwise, prove that if an ellipsoidal conductor whose principal semi-axes are a, b, c is charged so that its potential is V_0, the potential on S_λ is

$$V_0\int_\lambda^\infty \frac{d\theta}{\sqrt{\{(a^2+\theta)(b^2+\theta)(c^2+\theta)\}}}\Bigg/\int_0^\infty \frac{d\theta}{\sqrt{\{(a^2+\theta)(b^2+\theta)(c^2+\theta)\}}}.$$

8.2 The method of images

1. The technique of the method

Here we describe the method of images in the context of electrostatics but the method is applicable to magnetostatics, steady current flow and hydrodynamics [see *Elementary Classical Hydrodynamics*, Commonwealth Library, Pergamon Press, pp. 81, 120, 140]. Consider an electrostatic field filling all space due to a finite system of fixed point charges, volume charges, charged conductors and dielectrics. The equipotential surfaces must be closed surfaces (apart from the equipotential surface $V = 0$ in the case when the field is of infinite extent). If any equipotential surface $V = V_0$ (constant) is replaced by a conducting surface S at potential V_0 the uniqueness theorem implies that the field on each side of S remains unchanged. This suggests a possible method of solving the problem of an electrostatic system on one side (2) of a conductor S as follows:

Seek a system of charges, etc., on the *other side* (1) of the conductor S which makes S an equipotential surface (at the correct potential if this is prescribed). Then the field on side (2) due to *all the charges, etc.*, satisfies all the conditions of the uniqueness theorem and gives the solution of the problem. The system of charges, etc., on side (1) is called the *image system* in S of the charges, etc., on side (2).

2. Electrostatic images of point charges *in vacuo*

Suppose that S is an equipotential surface $V = V_0$ in the field of the point charges Q_i at A_i $(i = 1, 2, \ldots, m)$, Q'_j at A'_j $(j = 1, 2, \ldots, n)$, where the unprimed points lie on side (2) of S and the primed points lie on the other

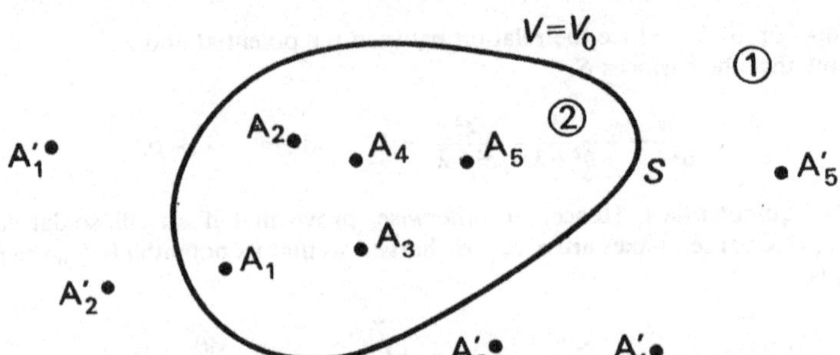

Fig. 8.2

side (1), Fig. 8.2. Then the potential within S remains unaltered when S is replaced by a conductor at potential V_0. Hence the potential within S when it is maintained at potential V_0 and contains the specified charges at the unprimed points is identical with that of the $m+n$ charges at A_i, A'_j. The set of charges Q'_j forms the image system in S of the set Q_i. Conversely the system Q_i forms the image system in S (maintained at potential V_0) of the system Q'_j.

The solid angle theorem

We derive now a theorem which can be used to determine the induced charge on a finite area of side (2) of the conductor S described above. Let $\boldsymbol{r}_1$ be the position vector, referred to A_1, of a point P of S and let $\hat{\boldsymbol{n}}$ be the unit normal to S at P directed from inside to outside S. Then the field at P due to Q_1 is $\boldsymbol{E}_1 = Q_1\boldsymbol{r}_1/(4\pi\varepsilon_0 r_1^3)$ and the induced charge density at P_1 on the ***inside surface*** of S is given by $\sigma_1 = -\varepsilon_0\hat{\boldsymbol{n}}\cdot\boldsymbol{E}_1$. It follows that the charge induced on the element dS of the inner surface of S is

$$-\varepsilon_0\hat{\boldsymbol{n}}\cdot\boldsymbol{E}_1\, \mathrm{d}S = -(Q_1\, \mathrm{d}S\cos\theta)/(4\pi r_1^2) = -Q_1\, \mathrm{d}\omega_1/(4\pi)$$

where $\mathrm{d}\omega_1$ is the element of solid angle subtended at A_1 by dS. It follows that the charge induced on the area T of the inner surface of S due to the field of Q_1 is $-Q_1\Omega_1/(4\pi)$, where Ω_1 is the solid angle subtended by T at A_1. Similar results hold for other internal charges. For external (primed) charges, however, the field is directed through S from side (1) into side (2) and so the contribution of Q'_1 to the total charge is $+Q'_1\Omega'_1/(4\pi)$. By addition we obtain the total induced charge on T as

$$Q = \frac{1}{4\pi}\left(\sum_{j=1}^{n} Q'_j\Omega'_j - \sum_{i=1}^{m} Q_i\Omega_i\right). \tag{8.7}$$

We use this theorem in the following examples. In each case appeal should strictly be made to the uniqueness theorem, although for brevity we do not explicitly do this in the worked examples.

Example 1. **A point charge Q is at a point A distant a from a semi-infinite earthed solid conductor with a plane face S. Show that the surface density of charge at a point P on S is equal to**

$$-Qa/\{2\pi(a^2+x^2)^{3/2}\},$$

where x is the distance from P to the foot of the perpendicular from A onto S. Verify by integration that the total induced charge on the plane is $-Q$.

Show that the lines of force, which start on the point charge and make an angle ϕ or less with the normal from A to the plane (not with the normal produced), end on a circular region of the plane whose radius subtends an angle θ at A, where

$$\sin\tfrac{1}{2}\phi = 2^{1/2}\sin\tfrac{1}{2}\theta.$$

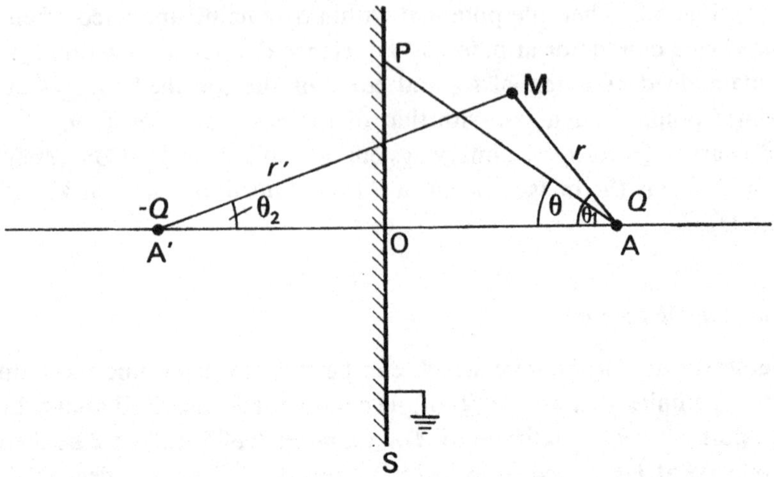

FIG. 8.3

Consider the field to the right of S due to a point charge Q at A and a point charge $-Q$ at the optical image A' of A in S, Fig. 8.3. [In this case AOA' is perpendicular to S and O is the mid-point of AA'.] Then the potential at a field point M is

$$V = \frac{1}{4\pi\varepsilon_0}\left(\frac{Q}{r} - \frac{Q}{r'}\right), \tag{1}$$

where r, r' are respectively the distances AM, $A'M$. But this potential function is such that $V = 0$ on S (where $r = r'$) satisfies Laplace's equation to the right of S, except at A ($r = 0$) where $V = Q/(4\pi\varepsilon_0 r) + O(1)$ in the neighbourhood, and satisfies the standard boundary conditions at infinity. Hence, by the uniqueness theorem, V is the unique potential for the field of the charge Q in front of the earthed conductor. In fact, the image of a point charge in an infinite earthed plane is an equal and opposite charge at the optical image point.

At P, where angle $OAP = \theta$ and $OP = x$, the electric field is normal to the plane, is directed into the plane and is of magnitude $(2Q\cos\theta)/(4\pi\varepsilon_0 AP^2)$. Hence the surface density of induced charge is

$$\sigma = -\varepsilon_0 E_n = \frac{-Q\cdot OA}{2\pi AP^3} = \frac{-Qa}{2\pi(a^2+x^2)^{3/2}}$$

as required. The total charge Q_1 on the plane S is obtained by calculating the charge on a ring of radius x and width δx, i.e. $2\pi x\sigma\,\delta x$, and integrating w.r. to x from 0 to ∞. We find

$$Q_1 = -Qa\int_0^\infty \frac{x\,dx}{(a^2+x^2)^{3/2}} = \left[\frac{Qa}{\sqrt{(a^2+x^2)}}\right]_0^\infty = -Q$$

as required. This result is obvious by eqn. (8.7).

With the usual notation (Vol. 1, p. 47) the equation of a line of force is

$$Q\cos\theta_2 - Q\cos(\pi-\theta_1) = \text{constant}.$$

For the lines which leave A at the angle ϕ the right-hand side is $Q(1+\cos\phi)$. Where the line of force reaches the plane $\theta_1 = \theta_2$, i.e.

$$2\cos\theta_1 = (1+\cos\phi)$$

or

$$2\sin^2(\theta_1/2) = \sin^2(\phi/2).$$

The stated result follows. The force exerted on the charge at A is that due to a charge $-Q$ at A', i.e. an attraction $Q^2/(16\pi\varepsilon_0 a^2)$.

Example 2. An earthed conductor is formed of two planes inclined at an angle $\pi/2$. The planes are bounded by their line of intersection and are of very great area. A point charge Q is placed at P between the two planes such that the perpendicular from P to the line of intersection of the planes makes angles $\pi/6$, $\pi/3$ with the planes. Show that the charges induced on the planes are $-2Q/3$, $-Q/3$ respectively.

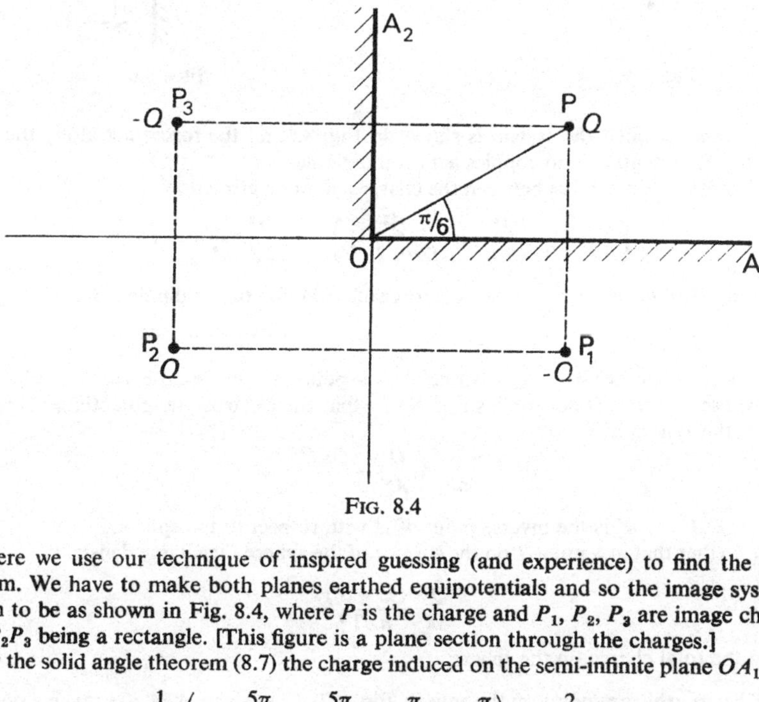

FIG. 8.4

Here we use our technique of inspired guessing (and experience) to find the image system. We have to make both planes earthed equipotentials and so the image system is taken to be as shown in Fig. 8.4, where P is the charge and P_1, P_2, P_3 are image charges, $PP_1P_2P_3$ being a rectangle. [This figure is a plane section through the charges.]

By the solid angle theorem (8.7) the charge induced on the semi-infinite plane OA_1 is

$$\frac{1}{4\pi}\left(-Q\frac{5\pi}{3}-Q\frac{5\pi}{3}+Q\frac{\pi}{3}+Q\frac{\pi}{3}\right) = -\frac{2}{3}Q,$$

the various terms being the contributions from P, P_1, P_2, P_3 in that order. Similarly the charge induced on the other half-plane is $-Q/3$. [The total induced charge must be equal and opposite to the exciting charge at P since all tubes of force leaving P end on the earthed planes.]

Example 3. An electric dipole of moment $\boldsymbol{m}$ is placed at a perpendicular distance a from an infinite plane conducting sheet, with its axis normal to the plane. Show that the dipole is attracted to the plane by a force of magnitude $3m^2/(32\pi\varepsilon_0 a^4)$.

We consider the general case, and regard a dipole as the limiting combination of two point charges $\pm Q$ separated by a distance l. Using the results of example 1, Fig. 8.5, we see that the image dipole has the same scalar moment, but if the original dipole makes the angle α with the normal to the plane, the image dipole makes the angle α on the opposite side of the normal, i.e. the image dipole makes the angle 2α with the real dipole.

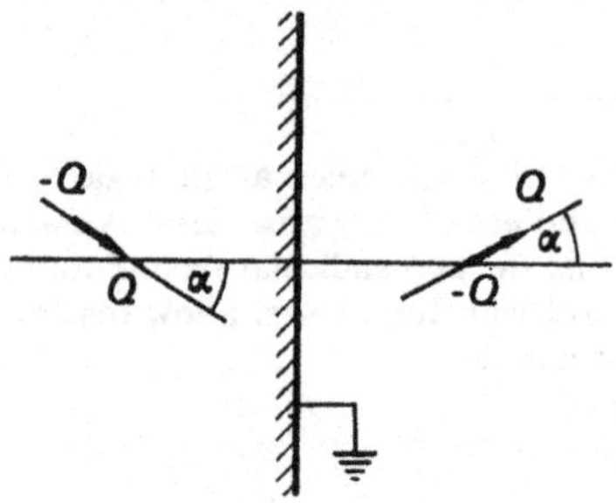

FIG. 8.5

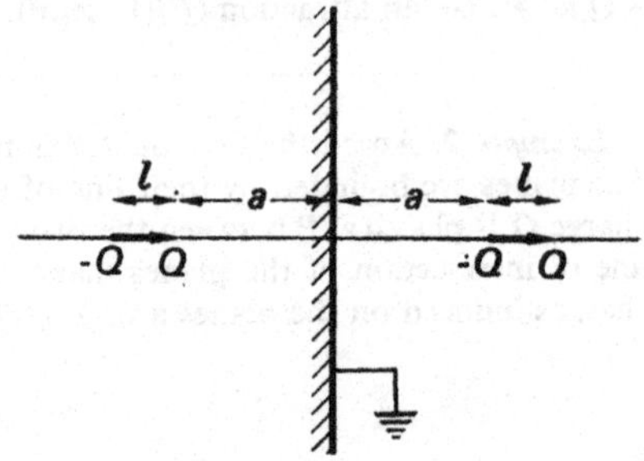

FIG. 8.6

In the case specified the system is shown in Fig. 8.6, all the forces act along the line joining the dipoles and so no couples act on the dipoles.

Consideration of the force between the charges gives an attraction

$$\frac{1}{4\pi\varepsilon_0}\left\{\frac{Q^2}{4a^2}-\frac{2Q^2}{(2a+l)^2}+\frac{Q^2}{(2a+2l)^2}\right\} = \frac{3l^2Q^2}{32\pi\varepsilon_0 a^4}+O(Q^2l^3).$$

In the limit $lQ \to m$ and $Q^2l^3 \to 0$ and the force of attraction on the dipole is $3m^2/(32\pi\varepsilon_0 a^4)$.

Example 4. A point charge Q is situated at a point A outside an earthed conducting sphere whose centre is O and radius is a. Show that the electrostatic potential at a point P outside the sphere is

$$V = \frac{1}{4\pi\varepsilon_0}\left(\frac{Q}{AP}-\frac{Qaf^{-1}}{A'P}\right),$$

where $f = OA$ and A' is the inverse point of A with respect to the sphere.

Show further that at a point T on the surface of the sphere the charge density is

$$-\frac{Q}{4\pi}\frac{(f^2-a^2)}{aAT^3}$$

and find the total charge on the sphere.

Let T be an arbitrary point on the sphere, Fig. 8.7. Then since A, A' are inverse points $OA \cdot OA' = a^2$ and so $OA'/OT = OT/OA$. It follows that the triangles $OA'T$, OTA are similar and therefore

$$A'T/AT = OT/OA = a/f. \tag{1}$$

Therefore the sphere is an equipotential surface $V = 0$ when a charge Q is placed at A and a charge $-aQ/f$ is placed at A'. In fact, the charge $-aQ/f$ at A' is the image in the sphere of the charge Q at A. The potential at an arbitrary external point P is

$$V(=V_1) = \frac{1}{4\pi\varepsilon_0}\left(\frac{Q}{r}-\frac{Qa}{fr'}\right), \tag{2}$$

where $r = AP$, $r' = A'P$. On and inside the sphere $V = 0$.

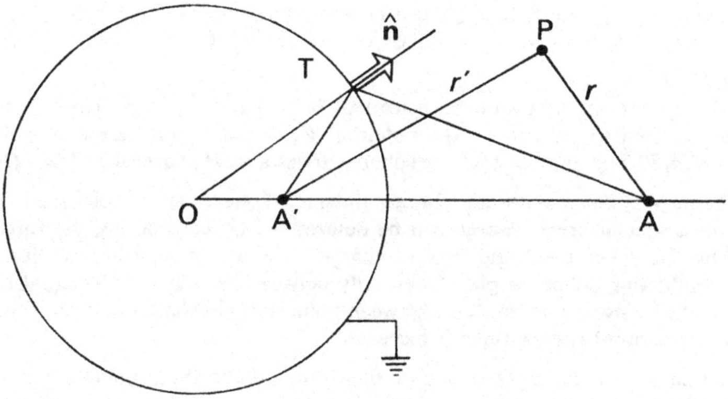

FIG. 8.7

The electrostatic field at T due to the charges is

$$\boldsymbol{E}(T) = \frac{Q}{4\pi\varepsilon_0}\left(\frac{\overrightarrow{AT}}{AT^3} - \frac{a}{f}\frac{\overrightarrow{A'T}}{A'T^3}\right)$$

and, using the relations

$$\overrightarrow{AT} = \overrightarrow{AO} + \overrightarrow{OT}, \qquad \overrightarrow{A'T} = \overrightarrow{A'O} + \overrightarrow{OT},$$

this becomes

$$\boldsymbol{E}(T) = \frac{Q}{4\pi\varepsilon_0}\left[\left(\frac{1}{AT^3} - \frac{a}{fA'T^3}\right)\overrightarrow{OT} + \left(\frac{\overrightarrow{AO}}{AT^3} - \frac{a}{f}\frac{\overrightarrow{A'O}}{A'T^3}\right)\right]. \tag{3}$$

But since the sphere is an equipotential the field at T is purely radial, i.e. the field acts along $\overrightarrow{OT}$ and so the components along OA on the right-hand side of (3) cancel one another. Therefore the surface density of induced charge at T is given by

$$\sigma = \varepsilon_0 \hat{\boldsymbol{n}}\cdot\boldsymbol{E}(T) = \frac{Qa}{4\pi}\left(\frac{1}{AT^3} - \frac{a}{fA'T^3}\right),$$

$\hat{\boldsymbol{n}}$ being the unit outward normal at T. Using (1) we find

$$\sigma = \frac{-Q(f^2-a^2)}{4\pi a AT^3} \tag{4}$$

as required. Note that, if $Q > 0$, σ is everywhere negative. Also AT increases as the angle AOT increases and so $|\sigma|$ is greatest when T lies on AO, least when T lies on OA produced. By the solid angle theorem (8.7) the charged induced on the sphere is $-aQ/f$.

Notes

1. We can superpose the above solution and that of the sphere when it is insulated and charged. Note, however, that, when earthed, the sphere carries a charge $-aQ/f$ and so, when the sphere is insulated and carries a total charge Q_1, we must add to V_1 the potential V_2 of a charge Q_1+Qa/f at O, i.e. $V_2 = (Q_1+Qa/f)/(4\pi\varepsilon_0 R)$ where $R = OP$. In this case the surface density of charge is

$$\sigma_1 = \frac{(Q_1+Qa/f)}{4\pi a^2} - \frac{Q(f^2-a^2)}{4\pi a AT^3}. \tag{5}$$

Since $f-a \leqslant AT \leqslant f+a$, it follows that σ_1 is everywhere positive provided that $Q_1 > Qa^2(3f-a)/\{f(f-a)^2\}$. Furthermore, σ_1 is everywhere negative provided that

$Q_1 < -Qa^2(3f+a)/\{f(f+a)^2\}$. If Q_1 lies between these limits, σ_1 will be partly positive and partly negative, the dividing curve on the sphere being a ring (small circle) of points where σ_1 is zero.

2. If the sphere of the example, instead of being earthed, is insulated and raised to potential V_0, then we add to V_1 the potential V_2 of the sphere raised to potential V_0, i.e. $V_2 = V_0a/R$. This implies that the sphere carries a total charge $4\pi\varepsilon_0 V_0 a - Qa/f$.

3. The image systems of several charges outside a sphere can be combined. The forces exerted on a specific (real) charge can be determined by considering the forces exerted upon it by the other (real and image) charges. For example, when an insulated uncharged conducting sphere is placed centrally between two charges of equal magnitude, if they are of like signs, the repulsion between them is diminished, but if they are of unlike signs, the attraction between them is increased.

4. The image of a charge Q' at the point A *inside* an earthed spherical conductor and distance f' from the centre of the conductor is a charge $-Q'a/f'$ at the inverse of A.

Example 5. An earthed conducting sphere of radius a has its centre at O. An electric dipole of moment $\boldsymbol{M}$ is held at a point P such that the position vector of P relative to O is $\boldsymbol{c}(c^2 > a^2)$. Show that the image system consists of a dipole of moment

$$\frac{a^3}{c^5}[2\boldsymbol{c}(\boldsymbol{c}\cdot\boldsymbol{M})-c^2\boldsymbol{M}] \quad (c = |\boldsymbol{c}|),$$

together with a certain point charge, at a certain point.

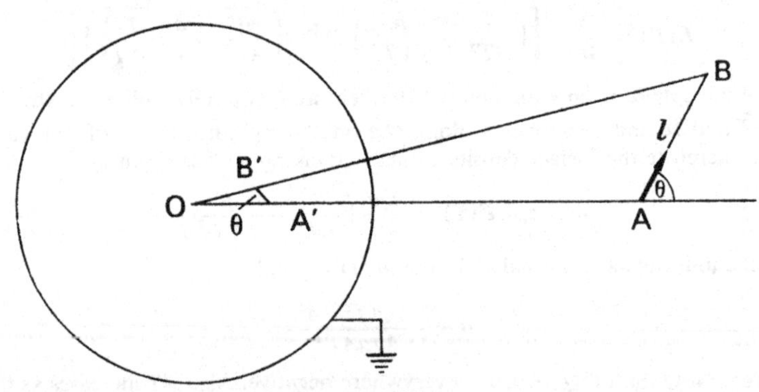

FIG. 8.8

Let the dipole be constituted by the charges $-Q$, Q at A, B respectively, Fig. 8.8, where $AB = l$ and $\lim Ql = M$. Then the image system consists of charges Qa/c, $Q' = -Qa/OB$ at A', B' respectively, where primes denote inverse points. The triangles $OA'B'$, OBA are similar, so that

$$A'B'/AB = OA'/OB = a^2/(cOB).$$

Also $$OB = \sqrt{(OA^2+AB^2+2OA\cdot OB\cos\theta)} = c\left\{1+\frac{l\cos\theta}{c}+O(l^2)\right\}.$$

Hence, neglecting terms of $O(l^2)$,

$$A'B' = a^2l/c^2, \quad Q' = -Qa/c+(Qal\cos\theta)/c^2.$$

It follows that the image system consists of a dipole at A' of scalar moment a^3M/c^3 directed along $B'A'$ together with a point charge $q = (Ma \cos\theta)/c^2$ at A'. The point charge q can be expressed in terms of the vectors thus,

$$q = \frac{a}{c^3} \boldsymbol{M} \cdot \boldsymbol{c}.$$

The vector moment of the image dipole is $a^3M\hat{\boldsymbol{k}}/c^3$ where $\hat{\boldsymbol{k}}$ is a unit vector along $B'A'$. But a unit vector directed along $B'A'$ is equivalent to a unit vector, $-\boldsymbol{M}/M$, directed along BA plus twice the projection along OA, $2(\boldsymbol{M}\cdot\boldsymbol{c})\boldsymbol{c}/(Mc^2)$, of a unit vector directed along AB, i.e. $\hat{\boldsymbol{k}} = -\boldsymbol{M}/M + 2\boldsymbol{c}(\boldsymbol{M}\cdot\boldsymbol{c})/(Mc^2)$. The moment of the dipole is therefore

$$\frac{a^3}{c^5}\{2\boldsymbol{c}(\boldsymbol{M}\cdot\boldsymbol{c}) - c^2\boldsymbol{M}\},$$

as required.

Example 6. An infinite series of images

A charge Q is placed at the point O midway between two parallel infinite conducting earthed planes P_1, P_1' separated by distance $2a$, Fig. 8.9.

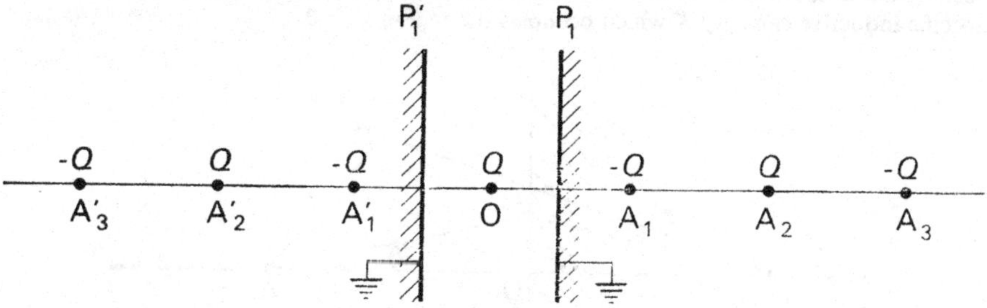

FIG. 8.9

To ensure that plane P_1 is an earthed equipotential we insert an image charge $-Q$ at A_1 where $OA_1 = 2a$. Similarly the image charge $-Q$ at A_1', where $OA_1' = 2a$, makes P_1' an earthed equipotential. But we must now insert the images of A_1' in P_1 and A_1 in P_1', i.e. charges Q, at A_2, A_2' where $OA_2 = 4a = OA_2'$. Then further image charges $-Q$, $-Q$ must be placed at A_3, A_3' where $OA_3 = 6a = OA_3'$, etc. In fact, there is an infinite set of image charges. However, the field can be determined to a reasonable degree of accuracy by using the nearer images. A simple formula for the potential in closed form is not possible in this case but can be found for the analogous two-dimensional problem (see p. 311) where conformal transformations can be used.

This method can be generalized for point charges within conductors with plane boundaries. For example, the image system for a point charge within a cuboid with earthed sides consists of a triply infinite system of charges.

Note that, as before, we can superpose on the above (image) fields the simpler fields obtained earlier, e.g. that of the field between two infinite plane conductors maintained at distinct potentials.

---

3. Images when Dielectrics are Present

The method of images, as described on p. 270, is not used in exactly that form when dielectrics are present. In certain cases, when the surface of the dielectric has a convenient shape, a distribution of electric charges can be devised which gives rise to fields satisfying the boundary conditions, p. 263. In these cases, the distributions of charge can be called images. In devising suitable arrangements of charges we obtain guidance from the image systems in conducting surfaces, but the process is, in general, rather more difficult than the image processes already discussed.

The following example illustrates one of the available techniques.

Example. A point charge q is placed at a point A with coordinates $(a, 0, 0)$ referred to rectangular axes $Oxyz$ in front of the plane face, $x = 0$, of an infinite slab of dielectric of specific inductive capacity K which occupies the region $x < 0$.

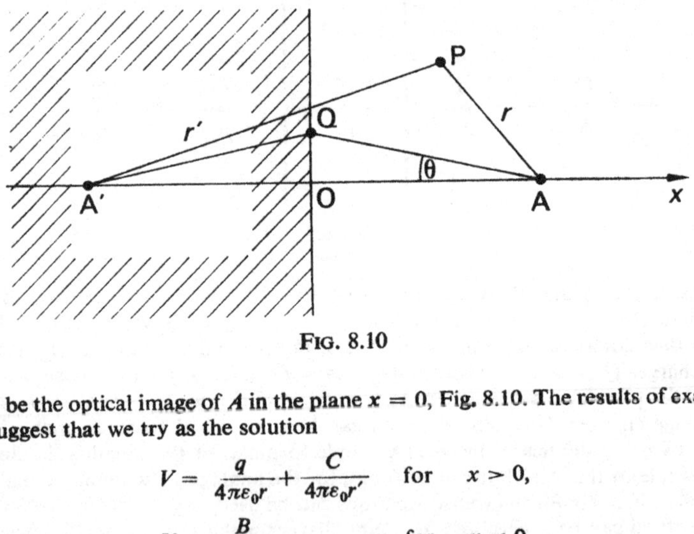

FIG. 8.10

Let A' be the optical image of A in the plane $x = 0$, Fig. 8.10. The results of example 1, p. 271, suggest that we try as the solution

$$V = \frac{q}{4\pi\varepsilon_0 r} + \frac{C}{4\pi\varepsilon_0 r'} \quad \text{for} \quad x > 0, \tag{1}$$

$$V = \frac{B}{4\pi\varepsilon_0 r} \quad \text{for} \quad x < 0, \tag{2}$$

where B, C are constants to be determined and $r = AP$, $r' = A'P$, P being a field point. The only singularity in the assumed potential occurs at A where the potential is $q/(4\pi\varepsilon_0 r) + O(1)$ as is required by the uniqueness theorem. Further, V satisfies the standard boundary conditions at infinity. The conditions that V and D_n are continuous where $r = r'$ at an arbitrary point Q on the plane $x = 0$, the surface of discontinuity of K, are

$$\frac{q}{4\pi\varepsilon_0 r} + \frac{C}{4\pi\varepsilon_0 r} = \frac{B}{4\pi\varepsilon_0 r}, \tag{3}$$

$$\frac{1}{4\pi}\left(\frac{q}{r^2} - \frac{C}{r^2}\right)\cos\theta = \frac{KB\cos\theta}{4\pi r^2}. \tag{4}$$

Solution of eqns. (3), (4), after cancelling common factors $1/r$, $1/r^2$, $\cos\theta$, gives

$$C = \frac{-(K-1)q}{(K+1)}, \quad B = \frac{2q}{(K+1)},$$

and, by the uniqueness theorem, (1), (2) now give the solution of the problem. In fact, the potential for $x > 0$ is the same as if the dielectric were removed and an additional charge $-(K-1)\,q/(K+1)$ were placed at A'; the potential for $x < 0$ is the same as if the dielectric were removed and the magnitude of the charge at A were reduced to $2q/(K+1)$.

Notes

1. We regain the solution of a point charge q in front of an infinite conducting slab, example 1, p. 271, by letting $K \to \infty$. [This method of letting $K \to \infty$ provides some check on the correctness of a result when dielectrics are present.] Further, when $K = 1$ the field reduces to that of a point charge at A in otherwise empty space.

2. The force exerted on the charge at A is the same as that which is exerted by a charge $-(K-1)q/(K+1)$ at A', i.e. an attraction $(K-1)q^2/\{16\pi a^2\varepsilon_0(K+1)\}$.

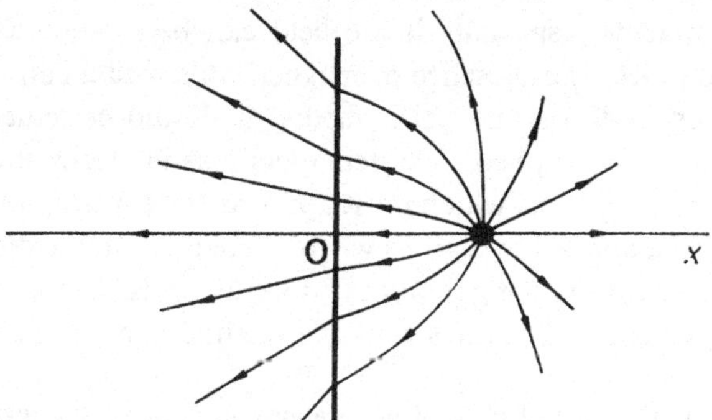

FIG. 8.11

3. The lines of force are shown in Fig. 8.11. Note that they are straight for $x < 0$. It is easy to show that, if a line of force starts from A in a direction making an angle α with the normal to the plane face, it will make a constant angle β with the normal in the dielectric, where

$$\sin\left(\frac{1}{2}\beta\right) = \left(\frac{1+K}{2K}\right)^{1/2} \sin\left(\frac{1}{2}\alpha\right).$$

4. As an exercise the reader should show that a dipole of moment $\boldsymbol{M}$ directed along the x-axis is attracted by a force $3(K-1)M^2/\{32\pi\varepsilon_0 a^4(K+1)\}$.

---

4. Images in magnetostatics

The discussion given in Chapter 6 started from linear currents as the basic source of a magnetic field. In those cases where the current is confined strictly to linear circuits the field equation $\textbf{curl}\ \boldsymbol{H} = \boldsymbol{J}$ reduces to

$$\textbf{curl}\ (\boldsymbol{B}/\mu_0) = \boldsymbol{0}, \quad \text{i.e.} \quad \textbf{curl}\ \boldsymbol{B} = \boldsymbol{0},$$

for the region (usually multiply connected) outside the conducting wires. Hence it is possible to introduce a magnetostatic potential Ω such that

$$\boldsymbol{B}/\mu_0 = \boldsymbol{H} = -\mathbf{grad}\,\Omega.$$

In many cases, e.g. a small circuit, we can regard Ω as analogous to the electrostatic potential V, provided we regard magnetic dipoles (or particles) as arising from combinations of magnetic poles, analogous to electric point charges. The field-lines of $\boldsymbol{H}$ in the magnetic case are orthogonal to "equipotential surfaces" $\Omega = \text{const}$. The method of images in electrostatics relies on the application of the uniqueness theorem to regions in which $\nabla^2 V = 0$ bounded by an equipotential surface on which certain boundary conditions hold. There are some cases in magnetostatics in which $\nabla^2\Omega = 0$ and Ω satisfies similar boundary conditions. In these cases it is possible sometimes to devise image systems, especially if the field can be represented as arising from magnetic poles. The presence of magnetizable media can be treated in a manner corresponding to dielectric media. It should be remembered that the method cannot be applied so systematically as in electrostatics because of the boundary conditions which must apply at the surface, and in no case can the method be applied to regions where there is a volume distribution of current. In these regions the potential Ω does not exist (since $\mathbf{curl}\,\boldsymbol{H} \neq \boldsymbol{0}$). The following example illustrates how the method can be applied.

Example. An infinite slab of iron, of permeability μ, occupies the region for which $x > 0$ and has a plane face coinciding with the plane $x = 0$. If a small magnet, whose magnetic moment is M and whose axis is directed along the positive direction of the axis of x, is placed with its centre at the point $(-a, 0, 0)$, prove that it will be attracted to the iron by a force

$$\frac{3\mu_0(\mu-\mu_0)M^2}{32\pi(\mu+\mu_0)a^4}.$$

For this we use the method, adapted slightly, of the example on p. 278 of § 8.2 (3) which gives the images used in the presence of a dielectric.

We assume that the field in the vacuum is due to the given magnet of moment M and the contribution of the soft iron is given by that of another magnet of moment M' situated at the optical image $\{a\,0\,0\}$. The field inside the soft iron is assumed to be that of a magnet M'' situated at $\{-a\,0\,0\}$ in the absence of soft iron. The form taken by the magnetostatic potential is

$$\Omega \begin{cases} = M''(x+a)/(4\pi r^3), & x > 0, \\ = M'(x-a)/(4\pi r_1^3) + M(x+a)/(4\pi r^3), & x < 0. \end{cases}$$

Here r denotes the distance of the field point from $\{-a\,0\,0\}$ and r_1 the distance from $\{a\,0\,0\}$. The boundary conditions to be satisfied at $x = 0$ are

$$[\Omega] = 0 \quad \text{and} \quad [\mu\partial\Omega/\partial x] = 0.$$

For field points on the surface $x = 0$, at distance y from the origin, $r = r_1 = (a^2+y^2)^{1/2}$.

The condition $[\Omega] = 0$ gives

$$M'' = M - M'. \tag{1}$$

Also

$$\frac{\partial \Omega}{\partial x} = \begin{cases} \dfrac{M''}{4\pi}\left[\dfrac{1}{r^3} - \dfrac{3(x+a)^2}{r^5}\right], & x > 0, \\ \dfrac{M'}{4\pi}\left[\dfrac{1}{r_1^3} - \dfrac{3(x-a)^2}{r_1^5}\right] + \dfrac{M}{4\pi}\left[\dfrac{1}{r^3} - \dfrac{3(x+a)^2}{r^5}\right], & x < 0. \end{cases}$$

Hence on $x = 0$, the factors in square brackets are all equal and the condition $[\mu\partial\Omega/\partial x] = 0$ gives

$$\mu M'' = \mu_0(M' + M). \tag{2}$$

Solution of (1) and (2) gives

$$M' = \frac{\mu - \mu_0}{\mu + \mu_0} M, \qquad M'' = \frac{2\mu_0}{\mu + \mu_0} M.$$

(This result would also be obtained by the same method applied to magnetic poles which are combined to give, in the limit, a magnetic dipole.)

Since the effect of the soft iron in the region $x < 0$ is given by that of the magnetic moment M', the force exerted on M by the iron is given by $\boldsymbol{F} = (\boldsymbol{m}\cdot\mathbf{grad})\boldsymbol{B}$ where $\boldsymbol{B}$ is the flux density produced in the vicinity of M by the soft iron, i.e. by M'. Hence $\boldsymbol{B} = \{B\ 0\ 0\}$ where

$$B = -\frac{\mu_0 M'}{4\pi} \frac{\partial}{\partial x}\left\{\frac{x-a}{r_1^3}\right\} = -\frac{\mu_0 M'}{4\pi} \cdot \frac{2}{(x-a)^3},$$

since $r_1 = -(x-a)$ when $x < a$. The force $\boldsymbol{F}$ therefore has only an x-component given by

$$F_x = M\left[\frac{\partial B}{\partial x}\right]_{x=-a} = -\frac{\mu_0 MM'}{4\pi} \frac{\partial}{\partial x}\left\{\frac{2}{(x-a)^3}\right\}_{x=-a}$$

$$= \frac{\mu_0 MM'}{4\pi}\left[\frac{6}{(x-a)^4}\right]_{x=-a} = \frac{3\mu_0(\mu-\mu_0)M^2}{32\pi(\mu+\mu_0)a^4}.$$

---

5. Images in steady current flow problems

For brevity here we state two image systems for an anode emitting current I in a semi-infinite uniform conductor bounded by an infinite plane. The student should verify that the boundary conditions are satisfied.

(a) When the plane boundary is a free surface, the image is an equal anode of the same shape and size also emitting current I.

(b) When the plane boundary is itself a perfectly conducting electrode, the image is an equal cathode absorbing current I.

Example. **Two spherical electrodes each of radius δ are immersed in the sea at a distance b apart, and each at a height h above the sea-bed. The conductivities of the sea and of the sea-bed are σ_1 and σ_2, respectively, and the depth of the sea is such that, in relation to h and b, it may be assumed infinite. A constant potential difference V_0 is maintained between the electrodes. Find the current that flows between them, up to terms of order δ/h; δ/b relative to the principal term.**

Denoting the properties of sea-water and the sea-bed by the suffixes 1, 2, respectively, we first find the potential due to an isolated anode of radius δ emitting current I at height h

above the sea-bed, Fig. 8.12 (i). Using the technique of p. 278, where the problem of a point charge in the vicinity of an infinite plane discontinuity in the dielectric constant was solved, we find that in regions (1), (2),

$$V_1 = \frac{I}{4\pi\sigma_1 r_1} + \frac{(\sigma_1-\sigma_2)I}{4\pi\sigma_1(\sigma_1+\sigma_2)r_2},$$

$$V_2 = \frac{2I}{4\pi(\sigma_1+\sigma_2)r_1}.$$

These potentials satisfy the boundary conditions $[V] = 0 = [j_n]$ at the floor of the ocean. We have neglected terms of order δ/h, because we have effectively taken the spheres to be

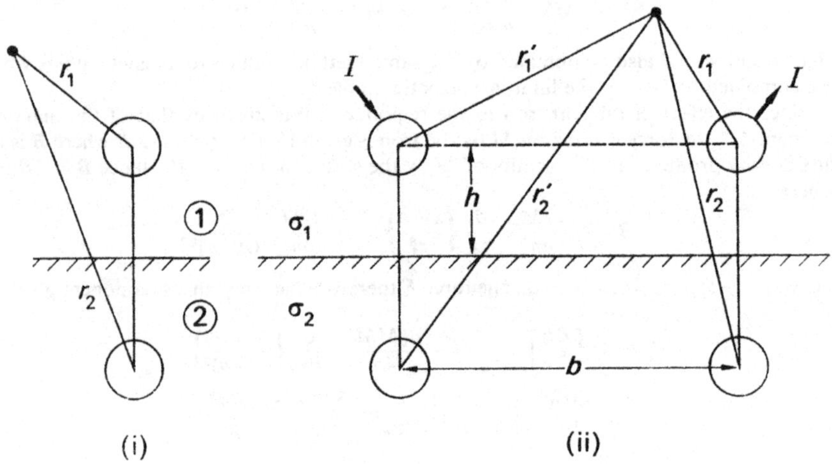

FIG. 8.12

point electrodes in the expression for V_1 and V_2. Then, with the same approximations throughout, the potential in region (1) for our problem is, see Fig. 8.12 (ii),

$$V = \frac{I}{4\pi\sigma_1}\left\{\frac{1}{r}-\frac{1}{r_1'}+\left(\frac{\sigma_1-\sigma_2}{\sigma_1+\sigma_2}\right)\left(\frac{1}{r_2}-\frac{1}{r_2'}\right)\right\}.$$

On the anode $r_1 = \delta$, $r_1' = b$, $r_2 = 2h$, $r_2' = \sqrt{(b^2+4h^2)}$.

$$\therefore \quad V(\text{anode}) = \frac{I}{4\pi\sigma_1}\left\{\frac{1}{\delta}-\frac{1}{b}+\left(\frac{\sigma_1-\sigma_2}{\sigma_1+\sigma_2}\right)\left(\frac{1}{2h}-\frac{1}{\sqrt{(b^2+4h^2)}}\right)\right\}.$$

Similarly V(cathode) $= -V$(anode) and so the potential difference between the electrodes is $2V$(anode).

But this is given to be V_0 and it follows that

$$I = 2\pi\sigma_1 V_0\bigg/\left\{\frac{1}{\delta}-\frac{1}{b}+\left(\frac{\sigma_1-\sigma_2}{\sigma_1+\sigma_2}\right)\left(\frac{1}{2h}-\frac{1}{\sqrt{(b^2+4h^2)}}\right)\right\}.$$

Note the special case when $\sigma_2 \to 0$, i.e. when the sea-bed is non-conducting. This case can be worked out from the image system stated on p. 272.

Exercises 8.2

1. Two point charges Q_1 and Q_2 are situated at points A and B which are a distance $2a$ apart and they lie to the same side of, and are each at a distance b from, an infinite conducting plane which is at zero potential. Describe a suitable system of images.

Show that the force acting on the charge Q_1 will act in the line joining the two charges if

$$Q_2 = -Q_1(1+a^2/b^2)^{3/2}.$$

When $a = b$, find the *magnitude* of the force acting on Q_1

(i) when $Q_1 = Q_2$;

(ii) when $Q_1 = -Q_2$.

Sketch the lines of force in the plane through AB perpendicular to the sheet when $Q_1 = Q_2$.

2. A point charge Q is situated at the point (a, a, a) in the positive octant, and the positive portions of the coordinate planes are occupied by three mutually perpendicular plane conductors maintained at zero potential. Prove that the point charge is acted on by a force towards the origin of amount

$$Q^2(2+6\sqrt{3}-3\sqrt{6})/(96\pi\varepsilon_0 a^2).$$

3. A thundercloud, which may be regarded as an electric doublet with its axis vertical, is moving uniformly along a horizontal straight line and directly approaching an observer on the ground who is recording the electric intensity close to the surface. Show that the rate of change of the electric intensity vanishes when the elevation of the cloud is $\tan^{-1}\frac{1}{2}$.

4. An electric dipole of strength m is placed at distance a from an earthed infinite conducting plane. Show that if the dipole is normal to the plane, the dipole is attracted towards the plane by a force $3m^2/(32\pi\varepsilon_0 a^4)$, but if the dipole is parallel to the plane, the force of attraction is $3m^2/(64\pi\varepsilon_0 a^4)$.

5. An electric dipole of moment m is held with its centre fixed at a distance a from an infinite conducting plane which is earthed. If the dipole is free to turn about its centre in a plane perpendicular to the conducting plane, find the position of stable equilibrium and show that the period of small oscillations about this position is $(8\pi/m)\sqrt{(2\pi\varepsilon_0 a^3 I)}$, where I is the moment of inertia of the dipole.

6. Point charges $+Q$, $-Q$ are *in vacuo* at a distance $4a$ apart. An insulated uncharged conducting sphere of radius a is placed with its centre midway between the point charges. Find the total charge induced on the hemisphere facing $+Q$.

7. Equal and opposite charges Q, $-Q$ are placed on opposite sides of a conducting sphere of radius a on the same diameter, each being at a distance $d > a$ from the centre of the sphere, which is earthed. Prove that each charge is attracted towards the sphere with force

$$\frac{Q^2}{4\pi\varepsilon_0}\left\{\frac{4a^3d^3}{(d^4-a^4)^2}+\frac{1}{4d^2}\right\}.$$

How is the result altered if the sphere is insulated and carries charge Q'?

8. A point charge Q is placed inside an otherwise empty hollow spherical conductor raised to a potential V_0. If the conductor has a radius a and the charge is at a distance c from its centre, find the potential at any point in the interior. [P.T.O.]

Show that the surface density of the induced charge at a point on the spherical conductor at a distance r from the charge Q is

$$-\frac{Q(a^2-c^2)}{4\pi ar^3},$$

and calculate the force which it exerts on the charge Q.

9. An electric dipole of strength M is placed at a point P at distance c from the centre O of an earthed conducting sphere of radius a where a is less than c. Find the potential at every point of space if (i) the dipole points along OP, (ii) the dipole is normal to OP, and show that the image system consists of a dipole and a charge in the first case, but of a dipole only in the second case.

10. A dipole of moment M is placed at a distance c ($> a$) from the centre of a fixed conducting sphere of radius a, and points in the radial direction. The sphere is insulated, and carries a charge Q. Show that, if the dipole is in equilibrium, then

$$Q = \frac{Ma^3(a^6-4a^4c^2+6a^2c^4-6c^6)}{c^2(c^2-a^2)^4}.$$

11. A point charge Q is placed outside a semi-infinite medium of dielectric constant K bounded by a plane face. Obtain the system of images that gives the potential at all points in space, and verify that the potentials so obtained do in fact satisfy all the conditions of the problem.

Two equal charges of magnitude Q are placed at a distance $2a$ apart, outside the semi-infinite medium, and at the same distance d from the surface. Find the force on each charge.

12. An infinite slab of homogeneous material of dielectric constant K fills the region $z < 0$, being separated from free space by the plane $z = 0$. A charge Q_1 lies at $(0, 0, a)$ $(a > 0)$, and another charge Q_2 is embedded in the material at $(0, 0, -a)$. Show how to choose suitable image charges at these points, and prove that the charge Q_1 is repelled from the slab by a force

$$\frac{1}{16\pi\varepsilon_0 a^3}\cdot\frac{2Q_1Q_2-(K-1)Q_1^2}{K+1}.$$

8.3 Three-dimensional problems with axial symmetry

In this section we solve a variety of three-dimensional potential problems with axial symmetry using the separable solutions of Laplace's equation. In particular, we use the Legendre polynomials.

Throughout, the justification for the validity of our results rests with the uniqueness theorems. Although, for brevity, this will not be explicitly stated on each occasion, nevertheless the reader should verify, for example, that the solutions of the electrostatics problems given do satisfy conditions (i)–(vi) of the general uniqueness theorem of electrostatics, p. 263.

Example 1. A charge Q is placed at an external point A distant c from the centre O of a conducting sphere, whose radius is a and whose total charge is Q', Fig. 8.13. Show that at a point T of the sphere the surface density of charge is

$$\frac{Q'}{4\pi a^2}-\frac{Q}{4\pi}\sum_{n=1}^{\infty}(2n+1)\frac{a^{n-1}}{c^{n+1}}P_n(\cos\theta),$$

where θ is the angle TOA. Find the force exerted on the charge Q.

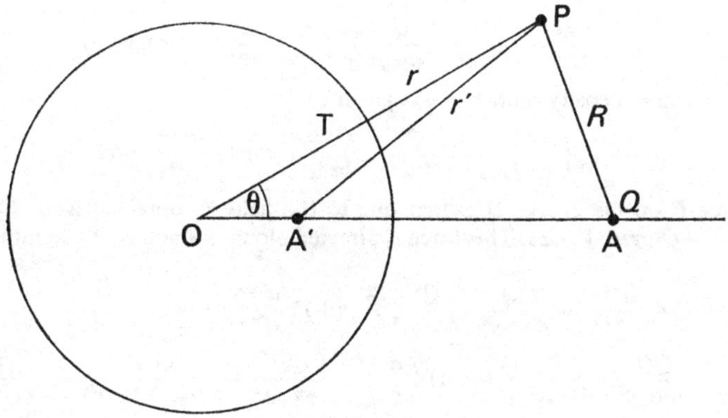

FIG. 8.13

The exciting potential, due to the charge at A, is

$$V_1 = \frac{Q}{4\pi\varepsilon_0 R} = \frac{Q}{4\pi\varepsilon_0 \sqrt{(r^2+c^2-2rc\cos\theta)}} = \frac{Q}{4\pi\varepsilon_0 c}\sum_0^\infty \left(\frac{r}{c}\right)^n P_n(\cos\theta)$$

or $r \leqslant c$. We suppose that the disturbance potential (due to the presence of the sphere) is

$$V_2 = \frac{Q}{4\pi\varepsilon_0 c}\sum_0^\infty A_n\left(\frac{a}{r}\right)^{n+1} P_n(\cos\theta) \tag{1}$$

for $r \geqslant a$, so that the total potential for $a \leqslant r \leqslant c$ is

$$V = V_1+V_2 = \frac{Q}{4\pi\varepsilon_0 c}\sum_0^\infty \left\{\left(\frac{r}{c}\right)^n + A_n\left(\frac{a}{r}\right)^{n+1}\right\}P_n(\cos\theta), \tag{2}$$

where the A_n are constants to be determined.

Note that V_2 consists of external harmonics satisfying Laplace's equation.

In order that the sphere $r = a$ should be an equipotential, we must have

$$\frac{Q}{4\pi\varepsilon_0 c}\sum_0^\infty \left\{\left(\frac{a}{c}\right)^n + A_n\right\}P_n(\cos\theta) = \text{constant}. \tag{3}$$

Since the P_n form a complete set of orthogonal functions, it follows from (3) that $(a/c)^n + A_n = 0$ for $n > 0$. Since the total charge on the sphere is Q', for points at a large distance from the origin the whole system behaves like a point charge $Q+Q'$ at the origin. Hence the dominant term in the expression for the potential is $(Q+Q')/(4\pi\varepsilon_0 r)$ for large values of r. In this case we have for V_1,

$$4\pi\varepsilon_0 V_1 = \frac{Q}{R} = \frac{Q}{\sqrt{(r^2+c^2-2rc\cos\theta)}} = \frac{Q}{r}\sum_0^\infty \left(\frac{c}{r}\right)^n P_n(\cos\theta).$$

The dominant term in the expression for $V = V_1 + V_2$ is therefore

$$\frac{Q+Q'}{4\pi\varepsilon_0 r} = \frac{Q}{4\pi\varepsilon_0 r} + \frac{QA_0 a}{4\pi\varepsilon_0 cr}.$$

$$\therefore \quad \frac{Q}{r}\left(1+\frac{A_0 a}{c}\right) = \frac{Q}{r}\left(1+\frac{Q'}{Q}\right)$$

$$\therefore \quad A_0 = \frac{cQ'}{aQ}.$$

Therefore

$$V = \frac{Q'}{4\pi\varepsilon_0 r} + \frac{Q}{4\pi\varepsilon_0 c} + \frac{Q}{4\pi\varepsilon_0 c}\sum_1^\infty \left(\frac{r^n}{c^n} - \frac{a^{2n+1}}{r^{n+1}c^n}\right)P_n(\cos\theta).$$

The surface charge density σ on $r = a$ is given by

$$\sigma = -\varepsilon_0\left(\frac{\partial V}{\partial r}\right)_{r=a} = \frac{Q'}{4\pi a^2} - \frac{Q}{4\pi}\sum_1^\infty (2n+1)\frac{a^{n-1}}{c^n}P_n(\cos\theta).$$

The force F on the charge Q is that due to the field E_2 obtained from V_2 only, i.e. $F = QE_2 = -Q(\mathbf{grad}\ V_2)_{\substack{r=c\\ \theta=0}}$. This force is directed along OA and is of magnitude

$$\frac{Q^2}{4\pi\varepsilon_0 c}\sum_0^\infty (n+1)A_n\frac{a^{n+1}}{c^{n+2}} = \frac{QQ'}{4\pi\varepsilon_0 c^2} - \frac{Q^2}{4\pi\varepsilon_0 c^2}\sum_1^\infty (n+1)\frac{a^{2n+1}}{c^{2n+1}}$$

$$= \frac{QQ'}{4\pi\varepsilon_0 c^2} - \frac{Q^2 a}{4\pi\varepsilon_0 c^3}\sum_1^\infty (n+1)\left(\frac{a^2}{c^2}\right)^n = \frac{QQ'}{4\pi\varepsilon_0 c^2} - \frac{Q^2a^3}{4\pi\varepsilon_0 c^3(c^2-a^2)} - \frac{Q^2a^3}{4\pi\varepsilon_0 c(c^2-a^2)^2}.$$

Here we have used the result

$$\sum_1^\infty (n+1)x^n = \frac{d}{dx}\left(\sum_1^\infty x^{n+1}\right) = \frac{d}{dx}\left(\frac{x^2}{1-x}\right) \quad \text{for} \quad |x| < 1$$

in summing the series.

Note that when the sphere is earthed and so carries an unspecified charge the constant of eqn. (3) is zero and so $A_n = -(a/c)^n$ for $n = 0, 1, 2, 3, \ldots$.

In this case

$$V_2 = \frac{-Q}{4\pi\varepsilon_0 a}\sum_0^\infty \left(\frac{a^2}{cr}\right)^{n+1} P_n(\cos\theta) = \frac{-Qa/c}{4\pi\varepsilon_0\sqrt{\{r^2+(a^2/c)^2-2(a^2/c)r\cos\theta\}}} = \frac{-Qa/c}{4\pi\varepsilon_0 r'},$$

where r' is the distance of a field point P from A', the inverse of A in the sphere. This result reaffirms the image system obtained in example 4, p. 274.

Example 2. Obtain series expressions for the electrostatic potential inside and outside a uniform dielectric sphere of radius a due to a point charge Q distant c ($> a$) from its centre O. Show that the charge is attracted towards O by a force

$$\frac{(K-1)Q^2}{4\pi\varepsilon_0 c^2}\sum_1^\infty \frac{n(n+1)}{(1+n+nK)}\left(\frac{a}{c}\right)^{2n+1}.$$

With the same notation and figure (8.13) as in example 1, p. 285, except that the sphere is now made of uniform dielectric, the exciting potential due to the charge Q is

$$V_1 = \frac{Q}{4\pi\varepsilon_0 R} = \frac{Q}{4\pi\varepsilon_0\sqrt{(c^2+r^2-2rc\cos\theta)}}.$$

$$\therefore\quad V_1 = \frac{Q}{4\pi\varepsilon_0 c}\sum_0^\infty \left(\frac{r}{c}\right)^n P_n(\cos\theta) \quad \text{for} \quad r \leqslant c, \tag{1}$$

$$V_1 = \frac{Q}{4\pi\varepsilon_0 c}\sum_0^\infty \left(\frac{c}{r}\right)^{n+1} P_n(\cos\theta) \quad \text{for} \quad c \leqslant r. \tag{2}$$

We take the disturbance potential, due to the presence of the sphere, to be

$$V_2 = \frac{Q}{4\pi\varepsilon_0 c}\sum_0^\infty A_n\left(\frac{r}{a}\right)^n P_n(\cos\theta) \quad \text{for} \quad r \leqslant a, \tag{3}$$

$$V_2 = \frac{Q}{4\pi\varepsilon_0 c}\sum_0^\infty A_n\left(\frac{a}{r}\right)^{n+1} P_n(\cos\theta) \quad \text{for} \quad r \geqslant a. \tag{4}$$

Note that we have already satisfied the condition that V is continuous at $r = a$. The other boundary condition, $[D_r]_{r=a} = 0$, gives

$$K\varepsilon_0 \frac{Q}{4\pi\varepsilon_0 c}\sum_0^\infty \left(\frac{na^{n-1}}{c^n}+\frac{n}{a}A_n\right)P_n(\cos\theta) = \varepsilon_0 \frac{Q}{4\pi\varepsilon_0 c}\sum_0^\infty \left(\frac{na^{n-1}}{c^n}-\frac{(n+1)}{a}A_n\right)P_n(\cos\theta).$$

$$\therefore\quad A_n = \frac{-(K-1)na^n}{(n+1+Kn)c^n}. \tag{5}$$

The force on the charge Q is

$$F = Q(E_2)_A = -Q(\text{grad } V_2)_A = -Q\left(\frac{\partial V_2}{\partial r}\right)_A$$

directed along OA. This gives the required attraction towards O.

Example 3. Use the result of example 2 to find the electrostatic potential outside the sphere due to an electric dipole of magnitude m, distant c from the centre, and with its axis in a radial direction.

Show that there is no couple on the dipole and the force on it is a repulsion

$$\frac{m^2(K-1)}{4\pi\varepsilon_0 c^4}\sum_1^\infty \frac{n(n+1)^2(n+2)}{n+1+nK}\left(\frac{a}{c}\right)^{2n+1},$$

where K is the dielectric constant.

We obtain the potential by regarding the dipole as a combination of charges $-Q$ at c, $+Q$ at $c+\delta c$ and then proceeding to the limit in the usual manner. With the notation of example 2, and using primes to denote the potential when the dipole is present,

$$V' = \lim\{V(Q, c+\delta c)-V(Q, c)\} = \lim\left\{\frac{\partial V}{\partial c}\,\delta c\right\}$$
$$= \lim\left\{Q\,\delta c\,\frac{\partial(V/Q)}{\partial c}\right\} = m\,\frac{\partial(V/Q)}{\partial c}.$$

The required potential follows from equations (1)–(5) of example 2.

Since the forces on the charges $-Q$, $+Q$ are radial, there is no couple acting on the dipole ($\mathbf{m}$ and $\mathbf{E}$ are parallel). The outward (radial) force on the dipole is (see § 2.10) the radial component of $(\mathbf{m}\cdot\nabla)\mathbf{E}_2$, which equals

$$-m\left(\frac{\partial^2 V_2'}{\partial r^2}\right)_{\substack{r=c\\ \theta=0}} = \frac{-m^2}{4\pi\varepsilon_0}\left\{\frac{\partial^3}{\partial r^2\,\partial c}\left(\sum_0^\infty \frac{a^{n+1}A_n}{c^{n+1}r^{n+1}}\right)\right\}_{r=c}$$

$$= \frac{m^2}{4\pi\varepsilon_0}\left\{\frac{\partial^3}{\partial r^2\,\partial c}\left(\sum_0^\infty \frac{(K-1)na^{2n+1}}{(n+1+Kn)c^{n+1}r^{n+1}}\right)\right\}_{r=c} = -\frac{m^2(K-1)}{4\pi\varepsilon_0}\sum_0^\infty \frac{n(n+1)^2(n+2)a^{2n+1}}{(n+1+Kn)c^{2n+5}}$$

which is a force of repulsion as required.

Example 4. A condenser is formed from two conducting spherical shells S, T of radii a and b ($a < b$) with centres at O_1 and O_2 respectively, where O_1O_2 is of length c, and c is so small that powers of c/a above the first may be neglected. The conductor S is maintained at a potential V_0 and T is earthed. Verify that the potential in the space between the conductors is given approximately by

$$V = V_0\frac{ab}{b-a}\left\{\frac{1}{r}-\frac{1}{b}-\frac{c}{b^3-a^3}\left(r-\frac{a^3}{r^2}\right)\cos\theta\right\},$$

where r, θ are polar coordinates with O_1 as pole and O_2O_1 as initial line.

Deduce that the surface density at a point of the inner shell S is

$$\frac{Q}{4\pi a^2}\left(1+\frac{3a^2c\cos\theta}{b^3-a^3}\right),$$

where Q is the total charge on the inner shell.

Prove also that the mechanical force on S in the direction of c increasing is

$$\frac{4\pi V_0^2\varepsilon_0(ab)^2c}{(b-a)^2(b^3-a^3)},$$

and hence find the electrostatic energy of the system.

Deduce an expression for the capacity of the condenser.

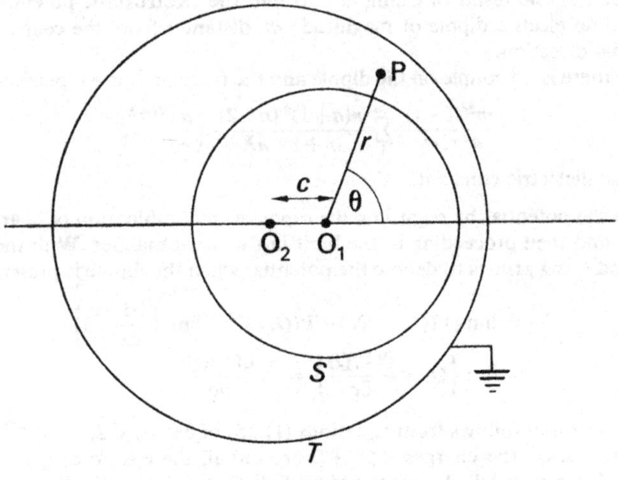

FIG. 8.14

We take the equations of the shells to be $r = a$, $r = b(1+\lambda\cos\theta)$ where $\lambda = -c/b$, Fig. 8.14. The form of the outer surface suggests that we try as our potential function

$$V = A+\frac{B}{r}+Cr\cos\theta+\frac{D\cos\theta}{r^2}, \tag{1}$$

where A, B, C, D are constants and C, D are each of order λ. Since $V = V_0$ on $r = a$,

$$A+\frac{B}{a}+\left(Ca+\frac{D}{a^2}\right)\cos\theta = V_0. \tag{2}$$

Also $V = 0$ on $r = b(1+\lambda\cos\theta)$ and therefore, neglecting terms of order λ^2,

$$A+\frac{B}{b}(1-\lambda\cos\theta)+\left(Cb+\frac{D}{b^2}\right)\cos\theta = 0. \tag{3}$$

Because eqns. (2), (3) must be satisfied for *all* values of θ we equate the coefficients of $\cos\theta$

to zero, and the remaining terms also must be equated. Therefore

$$A+\frac{B}{a} = V_0, \qquad Ca+\frac{D}{a^2} = 0,$$

$$A+\frac{B}{b} = 0, \qquad Cb+\frac{D}{b^2} = \frac{B\lambda}{b}.$$

$$\therefore \quad A = \frac{-aV_0}{b-a}, \qquad B = \frac{abV_0}{b-a},$$

$$C = \frac{ab^2\lambda V_0}{(b-a)(b^3-a^3)}, \qquad D = \frac{-a^4b^2\lambda V_0}{(b-a)(b^3-a^3)},$$

which gives the required result.

Note that in the case $\lambda = 0$ the result agrees with that obtained for the spherical condenser, Vol. 1, example 2, p. 86.

The surface density at (a, θ, ϕ) on the inner shell S is

$$\sigma = -\varepsilon_0\left(\frac{\partial V}{\partial r}\right)_{r=a} = \varepsilon_0\left\{\frac{B}{a^2}-\left(C-\frac{2D}{a^3}\right)\cos\theta\right\},$$

i.e.

$$\sigma = \frac{\varepsilon_0 bV_0}{a(b-a)}\left\{1+\frac{3a^2c\cos\theta}{b^3-a^3}\right\}. \tag{4}$$

But $\oint_{r=a} \sigma\, dS = Q$ and so

$$Q = 4\pi a^2 bV_0\varepsilon_0/\{a(b-a)\}. \tag{5}$$

It follows that the surface density on the inner shell S is as stated.

The normal outward force per unit area on $r = a$ is $\sigma^2/(2\varepsilon_0)$. By symmetry the resultant force must act along the direction $\theta = 0$ and is of magnitude

$$\frac{\pi a^2}{\varepsilon_0}\int_0^\pi \sigma^2 \sin\theta \cos\theta\, d\theta.$$

Using (4) and neglecting $(c/a)^2$, we find that this force is of magnitude

$$\frac{4\pi a^2b^2cV_0^2\varepsilon_0}{(b-a)^2(b^3-a^3)}+O\left(\frac{c^2}{a^2}\right) = F(c), \tag{6}$$

say, acting along $\theta = 0$. This force tends to move the inner shell along the line $\theta = 0$, thereby decreasing the distance between the shells. It follows that the equilibrium position is unstable.

The electrostatic energy of the condenser is given by $W = \frac{1}{2}Q^2/C$ (see § 3.4), so that when c increases to $c+\delta c$, W increases to $W+\delta W$ where $\delta W = -\frac{1}{2}(Q^2/C^2)\,\delta C$. But $-\delta W$ must equal the work done by the force $F(c)$, i.e.

$$\delta W = -F(c)\,\delta c.$$

It follows from (6) that

$$\frac{Q^2\,\delta C}{C^2} = 2F(c)\,\delta c = \left\{\frac{8\pi a^2b^2cV_0^2\varepsilon_0}{(b-a)^2(b^3-a^3)}+O\left(\frac{c^2}{a^2}\right)\right\}\delta c.$$

Integration gives

$$\frac{Q^2}{C} = \frac{Q^2}{C_0}-\frac{4\pi a^2b^2c^2V_0^2\varepsilon_0}{(b-a)^2(b^3-a^3)}+O\left(\frac{c^3}{a^2}\right),$$

where $C_0 = 4\pi\varepsilon_0 ab/(b-a)$ is the capacity when $c = 0$ (see § 3.2). Using (5), we obtain

$$\frac{1}{C} = \frac{b-a}{4\pi ab\varepsilon_0}\left\{1-\frac{abc^2}{(b-a)(b^3-a^3)}\right\}+O(c^3),$$

$$\therefore\ C = \frac{4\pi ab\varepsilon_0}{(b-a)}\left\{1+\frac{abc^2}{(b-a)(b^3-a^3)}\right\}+O(c^3).$$

Note that we have obtained W and C correct to the second order in c/a despite the fact that V and σ were calculated correct to the first order only.

Example 5. An almost spherical earthed conductor possessing axial symmetry is bounded by the surface whose equation in spherical polar coordinates is

$$r = a\{1+f(\theta)\},$$

where $f(\theta)$ is a continuous function in the range 0 to π whose absolute value does not exceed a small constant λ. Use Green's reciprocal theorem to show that if a charge Q is placed at a point (r, θ) outside the conductor the total charge induced on its surface, to the first order in λ, is

$$-\frac{Qa}{r}\left\{1+\sum_0^\infty \alpha_n\left(\frac{a}{r}\right)^n P_n(\cos\theta)\right\},$$

where
$$\alpha = (n+\tfrac{1}{2})\int_0^\pi f(\theta)\,P_n(\cos\theta)\sin\theta\,d\theta.$$

Here we regard the charge Q as located on a small sphere and we consider two cases:

(1) The given conductor is insulated and at unit potential, the small sphere being uncharged.

(2) The given conductor is earthed and carries charge Q', the small sphere carrying a charge Q.

Then the reciprocal theorem gives

$$0 = Q'+QV,$$

.e.
$$Q' = -QV, \tag{1}$$

where V is the potential at the point (r, θ, ϕ) when the given conductor is insulated and raised to unit potential.

To find V, we first express the equation of the given surface in the form

$$r = a\left\{1+\sum_0^\infty \alpha_n P_n(\cos\theta)\right\}, \tag{2}$$

where the α_n are constants. Then comparison with the given equation of the surface implies that

$$\sum_0^\infty \alpha_n P_n(\cos\theta) = f(\theta).$$

Since the $P_n(\cos\theta)$ form a set of orthogonal functions, multiplication by $P_n(\cos\theta)\sin\theta$ and integration w.r.t. to θ from 0 to π now gives

$$\frac{2\alpha_n}{2n+1} = \int_0^\pi f(\theta)\,P_n(\cos\theta)\sin\theta\,d\theta,$$

and it is clear that $\alpha_n = O(\lambda)$.

Now we assume that the potential outside the given conductor when raised to unit potential is

$$V = \sum_0^\infty A_n\left(\frac{a}{r}\right)^{n+1} P_n(\cos\theta),$$

where $A_n = O(\lambda)$ for $n > 0$. Using (2), we find, correct to the first order in λ,

$$1 = A_0\left\{1-\sum_0^\infty \alpha_n P_n(\cos\theta)\right\}+\sum_1^\infty A_n P_n(\cos\theta).$$

It follows that

$$A_0 = 1+\alpha_0, \quad A_n = A_0\alpha_n = \alpha_n \quad \text{for} \quad n>0,$$

correct to the first order.

$$\therefore\quad V = \frac{1}{r}+\sum_0^\infty \alpha_n\left(\frac{a}{r}\right)^{n+1} P_n(\cos\theta)$$

and the given result follows from (1).

Example 6. A conductor at potential V_0 is bounded by the nearly spherical surface $r = a(1+5\lambda\cos^3\theta)$, where λ is very small. Show that the surface density of charge at any point on the surface is approximately

$$\frac{V_0\varepsilon_0}{a}(1-6\lambda\cos\theta+10\lambda\cos^3\theta).$$

Since $\quad P_1 = \cos\theta, \quad P_3 = \tfrac{1}{2}(5\cos^3\theta-3\cos\theta),$

therefore $\quad 5\cos^3\theta = 3P_1+2P_3$

and the equation of the surface can be written

$$r = a(1+3\lambda P_1+2\lambda P_3).$$

We try as the potential outside the conductor

$$V = \frac{V_0 a}{r}+\sum_0^\infty A_n\left(\frac{a}{r}\right)^{n+1} P_n,$$

where each A_n is of order λ.

The boundary condition $V = V_0$ on the conductor gives

$$V_0 = V_0(1-3\lambda P_1-2\lambda P_3)+\sum_0^\infty A_n P_n.$$

$$\therefore\quad A_n = 0 \quad \text{unless} \quad n = 1 \quad \text{or} \quad 3, \quad A_1 = 3\lambda V_0, \quad A_3 = 2\lambda V_0.$$

$$\therefore\quad V = V_0\left(\frac{a}{r}+\frac{3\lambda a^2 P_1}{r^2}+\frac{2\lambda a^4 P_3}{r^4}\right).$$

At the surface of the conductor $\partial/\partial r = \partial/\partial n$, correct to the first order in λ, and so the surface density of charge is

$$\begin{aligned}\sigma &= -\varepsilon_0\left(\frac{\partial V}{\partial r}\right)_{r=a(1+3\lambda P_1+2\lambda P_3)}\\ &= \frac{V_0\varepsilon_0}{a}[\{1-2(3\lambda P_1+2\lambda P_3)\}+6\lambda P_1+8\lambda P_3]\\ &= \frac{V_0\varepsilon_0}{a}(1+4\lambda P_3) = \frac{V_0\varepsilon_0}{a}\{1+2\lambda(5\cos^3\theta-3\cos\theta)\}\\ &= \frac{V_0\varepsilon_0}{a}(1-6\lambda\cos\theta+10\lambda\cos^3\theta).\end{aligned}$$

Example 7. A charge Q is spread evenly over the surface of a circular disc of radius a. Show that at points on the axis of symmetry, at distance $|z|$ from the centre of the disc, the electrostatic potential is

$$\frac{Q}{2\pi a^2\varepsilon_0}\{\sqrt{(z^2+a^2)}-|z|\}.$$

Deduce expressions for the potential at any point which is a distance r from the centre of the disc.

To find the electrostatic potential at P, Fig. 8.15, we divide the disc into thin rings, a typical ring of radius $\tilde{\omega}$ carrying total charge $(2\pi\tilde{\omega}\,\delta\tilde{\omega}\,Q)/(\pi a^2)$. The potential at P of this ring of charge is

$$\delta V_P = \frac{Q\tilde{\omega}\,\delta\tilde{\omega}}{2\pi a^2\varepsilon_0\sqrt{(\tilde{\omega}^2+z^2)}}.$$

Integration gives

$$V_P = \frac{Q}{2\pi a^2\varepsilon_0}\int_0^a \frac{\tilde{\omega}\,d\tilde{\omega}}{\sqrt{(\tilde{\omega}^2+z^2)}} = \frac{Q}{2\pi a^2\varepsilon_0}\{\sqrt{(z^2+a^2)}-|z|\}. \tag{1}$$

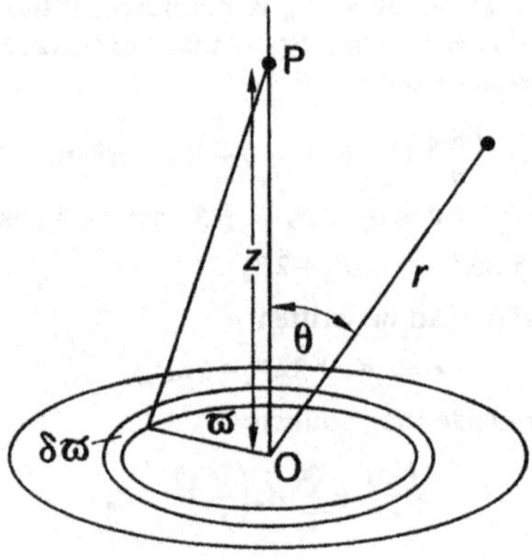

FIG. 8.15

Equation (1) gives the potential at points on the axis of the disc. To obtain the potential at points not on the axis we proceed as follows:

Since the system is axially symmetric about Oz we suppose that the electrostatic potential $V(r, \theta)$ can be expressed in spherical harmonics in the form

$$V = V_0 = \sum_{n=0}^{\infty} A_n\left(\frac{a}{r}\right)^{n+1} P_n(\cos\theta) \quad \text{for} \quad r \geqslant a, \tag{2}$$

$$V = V_1 = \sum_{n=0}^{\infty} B_n\left(\frac{r}{a}\right)^{n} P_n(\cos\theta) \quad \text{for} \quad r \leqslant a, \tag{3}$$

and we choose the constants A_n, B_n to give the correct expression for V at arbitrary points on the axis Oz.

Considering first a point on the axis where $0 \leqslant z \leqslant a$, equation (1) gives

$$V(z, 0) = \frac{Q}{2\pi a^2\varepsilon_0}\{\sqrt{(a^2+z^2)}-z\}$$

$$= \frac{Q}{2\pi a\varepsilon_0}\left\{\left(1+\frac{z^2}{a^2}\right)^{1/2}-\frac{z}{a}\right\},$$

i.e.

$$V(z, 0) = \frac{Q}{2\pi a\varepsilon_0}\left\{1-\frac{z}{a}+\frac{1}{2}\frac{z^2}{a^2}-\frac{1}{2\cdot 4}\frac{z^4}{a^4}+\ldots\right.$$

$$\left.+(-1)^{n-1}\frac{1\cdot 3\cdot 5\ldots(2n-3)}{2\cdot 4\ldots 2n}\frac{z^{2n}}{a^{2n}}+\ldots\right\} \tag{4}$$

on expanding $(1+z^2/a^2)^{1/2}$ by the binomial theorem. Since the expressions (3) and (4) are identical when $z = r$, $\theta = 0$ [so that $P_n(\cos\theta) = P_n(1) = 1$], we must have

$$B_0 = \frac{Q}{2\pi a\varepsilon_0}, \qquad B_1 = -\frac{Q}{2\pi a\varepsilon_0},$$

$$B_{2p+1} = 0 \quad \text{for} \quad p > 0, \qquad B_{2p} = (-1)^{p-1}\frac{1\cdot 3\cdot 5\ldots(2n-3)Q}{2^{n+1}n!\pi a\varepsilon_0}.$$

This gives the potential for $r \leqslant a$, $0 \leqslant \theta \leqslant \pi/2$. Symmetry about the plane $\theta = \pi/2$ implies that for $r \leqslant a$, $\pi/2 \leqslant \theta \leqslant \pi$ the same result holds except that B_1 must now take the value $+Q/(2\pi a\varepsilon_0)$.

Considering now points for which $z > a$, we expand $\sqrt{(a^2+z^2)}$ as a power series in $(a/z)^2$ and find

$$V(z, 0) = \frac{Qz}{2\pi a^2\varepsilon_0}\left\{\frac{1}{2}\frac{a^2}{z^2} - \frac{1}{2\cdot 4}\frac{a^4}{z^4} + \ldots + \frac{(-1)^{n-1}1\cdot 3\ldots(2n-3)}{2\ldots 4\ldots 2n}\left(\frac{a}{z}\right)^{2n} + \ldots\right\}. \quad (5)$$

Since the expressions (2) and (5) are identical when $z = r$, $\theta = 0$ we obtain

$$A_0 = \frac{Q}{4\pi\varepsilon_0}, \qquad A_{2p+1} = 0, \qquad A_{2p} = \frac{(-1)^p 1\cdot 3\ldots(2p-1)}{2^{p+2}(p+1)!\pi a\varepsilon_0}$$

and by symmetry this holds for $0 \leqslant \theta \leqslant \pi$.

Example 8. A distribution of electric charge is rigidly fixed to the surface of a spherical shell of radius a, the charge density at a point Q being $\sum\limits_n A_n P_n(\cos\theta)$, where θ is the colatitude angle of Q referred to an axis ON through the centre O of the sphere and the A_n are constants. The spherical shell rotates with constant angular velocity ω about ON in a right-handed sense. Show that the resulting density of surface current at Q can be expressed as

$$\omega a \sum_n B_n \frac{d}{d\theta} P_n(\cos\theta),$$

where

$$B_n = \frac{A_{n+1}}{2n+3} - \frac{A_{n-1}}{2n-1}, \qquad (A_{-1} = 0).$$

Show that the magnetic field at any point P inside the shell can be derived from the scalar potential

$$\Omega = \omega a^2 \sum_n \frac{n+1}{2n+1} B_n \left(\frac{r}{a}\right)^n P_n(\cos\alpha),$$

where r is the distance of P from O and α the colatitude of P. Find also the potential of the magnetic field outside the shell. [The relation $(2n+1)P_n(\mu) = P'_{n+1}(\mu) - P'_{n-1}(\mu)$ may be assumed.]

When the sphere rotates the surface current is $\sigma\omega a \sin\theta$.

$$\therefore\ I = \omega a \sum_0^\infty A_n \sin\theta\, P_n(\cos\theta).$$

But

$$\sin\theta\, P_n(\cos\theta) = \frac{1}{2n+1}(\sin\theta\, P'_{n+1} - \sin\theta\, P'_{n-1})$$

$$= \frac{1}{2n+1}\left(\frac{dP_{n-1}}{d\theta} - \frac{dP_{n+1}}{d\theta}\right) = \frac{\dot P_{n-1} - \dot P_{n+1}}{2n+1},$$

so that
$$I = \omega a \sum_0^\infty \frac{A_n}{2n+1}(\dot{P}_{n-1} - \dot{P}_{n+1})$$
$$= \omega a \sum_0^\infty \left(\frac{A_{n+1}}{2n+3} - \frac{A_{n-1}}{2n-1}\right)\dot{P}_n = \omega a \sum_0^\infty B_n \dot{P}_n,$$
where
$$\dot{P}_n = \frac{\mathrm{d}}{\mathrm{d}\theta} P_n(\cos\theta).$$

Since Ω must satisfy Laplace's equation inside and outside the sphere we assume
$$\Omega_+ = \frac{1}{r}\sum_0^\infty C_n\left(\frac{a}{r}\right)^n P_n(\cos\theta), \quad r \geqslant a,$$
$$\Omega_- = \frac{1}{a}\sum_0^\infty D_n\left(\frac{r}{a}\right)^n P_n(\cos\theta), \quad r \leqslant a,$$
where C_n, D_n are constants to be determined.

On the sphere the tangential components of H satisfy
$$\hat{n}\times(H_+ - H_-) = K.$$
$$\therefore \quad \frac{1}{a}\left(\frac{\partial\Omega_-}{\partial\theta}\right)_{r=a} - \frac{1}{a}\left(\frac{\partial\Omega_+}{\partial\theta}\right)_{r=a} = I,$$
i.e.
$$\frac{1}{a^2}\sum_0^\infty D_n\dot{P}_n - \frac{1}{a^2}\sum_0^\infty C_n\dot{P}_n = \omega a \sum_0^\infty B_n\dot{P}_n.$$
Hence we deduce that $D_n - C_n = \omega a^3 B_n$. Also on the sphere $\hat{n}\cdot(B_+ - B_-) = 0$, which gives
$$\left(\frac{\partial\Omega_+}{\partial r}\right)_{r=a} = \left(\frac{\partial\Omega_-}{\partial r}\right)_{r=a},$$
i.e.
$$\sum_0^\infty -(n+1)C_nP_n = \sum_0^\infty nD_nP_n.$$
$$\therefore \quad C_n = -\frac{n}{n+1}D_n, \quad D_n\left(1+\frac{n}{n+1}\right) = \omega a^3 B_n.$$
$$\therefore \quad D_n = \left(\frac{n+1}{2n+1}\right)\omega a^3 B_n$$
and
$$\Omega_- = \omega a^2 \sum_0^\infty \frac{n+1}{2n+1} B_n\left(\frac{r}{a}\right)^n P_n(\cos\theta).$$
Since $C_n = -nD_n/(n+1)$, we deduce that
$$\Omega_+ = -\frac{\omega a^3}{r}\sum_0^\infty \frac{n}{2n+1} B_n\left(\frac{a}{r}\right)^n P_n(\cos\theta).$$

Note. The boundary conditions on B, H taken together imply that Ω is *not* continuous across a surface current, unlike electrostatic potential across a surface charge.

Example 9. Show that for a current I flowing in a circular wire of radius a the scalar potential can be written in the form
$$\frac{I}{2}\left\{\left(\frac{r}{a}\right)P_1(\cos\theta) - \frac{1}{2}\left(\frac{r}{a}\right)^3 P_3(\cos\theta) + \frac{1\cdot 3}{2\cdot 4}\left(\frac{r}{a}\right)^5 P_5(\cos\theta) - \ldots\right\}$$
for $r < a$, where r, θ are measured from the centre and from the axis of the wire.

A second circular wire of radius b ($b \ll a$) is concentric with and has its plane perpendicular to that of the first wire. If this wire also carries a current I show that it will experience a couple of magnitude
$$\frac{\pi\mu_0 I^2 b^2}{2a}\left\{1 + O\left(\frac{b^2}{a^2}\right)\right\}.$$

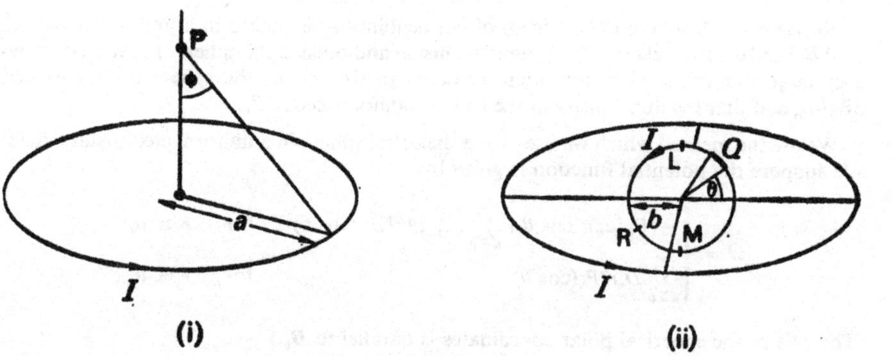

FIG. 8.16

We use the method of generalizing the expression for Ω on the axis to obtain the expression as a series of Legendre polynomials.

The solid angle subtended at P by the circuit, Fig. 8.16 (i), is $2\pi(1-\cos\phi)$ and so

$$\Omega = -\frac{I}{4\pi}2\pi(1-\cos\phi) = -\frac{I}{2}\left\{1-\frac{z}{(z^2+a^2)^{1/2}}\right\}.$$

When $|z| < a$,
$$\Omega = -\frac{I}{2}\left\{1-\frac{z}{a}\left(1+\frac{z^2}{a^2}\right)^{-1/2}\right\}$$

$$= -\frac{I}{2}\left\{1-\frac{z}{a}+\frac{1}{2}\left(\frac{z}{a}\right)^3-\frac{1\cdot3}{2\cdot4}\left(\frac{z}{a}\right)^5+\dots\right\}.$$

We replace z^n by the harmonic function $r^n P_n(\cos\theta)$, which reduces to z when $\theta = 0$ and obtain

$$\Omega(r,\theta) = -\frac{I}{2}\left[1-\left(\frac{r}{a}\right)P_1(\cos\theta)+\frac{1}{2}\left(\frac{r}{a}\right)^3P_3(\cos\theta)-\frac{1\cdot3}{2\cdot4}\left(\frac{r}{a}\right)^5P_5(\cos\theta)+\dots\right],$$

which is equivalent to the stated result since the constant term can be discarded.

To find the force on the small circle we need to know the radial component B_r of the field on the circle $r = b$, viz.

$$B_r = \mu_0 H_r = -\mu_0\left(\frac{\partial\Omega}{\partial r}\right)_{r=b} = \frac{\mu_0 I}{2a}\left\{P_1-\frac{3}{2}\left(\frac{b}{a}\right)^2P_3+\frac{1\cdot3\cdot5}{2\cdot4}\left(\frac{b}{a}\right)^4P_5-\dots\right\}.$$

Because of this field the force on an element of length $b\,d\theta$ at Q, Fig. 8.16 (ii), is perpendicular to the plane, is downwards and of magnitude $b\,d\theta IB_r$. This force is

$$b\,d\theta\cdot\mu_0\frac{I^2}{2a}\left\{P_1-\frac{3}{2}\left(\frac{b}{a}\right)^2P_3+\dots\right\}.$$

An equal force acts at R, and the two have a moment about the diameter LM given by

$$d\Gamma = 2b\cos\theta\, b\,d\theta\mu_0\frac{I^2}{2a}\left\{P_1-\frac{3}{2}\left(\frac{b}{a}\right)^2P_3+\dots\right\}.$$

The resultant couple is therefore (since the moment about the initial line integrates to zero)

$$\Gamma = \frac{\mu_0 b^2 I^2}{a}\int_0^\pi \cos\theta\left[P_1-\frac{3}{2}\left(\frac{b}{a}\right)^2P_3+\dots\right]d\theta$$

$$= \frac{\mu_0 b^2 I^2}{a}\left\{\int_0^\pi \cos^2\theta\,d\theta+O\left(\frac{b^2}{a^2}\right)\right\} = \frac{\pi\mu_0 b^2 I^2}{2a}\left\{1+O\left(\frac{b^2}{a^2}\right)\right\}.$$

Example 10. A sphere of soft iron, of permeability μ, is placed in a uniform magnetic field $\boldsymbol{B}$. Find the formulae for the potentials inside and outside the sphere. Prove that, however large the value of μ, the intensity of magnetization of the sphere cannot exceed $3\boldsymbol{B}_0/\mu_0$, and that the flux density in the sphere cannot exceed $3\boldsymbol{B}_0$.

We use the method which we used for a dielectric sphere in a uniform electrostatic field. We suppose the potential function is given by

$$\Omega = \begin{cases} -(B_0/\mu_0)r\cos\theta + \sum_{n=0}^{\infty}(C_n/r^{n+1})P_n(\cos\theta) & \text{for} \quad r \geqslant a, \\ \sum_{n=0}^{\infty} D_n r^n P_n(\cos\theta) & \text{for} \quad r \leqslant a. \end{cases}$$

(The axis of the spherical polar coordinates is parallel to $\boldsymbol{B}_0$.)

The boundary conditions to be satisfied at $r = a$ are

$$[\Omega] = 0, \quad [B_r] = \left[-\mu\frac{\partial\Omega}{\partial r}\right] = 0.$$

$$\therefore \quad -(B_0/\mu_0)aP_1 + \sum_{n=0}^{\infty}(C_n/a^{n+1})P_n = \sum_{n=0}^{\infty} D_n a^n P_n,$$

$$-\mu_0(B_0/\mu_0)P_1 - \mu_0\sum_{n=0}^{\infty}(n+1)(C_n/a^{n+2})P_n = \mu\sum_{n=0}^{\infty} nD_n a^{n-1}P_n.$$

We equate the coefficients of the various P_n in these separately:

$$\frac{C_0}{a} = D_0; \quad -\frac{B_0 a}{\mu_0} + \frac{C_1}{a^2} = D_1 a; \quad \frac{C_n}{a^{n+1}} = D_n a^n \quad (n \geqslant 2):$$

$$-\frac{C_0}{a^2} = 0; \quad -B_0 - \frac{2\mu_0 C_1}{a^3} = \mu D_1; \quad -\frac{\mu_0(n+1)C_n}{a^{n+2}} = \mu n D_n a^{n-1} \quad (n \geqslant 2).$$

Only the centre pair of equations have a non-trivial solutio n. (This is usually assumed a the start of the solution of these problems in which the undisturbed field is uniform.)

The solution of the central pair gives

$$C_1 = \frac{a^3(\mu-\mu_0)}{\mu_0(\mu+2\mu_0)}B_0, \quad D_1 = -\frac{3B_0}{\mu+2\mu_0}.$$

The field inside the sphere is given by the potential

$$\Omega = -\frac{3B_0}{\mu+2\mu_0}r\cos\theta.$$

$$\therefore \quad \boldsymbol{B}/\mu = \boldsymbol{H} = -\textbf{grad}\,\Omega = +\frac{3\boldsymbol{B}_0}{\mu+2\mu_0}.$$

$$\therefore \quad \boldsymbol{B} = \frac{3\mu}{\mu+2\mu_0}\boldsymbol{B}_0.$$

We find the magnetization, from the relation $\boldsymbol{B} = \mu_0(\boldsymbol{H}+\boldsymbol{M})$, to be

$$\boldsymbol{M} = \frac{3(\mu-\mu_0)}{\mu+2\mu_0}\frac{\boldsymbol{B}_0}{\mu_0}.$$

Thus, if μ is very large,

$$\boldsymbol{B} \approx 3\boldsymbol{B}_0, \quad \boldsymbol{M} \approx 3\boldsymbol{B}_0/\mu_0.$$

(The electrostatic case is similar to this with ε, ε_0 replacing μ, μ_0. The limiting situation $\mu \to \infty$ corresponds, in electrostatics, to a conducting sphere placed in a uniform field.)

Exercises 8.3

1. A nearly spherical conductor is bounded by the surface

$$r = a(1+\lambda \cos \theta),$$

where λ is small. It is insulated and uncharged and placed in a field of potential $-Fr \cos \theta$. Show that, if λ^2 and higher powers of λ are neglected, the potential of the disturbed field is

$$V = -Fr\left(1-\frac{a^3}{r^3}\right) \cos \theta+\frac{\lambda F a^4}{r^3} (3 \cos^2 \theta-1).$$

2. Current flows through a medium of uniform conductivity σ between two nearly concentric spheres of radii b, a $(b > a)$ whose centres are a small distance $\varkappa a$ apart. The potential difference between the electrodes is V_0. Prove that the current density at the outer electrode is

$$\frac{\sigma V_0 a}{(b-a)b}\left[1+\frac{3ab^2\varkappa \cos \theta}{b^3-a^3}\right],$$

where θ is the angle between the line of centres and the radius vector to the point at which the current density is specified.

3. A point charge Q is placed at a distance c from the centre of an earthed conducting sphere of radius a, coated with a layer of dielectric of constant K, whose outer surface is a concentric sphere of radius b $(< c)$. Show that the potential at a point P in this layer is

$$\frac{Q}{4\pi\varepsilon_0 c}\sum_{n=0}^{\infty}\frac{(2n+1)b^{2n+1}(r^n-a^{2n+1}r^{-n-1})}{c^n\{(Kn+n+1)b^{2n+1}+(n+1)(K-1)a^{2n+1}\}}P_n(\cos \theta),$$

where r is the radius vector drawn from the centre of the sphere to P, and θ is the angle made by this radius vector with the line joining the charge to the centre of the sphere.

4. If U denotes the potential at a point P when a conductor is raised to unit potential, prove that, if a charge Q is placed at P and the conductor earthed, then the charge which is induced on the conductor is $-QU$.

The surface of a nearly spherical earthed conductor is given by $r = a+\lambda P_2(\cos \theta)$ where λ is small. Show that, neglecting squares and higher powers of λ, the charge induced by a point charge Q placed at the point $(r = c,\ \theta = 0,\ \phi = 0)$ is

$$-Q\left(\frac{a}{c}+\frac{\lambda a^2}{c^3}\right).$$

5. A point charge e is placed at a distance c from the centre O of a spherical cavity of radius a $(c < a)$, in a uniform and otherwise unbounded dielectric of specific inductive capacity k. Show that the part of the potential in the cavity due to the presence of the dielectric is

$$\frac{Q(1-k)}{4\pi\varepsilon_0 a}\sum_{n=1}^{\infty}\frac{n+1}{n+k+nk}\left(\frac{cr}{a^2}\right)^n P_n(\cos \theta),$$

where (r, θ, ϕ) are the spherical polar coordinates of a field point, r being measured from O as origin, and θ being measured from the line joining O to the charge.

Show further that the force on the point charge is away from O and of amount

$$\frac{(k-1)Q^2}{4\pi\varepsilon_0 a^2}\sum_{n=1}^{\infty}\frac{n(n+1)}{n+k+nk}\left(\frac{c}{a}\right)^{2n-1}.$$

6. An earthed conducting sphere of radius a is introduced into a permanent electric field whose potential at any point before the introduction of the sphere can be written in terms of integral spherical harmonics with centre at the centre of the sphere, in the form

$$V = \sum_0^\infty V_n.$$

Show that the surface density of the electrification induced on the sphere is

$$-\frac{\varepsilon_0}{4\pi a}\sum_0^\infty (2n+1)V_n.$$

7. The centre O of an earthed conducting sphere of radius a is at the centre of a thin circular wire of radius b which surrounds the sphere and carries a charge Q uniformly distributed round its circumference. Prove that the potential at any point P outside the sphere due to the charge induced on the sphere is

$$-\frac{Qa}{4\pi\varepsilon_0 br}\left[1+\sum_{n=1}^\infty \frac{1\cdot 3\cdot 5\ldots(2n-1)}{2\cdot 4\cdot 6\ldots 2n}\left(-\frac{a^4}{b^2r^2}\right)^n P_{2n}(\cos\theta)\right],$$

where r is the distance OP and θ is the angle between OP and the normal Oz to the plane of the wire. Using this result, or otherwise, prove that the potential due to the whole charge at a point on Oz outside the sphere and at a distance r from O is

$$\{Q(b^2+r^2)^{-1/2} - Qa(b^2r^2+a^4)^{-1/2}\}/(4\pi\varepsilon_0).$$

8. A soft iron shell of relative permeability μ is bounded by concentric spheres of radii a, b $(b > a)$. A magnetic dipole of moment $\boldsymbol{M}$ is at the common centre O. Show that the field outside the shell is identical with the field of a dipole $\boldsymbol{M}'$ at O in the absence of the shell, where

$$\boldsymbol{M}' = \boldsymbol{M}\left[1+\frac{2}{9}\,\frac{(\mu-1)^2}{\mu}\left(1-\frac{a^3}{b^3}\right)\right]^{-1}.$$

9. A spherical specimen of diamagnetic material (permeability $\mu_1 < \mu_0$) of radius λa is surrounded by a spherical shell of paramagnetic material (permeability $\mu_2 > \mu_0$) of inner radius λa and outer radius a. When the whole system is placed in a uniform external magnetic field in free space, it is found that the magnetic intensity in the inner sphere is equal to that of the uniform field. Show that

$$3\mu_2(\mu_0-\mu_1) = 2(1-\lambda^3)(\mu_2-\mu_0)(\mu_2-\mu_1).$$

8.4 Two-dimensional problems

Electric and magnetic fields are said to be two-dimensional when the field vectors at any point are always parallel to a fixed plane Oxy and are independent (in magnitude and direction) of the distance of the field point from this plane, i.e. the components of $\boldsymbol{D}$ and $\boldsymbol{B}$ are, in general, functions only of two space variables and the time. Briefly, the field in all planes parallel to Oxy is the same. For purposes of calculating forces and energy densities a layer of unit thickness is considered. Instead of, say, a point charge in three dimensions we now have a line-charge consisting of charge distributed continuously and uniformly along an infinite line perpendicular to Oxy; the line-charge is represented on a diagram by the point in which the line cuts Oxy and its strength is measured by the charge per unit length. The conductors in the field are cylinders with generators perpendicular to Oxy and are represented on a diagram by the curves in which they intersect the plane Oxy.

We can use any suitable coordinate system in the plane Oxy. In general, the most suitable coordinates are either plane polar coordinates [see (1) below] or rectangular cartesian coordinates which in most cases will be combined to form the real and imaginary parts of a single complex variable. The complex variable method is extremely powerful and is considered in detail on pp. 305–314.

1. Circular harmonics and their applications

The standard method of solving Laplace's equation by separation of variables can be applied to two-dimensional electromagnetic potential fields. When we use polar coordinates (r, θ) in the plane of the field, Laplace's equation is

$$\frac{\partial^2 V}{\partial x^2}+\frac{\partial^2 V}{\partial y^2} = \nabla^2 V = \frac{\partial^2 V}{\partial r^2}+\frac{1}{r}\frac{\partial V}{\partial r}+\frac{1}{r^2}\frac{\partial^2 V}{\partial \theta^2} = 0.$$

Writing $V = R(r)P(\theta)$ we find

$$\frac{r^2 R''(r)}{R(r)}+\frac{rR'(r)}{R(r)}+\frac{P''(\theta)}{P(\theta)} = 0.$$

If the field is electrostatic or magnetostatic in character the function V must be a single-valued function of position. However, for a magnetic field caused by currents flowing in long parallel wires or other conductors, the plane is a multiply connected region and V may contain cyclic functions.

(a) *Single connection*

In this case $P''/P = -m^2$, where $m = 1, 2, \ldots$. Therefore

$$P'' = -m^2 P, \quad P = A\cos m\theta + B\sin m\theta,$$

$$r^2 R'' + rR' - m^2 R = 0, \quad R = Cr^m + Dr^{-m}.$$

In the above A, B, C, D are arbitrary constants and we therefore obtain four types of solution of Laplace's equation, viz.

$$r^m\cos m\theta, \quad r^m\sin m\theta, \quad r^{-m}\cos m\theta, \quad r^{-m}\sin m\theta.$$

The first two are ***internal harmonics*** which vanish at the origin; the last two are ***external harmonics*** which vanish at infinity.

(b) *Multiple connection*

When the origin is excluded (e.g. by the presence of a wire carrying current) the magnetostatic field is doubly connected and we need cyclic solutions of Laplace's equation. The simplest one, corresponding to $m = 0$, above, is given by $P = A\theta + B$, $R = C \ln r + D$.

In the problems examined in this section the usual uniqueness theorems are assumed to hold. The uniqueness theorems are considered in (3) on p. 307.

Example 1. The potential of a line charge q at the origin O in otherwise empty space.

In this case the field is symmetrical about O, i.e. independent of θ, so that $V = V(r)$, $\mathbf{D} = (D_r, 0)$. Then Gauss's theorem, applied to a circle of radius r and centre O, gives

$$2\pi r D_r = 2\pi r \varepsilon_0 E_r = -2\pi r \varepsilon_0 \frac{dV}{dr} = q.$$

$$\therefore \quad V = -\frac{q}{2\pi\varepsilon_0} \ln r + \text{constant}.$$

[Note that every non-cyclic two-dimensional field symmetric about O must have a potential of the form $C \ln r + D$.]

Example 2. An infinite conducting earthed circular cylinder of radius a is surrounded by a concentric layer of dielectric of thickness $b-a$ and dielectric constant K. The cylinder is set with its axis perpendicular to a uniform electric field which at great distances from the cylinder is of strength E_0 directed along Ox, Fig. 8.17.

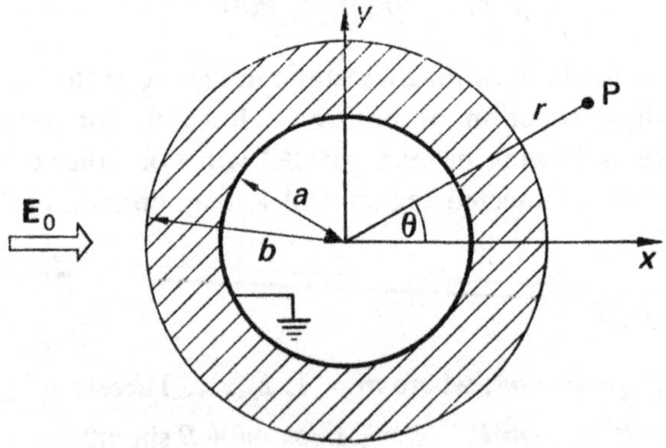

FIG. 8.17

The exciting potential can be taken as $= -E_0 x = -E_0 r \cos\theta$, and suggests that we assume for the total potential

$$V = -E_0 r \cos\theta + \frac{A\cos\theta}{r} \qquad \text{for} \quad r \geqslant b,$$

$$V = Br\cos\theta + \frac{C\cos\theta}{r} \qquad \text{for} \quad b \geqslant r \geqslant a,$$

$$V = 0 \qquad \text{for} \quad r \leqslant a,$$

where A, B, C are constants to be determined.

The boundary conditions are

$$V = 0 \quad \text{at} \quad r = a, \quad \text{giving}$$

$$Ba + C/a = 0, \tag{1}$$

$$[V]_{r=b} = 0, \quad \text{giving}$$

$$-E_0 b + A/b = Bb + C/b, \tag{2}$$

$$[D_r]_{r=b} = 0, \quad \text{giving}$$

$$\varepsilon_0(E_0 + A/a^2) = \varepsilon_0 K(-B + C/a^2). \tag{3}$$

Solution of eqns. (1), (2), (3) gives

$$\frac{A}{(K+1)a^2+(K-1)b^2} = \frac{B}{-2} = \frac{C}{2a^2} = \frac{E_0 b^2}{(K+1)b^2+(K-1)a^2}.$$

This gives the unique solution (see p. 307).

Special cases

1. When $K = 1$ we have the solution for an earthed cylinder of radius a in an otherwise uniform field, i.e.

$$V = -E_0\left(r - \frac{a^2}{r}\right)\cos\theta \quad \text{for} \quad r \geqslant a,$$

$$V = 0 \quad \text{for} \quad r \leqslant a.$$

In this case, the lines of force are as sketched in Fig. 8.18 (i). The disturbance field is the same as that of a dipole of moment $E_0 a^2 \boldsymbol{i}/(2\pi\varepsilon_0)$, where $\boldsymbol{i}$ is a unit vector directed along Ox.

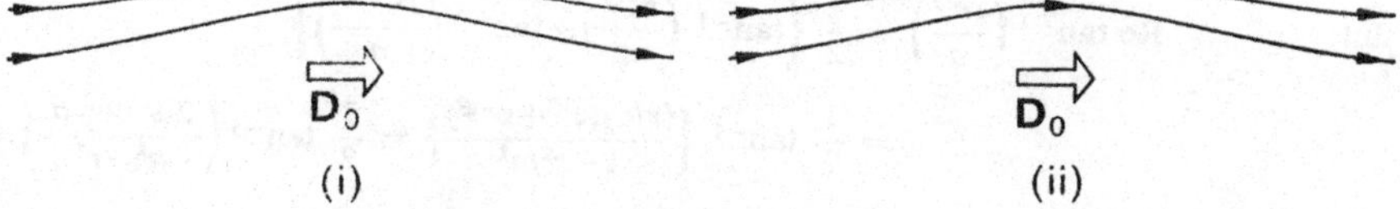

FIG. 8.18

2. When $a = 0$, we have the solution for a dielectric cylinder of radius b in an otherwise uniform field, i.e.

$$V = -E_0 r\cos\theta + \frac{(K-1)E_0 b^2\cos\theta}{(K+1)r} \quad \text{for} \quad r \geqslant b,$$

$$V = -\frac{2E_0 r\cos\theta}{(K+1)} \quad \text{for} \quad r \leqslant b.$$

Note that the field is uniform within the cylinder, being parallel to Ox and of strength $2E_0/(K+1)$. In this case the lines of force are as sketched in Fig. 8.18 (ii).

Example 3. A long hollow cylindrical conductor is divided into two parts by a plane through the axis, and the parts are separated by a small interval. If the two parts are kept at potentials V_1 and V_2, show that the potential at any point within the cylinder is

$$\frac{V_1+V_2}{2}+\frac{V_1-V_2}{\pi}\tan^{-1}\left(\frac{2ar\cos\theta}{a^2-r^2}\right),$$

where the origin is the centre of the cylinder and

$$V = V_1 \quad \text{for} \quad r = a \quad \text{and} \quad -\pi/2 < \theta < \pi/2,$$
$$V = V_2 \quad \text{for} \quad r = a \quad \text{and} \quad \pi/2 < \theta < \pi \quad \text{or} \quad -\pi < \theta < -\pi/2.$$

Since the potential on the conductor is an even function of θ, we try the non-singular solution

$$V = \sum_0^\infty A_n\left(\frac{r}{a}\right)^n \cos n\theta \quad \text{for} \quad r \leqslant a.$$

The boundary condition gives

$$\sum_0^\infty A_n \cos n\theta = \begin{cases} V_1 & \text{for} \quad 0 \leqslant |\theta| < \pi/2, \\ V_2 & \text{for} \quad \pi/2 < |\theta| < \pi, \end{cases} \tag{1}$$

and so the A_n's are the coefficients of the Fourier expansion of the right-hand side of (1). Hence

$$A_0 = \frac{V_1+V_2}{2}, \qquad A_n = \frac{2(V_1-V_2)}{\pi n}\sin\left(\frac{n\pi}{2}\right).$$

Therefore

$$V = \frac{V_1+V_2}{2}+\frac{2(V_1-V_2)}{\pi}\left\{\left(\frac{r}{a}\right)\cos\theta-\left(\frac{r}{a}\right)^3\frac{\cos 3\theta}{3}\right.$$
$$\left.+\ldots+(-1)^n\left(\frac{r}{a}\right)^{2n+1}\frac{\cos(2n+1)\theta}{(2n+1)}+\ldots\right\}. \tag{2}$$

The series on the right-hand side of (2) can be summed as the real part of

$$\frac{re^{i\theta}}{a}-\frac{r^3e^{3i\theta}}{3a^3}+\ldots = \frac{re^{i\theta}}{a}-\frac{1}{3}\left(\frac{re^{i\theta}}{a}\right)^3+\ldots = \tan^{-1}(re^{i\theta}/a).$$

But

$$\operatorname{Re}\tan^{-1}\left(\frac{re^{i\theta}}{a}\right) = \frac{1}{2}\left\{\tan^{-1}\left(\frac{re^{i\theta}}{a}\right)+\tan^{-1}\left(\frac{re^{-i\theta}}{a}\right)\right\}$$
$$= \frac{1}{2}\tan^{-1}\left\{\frac{(r/a)(e^{i\theta}+e^{-i\theta})}{1-r^2/a^2}\right\} = \frac{1}{2}\tan^{-1}\left(\frac{2ar\cos\theta}{a^2-r^2}\right).$$

Substitution in (2) gives the required result.

---

2. The use of the method of images

The method of images applies equally well to two-dimensional problems. For example, the image of a line charge q in a parallel earthed plane, from which it is distance a, is the optical image line charge $-q$, the charge q being attracted to the plane by a force $q^2/(4\pi a\varepsilon_0)$ per unit length.

Example 1. An electrified line charge q per unit length is parallel to a circular cylinder of radius a and dielectric constant K, the distance of the wire from the centre of the cylinder being c. Show that the force per unit length of the wire is

$$\frac{(K-1)a^2q^2}{2\pi\varepsilon_0(K+1)\,c(c^2-a^2)}.$$

With the notation of Fig. 8.19, the exciting potential due to the line charge is

$$\begin{aligned}
V_0 &= -\frac{q}{2\pi\varepsilon_0}\ln r' + \text{constant}\\
&= -\frac{q}{2\pi\varepsilon_0}\ln\sqrt{(c^2+r^2-2cr\cos\theta)} + \text{constant}\\
&= -\frac{q}{4\pi\varepsilon_0}\ln\{(c-re^{i\theta})(c-re^{-i\theta})\} + \text{constant}\\
&= -\frac{q}{4\pi\varepsilon_0}\left\{\log\left(1-\frac{r}{c}e^{i\theta}\right)+\log\left(1-\frac{r}{c}e^{-i\theta}\right)\right\} + \text{constant}\\
&= \frac{q}{2\pi\varepsilon_0}\sum_1^\infty\left(\frac{r}{c}\right)^n\frac{\cos n\theta}{n} + \text{constant} \quad \text{for} \quad r \leqslant c.
\end{aligned}$$

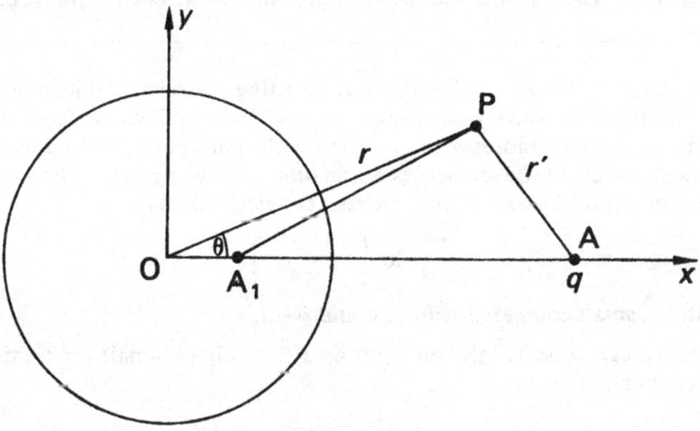

FIG. 8.19

We assume therefore that the non-singular disturbance potential is

$$V_1\begin{cases} = \dfrac{q}{2\pi\varepsilon_0}\sum_1^\infty A_n\left(\dfrac{a}{r}\right)^n\cos n\theta & \text{for} \quad r \geqslant a,\\[2ex] = \dfrac{q}{2\pi\varepsilon_0}\sum_1^\infty B_n\left(\dfrac{r}{a}\right)^n\cos n\theta & \text{for} \quad r \leqslant a,\end{cases}$$

so that $V = V_0+V_1$. The continuity of V and D_r at $r = a$ gives

$$A_n = B_n, \quad K\left[\frac{a^{n-1}}{c^n}+\frac{nB_n}{a}\right] = \frac{a^{n-1}}{c^n}-\frac{nA_n}{a},$$

i.e.

$$A_n = B_n = -\frac{(K-1)a^n}{n(K+1)c^n}.$$

The field at the line charge due to the cylinder is $(-\mathbf{grad}\ V_1)_{r=c,\ \theta=0}$ and is therefore

$$\frac{q}{2\pi\varepsilon_0}\sum_1^\infty \frac{na^n A_n}{c^{n+1}} = -\frac{q(K-1)}{2\pi\varepsilon_0(K+1)}\sum_1^\infty \frac{a^{2n}}{c^{2n+1}} = -\frac{q(K-1)a^2}{2\pi\varepsilon_0(K+1)c(c^2-a^2)}$$

directed along OA. The attraction per unit length on the line charge is therefore as stated.

Note. If we let $K \to \infty$ we obtain the field of a line charge outside a fixed conducting insulated uncharged circular cylinder. We find the disturbance potential is

$$V_2 = -\frac{q}{2\pi\varepsilon_0}\sum_1^\infty \left(\frac{a^2}{rc}\right)^n \frac{\cos n\theta}{n}$$

$$= \frac{q}{4\pi\varepsilon_0}\left\{\log\left(1-\frac{a^2}{rc}e^{i\theta}\right)+\log\left(1-\frac{a^2}{rc}e^{-i\theta}\right)\right\}$$

$$= \frac{q}{4\pi\varepsilon_0}\left[\ln\left\{r^2+\left(\frac{a^2}{c}\right)^2-2r\frac{a^2}{c}\cos\theta\right\}-2\ln r\right]$$

$$= \frac{q}{2\pi\varepsilon_0}\ln r_1-\frac{q}{2\pi\varepsilon_0}\ln r,$$

where r_1 is the distance of the field point P from A_1 the inverse of A in the circle. Therefore the image system consists of a line charge $-q$ at A_1, where $OA_1 = a^2/c$, and a line charge q at O. The solution for the case when the cylinder is earthed can be obtained by removing the line charge at O. These results can be obtained directly as in example 4, p. 274.

Example 2. A thin plane sheet of uniform conducting material of thickness t and conductivity σ fills the half-plane $0 \leqslant x \leqslant \infty$, $-\infty \leqslant y \leqslant \infty$. Two perfectly conducting circular electrodes, each of radius δ, are inserted with centres at $(a, 0)$ and $(b, 0)$ $(b > a)$. Current enters at one electrode and leaves at the other. Show, by the method of images or otherwise, that the equivalent resistance between the electrodes is

$$\frac{1}{2\pi\sigma t}\ln\left[\frac{(b^2-a^2)^2}{4ab\delta^2}\right],$$

provided that δ is small compared with a, b and $b-a$.

In this case we can consider the problem as a (two-dimensional) problem of flow in a sheet with conductivity σt.

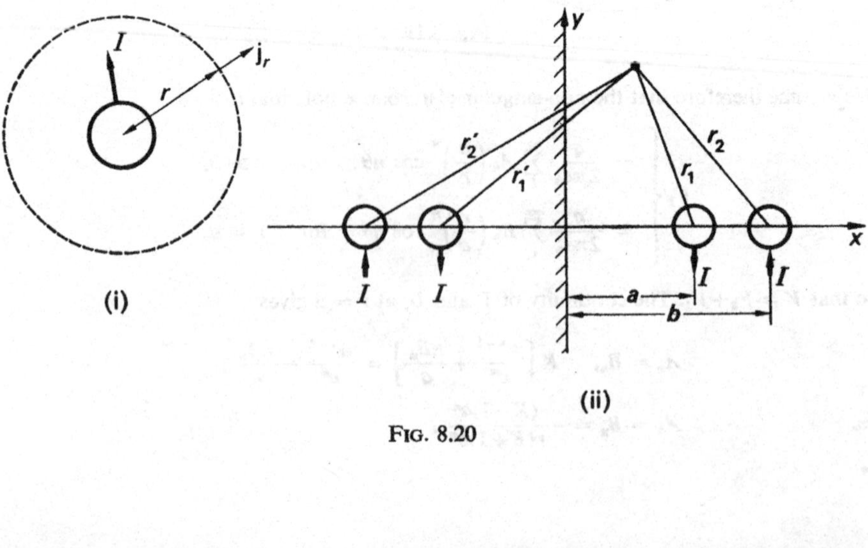

FIG. 8.20

Using two-dimensional polar coordinates, Fig. 8.20 (i), the radial current due to an isolated anode emitting current I is given by

$$2\pi r j_r = I.$$

$$\therefore\ j_r = \frac{I}{2\pi r}, \qquad V = -\frac{I}{2\pi\sigma t}\ln r + \text{constant}.$$

In the given problem the image system is as shown in Fig. 8.20 (ii). (Quantities of order δ/a, etc., have been neglected.) We can take the potential V as

$$-\frac{I}{2\pi\sigma t}[\ln r_1 - \ln r_2 + \ln r_1' - \ln r_2'] = -\frac{I}{2\pi\sigma t}\ln\left(\frac{r_1 r_1'}{r_2 r_2'}\right).$$

On the anode $V = V_a \approx -\dfrac{I}{2\pi\sigma t}\ln\left\{\dfrac{\delta\cdot 2a}{(b-a)(b+a)}\right\}$.

On the cathode $V = V_c \approx -\dfrac{I}{2\pi\sigma t}\ln\left\{\dfrac{(b-a)(b+a)}{\delta\cdot 2b}\right\}$.

The equivalent resistance is $(V_a - V_c)/I$, which gives the required result.

---

3. The use of conjugate functions

In two-dimensional potential fields of any kind, viz. electrostatic fields in the absence of volume charge, magnetic fields arising from a distribution of currents in wires, or the electric field in a conducting medium carrying a current, the potential satisfies Laplace's equation in two dimensions, viz.

$$\frac{\partial^2 V}{\partial x^2} + \frac{\partial^2 V}{\partial y^2} = 0. \qquad (8.8)$$

The use of functions of a complex variable $z(= x+iy)$ in such two-dimensional problems relies on the fact that an analytic function $w(z)$ satisfies the Cauchy–Riemann conditions, viz.

$$\frac{\partial u}{\partial x} = \frac{\partial v}{\partial y}, \qquad \frac{\partial u}{\partial y} = -\frac{\partial v}{\partial x},$$

where $w = u+iv$. Here u and v are called conjugate functions. These equations imply that both u and v satisfy Laplace's equation (8.8) and that the curves u = constant in the z-plane are orthogonal to the curves v = constant. We can therefore take u (or v) as the potential function for a field so that the lines v = constant (or u = constant) represent the field lines. The choice of u or v as the potential function is immaterial; in this book we shall usually take the real part u to be the potential and call the function $w(z)$ $(= u+iv)$ the complex potential of the field. In an electrostatic field any equipotential surface can be replaced by a conducting cylinder held at the appropriate potential, so that the field corresponding to an arbitrary function $w(z)$ can have any of the surfaces u = constant as its boundaries.

Since

$$E_x = -\frac{\partial u}{\partial x}, \quad E_y = -\frac{\partial u}{\partial y},$$

the Cauchy–Riemann conditions give

$$E_x - \mathrm{i}E_y = -\left(\frac{\partial u}{\partial x} - \mathrm{i}\frac{\partial u}{\partial y}\right) = -\left(\frac{\partial u}{\partial x} + \mathrm{i}\frac{\partial v}{\partial x}\right) = -\frac{\mathrm{d}w}{\mathrm{d}z}.$$

Hence, if the electric field has magnitude E in a direction making an angle θ with the real axis,

$$E_x - \mathrm{i}E_y = E\mathrm{e}^{-\mathrm{i}\theta} = -\frac{\mathrm{d}w}{\mathrm{d}z}.$$

$$\therefore \; E = \left|\frac{\mathrm{d}w}{\mathrm{d}z}\right|, \quad \theta = \pi - \arg\left(\frac{\mathrm{d}w}{\mathrm{d}z}\right). \tag{8.9}$$

If a line charge of strength q is situated at the origin, the potential function is $u = -(q \ln r)/(2\pi\varepsilon_0)$. [See p. 300.]

$$\therefore \; \frac{\partial u}{\partial x} = -\frac{qx}{2\pi\varepsilon_0 r^2} = \frac{\partial v}{\partial y}, \quad \frac{\partial u}{\partial y} = -\frac{qy}{2\pi\varepsilon_0 r^2} = -\frac{\partial v}{\partial x}.$$

$$\therefore \; v = -\frac{q}{2\pi\varepsilon_0}\tan^{-1}\left(\frac{y}{x}\right).$$

Hence the complex potential for the field is given by

$$u + \mathrm{i}v = -\frac{q}{2\pi\varepsilon_0}(\ln r + \mathrm{i}\theta),$$

i.e.

$$w(z) = -\frac{q}{2\pi\varepsilon_0}\log z. \tag{8.10}$$

Similarly, if the charge is at $z = z_0$, the complex potential is $-q/(2\pi\varepsilon_0)\log(z - z_0)$, and, for a number of charges q_s at z_s,

$$w = -\sum_s \{q_s/(2\pi\varepsilon_0)\}\log(z - z_s).$$

For the magnetic field arising from a current I at the origin the potential function (denoted by Ω in § 6.5) is

$$u = -\{I/(2\pi)\}\,\theta = -\{I/(2\pi)\}\tan^{-1}(y/x)$$

at the field point (x, y) [or (r, θ) in polar coordinates]. The corresponding complex potential is

$$u + \mathrm{i}v = w(z) = \frac{\mathrm{i}I}{2\pi}\log z. \tag{8.11}$$

Similarly the complex potential for the current flow from an electrode emitting current I into a medium of conductivity σ is

$$u+iv = w(z) = -\frac{I}{2\pi\sigma}\log z. \tag{8.12}$$

The relations corresponding to (8.9) for the magnetic field, magnitude B in direction θ, are

$$B = \mu\left|\frac{dw}{dz}\right|, \quad \theta = \pi - \arg\left(\frac{dw}{dz}\right); \tag{8.13}$$

and for the current density, magnitude J in direction θ, are

$$J = \sigma\left|\frac{dw}{dz}\right|, \quad \theta = \pi - \arg\left(\frac{dw}{dz}\right). \tag{8.14}$$

Similar generalizations for a number of sources can be made as in the electrostatic case.

Uniqueness theorem for two-dimensional problems

The uniqueness theorems of § 8.1 hold equally well for two-dimensional electromagnetic potential fields when the system is bounded externally; thus, in electrostatics if

(1) the potential of or the charge carried by each conductor is given,
(2) the strength and positions of all line charges and multiplets are given, the potential near a line charge q being $-(q \ln r)/(2\pi\varepsilon_0)+O(1)$, then the field is unique. Similar results hold for magnetostatic fields due to line magnets or line currents of known strengths.

However, when the fields extend to infinity in the complex plane some modifications must be made in the standard boundary conditions. We consider here two important cases. First, in electrostatics when there is a net line charge Q in the field we impose the condition that the potential at infinity behaves like $\{-Q/(2\pi\varepsilon_0)\} \ln R+O(1/R)$. Second, in magnetostatics when there is a net current flow I in the field we impose the condition that the potential at infinity behaves like $-I\theta/(2\pi)+O(1/R)$. These conditions make the surface integrals in the proofs of the uniqueness theorems vanish over the infinite boundaries.

When the electric field extends to infinity and is uniform there and of strength E_0 parallel to the x-axis, we impose the condition that the potential at infinity is $-E_0x+O(1/R)$.

Example 1. A charge of density q per unit length is distributed uniformly over the line $x = 0$, $y = a\ (a > 0)$. An infinite earthed conducting plane occupies the plane $y = 0$. Show that the potential in the region $y \geqslant 0$ is given by the real part of

$$\frac{-q}{2\pi\varepsilon_0}\log\left(\frac{z-ai}{z+ai}\right), \quad \text{where} \quad z = x+iy.$$

Show also that the force per unit length acting on the line charge is $q^2/(4\pi\varepsilon_0 a)$.

This complex potential gives the field due to a line charge of strength q through the point ai and $-q$ through $-ai$. The latter is outside the region $y \geqslant 0$. In fact, this is the potential of a charge and its image which has to satisfy the conditions:

(a) a singularity, the charge, is at $z = ia$;

(b) the plane $y = 0$ is a conductor.

We put $y = 0$ into the complex potential and obtain

$$-\frac{q}{2\pi\varepsilon_0}\log\left(\frac{x-ai}{x+ai}\right).$$

The fraction in brackets is the quotient of a number and its conjugate complex, and so for all values of x, when $y = 0$,

$$w = -\frac{q}{2\pi\varepsilon_0}\{\ln 1+2i\tan^{-1}(x/a)\} = u+iv.$$

$$\therefore \quad u = 0 \quad \text{for all values of } x.$$

This shows that the given potential corresponds to the stated conditions.

We calculate the force on the given charge as that arising from the attraction of the image charge.

Example 2. A complex potential w has as its real part the electrostatic potential u for a two-dimensional system of charged conductors. Show that the total force on any one conductor has components X and Y given by

$$X-iY = \frac{-i\varepsilon_0}{2}\int\left(\frac{dw}{dz}\right)^2 dz$$

where the contour integral is taken round the boundary of that conductor.

Two line charges of strength q and $-q$ are placed at the points c and ic respectively outside the circular cylinder $|z| = a$, which is an uncharged conductor. Show that the total force on the conductor is

$$\frac{\sqrt{2}q^2a^2c^3}{\pi\varepsilon_0(c^2-a^2)(c^4+a^4)}.$$

Consider an element PQ of the conductor (see Fig. 8.21) where the field strength has magnitude E and the charge density is $\sigma = \varepsilon_0 E$. The force per unit length on the conductor is $\sigma^2/(2\varepsilon_0) = \frac{1}{2}\varepsilon_0 E^2$ and acts in the direction of E. The components dX, dY of the force acting on this element are

$$dX = \tfrac{1}{2}\varepsilon_0 E^2\, ds \sin\psi, \quad dY = -\tfrac{1}{2}\varepsilon_0 E^2\, ds\cos\psi,$$

where ds is the length of the element PQ. Corresponding to the displacement PQ, $dz = ds\, e^{i\psi}$.

$$\therefore \quad dX - i\,dY = \tfrac{1}{2}\varepsilon_0 E^2(\sin\psi + i\cos\psi)\, ds = \tfrac{1}{2}i\varepsilon_0 E^2\, dz\, e^{-2i\psi}.$$

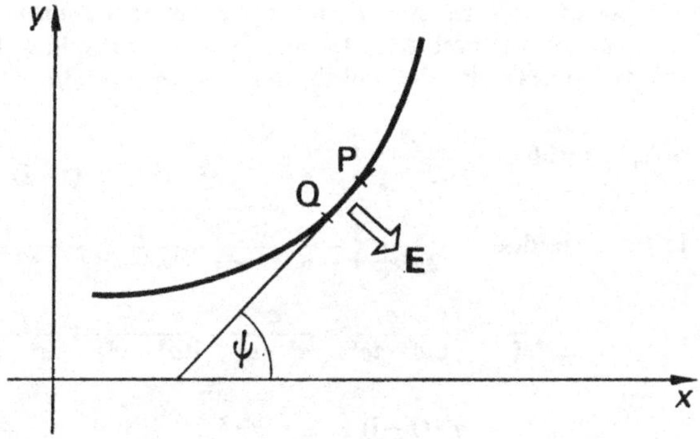

FIG. 8.21

But from eqn. (8.9), with $\theta = -\pi/2+\psi$, $E = -(\mathrm{d}w/\mathrm{d}z)\,\mathrm{e}^{\mathrm{i}(\psi-\pi/2)}$.

$$\therefore \quad E = \mathrm{i}\left(\frac{\mathrm{d}w}{\mathrm{d}z}\right)\mathrm{e}^{\mathrm{i}\psi}, \quad \text{and} \quad \mathrm{d}X-\mathrm{i}\,\mathrm{d}Y = -\frac{1}{2}\,\mathrm{i}\varepsilon_0\left(\frac{\mathrm{d}w}{\mathrm{d}z}\right)^2\mathrm{d}z.$$

The total force on the conductor is therefore as stated by integration around the boundary (cf. Blasius's theorem in hydrodynamics).

We can write down the complex potential for the special system given by using the image charges $-q$ at a^2/c and q at $\mathrm{i}a^2/c$, Fig. 8.22. The potential is

$$w(z) = -\frac{q}{2\pi\varepsilon_0}\{\log(z-c)-\log(z-\mathrm{i}c)-\log(z-a^2/c)+\log(z-\mathrm{i}a^2/c)\}.$$

$$\therefore \quad \frac{\mathrm{d}w}{\mathrm{d}z} = \frac{q}{2\pi\varepsilon_0}\left\{\frac{1}{z-\mathrm{i}c}-\frac{1}{z-c}+\frac{1}{z-a^2/c}+\frac{1}{z-\mathrm{i}a^2/c}\right\}.$$

$$\therefore \quad \left(\frac{\mathrm{d}w}{\mathrm{d}z}\right)^2 = \frac{q^2}{4\pi^2\varepsilon_0^2}\left\{\frac{1}{(z-\mathrm{i}c)^2}+\frac{1}{(z-c)^2}+\frac{1}{(z-a^2/c)^2}+\frac{1}{(z-\mathrm{i}a^2/c)^2}\right.$$
$$-\frac{2}{(z-\mathrm{i}c)(z-c)}+\frac{2}{(z-\mathrm{i}c)(z-a^2/c)}-\frac{2}{(z-\mathrm{i}c)(z-\mathrm{i}a^2/c)}$$
$$\left.-\frac{2}{(z-c)(z-a^2/c)}+\frac{2}{(z-c)(z-\mathrm{i}a^2/c)}-\frac{2}{(z-a^2/c)(z-\mathrm{i}a^2/c)}\right\}.$$

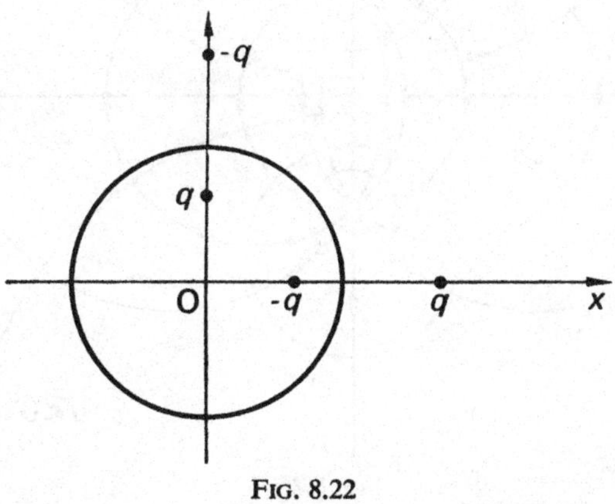

FIG. 8.22

To evaluate the forces in the form $X-iY$ we must evaluate a contour integral taken around the circle $|z| = a$. This is best done by the theorem of residues. In this case the integrand $(dw/dz)^2$ has two poles inside the contour at $z = a^2/c$ and ia^2/c. The residues are:

$$\text{at} \quad z = a^2/c, \quad \text{residue} \quad \frac{q^2}{2\pi^2\varepsilon_0^2}\left\{\frac{c}{a^2-ic^2}-\frac{c}{a^2-c^2}-\frac{c}{a^2(1-i)}\right\};$$

$$\text{at} \quad z = ia^2/c, \quad \text{residue} \quad \frac{q^2}{2\pi^2\varepsilon_0^2}\left\{-\frac{c}{i(a^2-c^2)}+\frac{c}{ia^2-c^2}-\frac{c}{a^2(i-1)}\right\}.$$

$$\therefore \quad X-iY = \frac{-i\varepsilon_0}{2}\cdot\frac{q^2}{2\pi^2\varepsilon_0^2}\cdot 2\pi i\left\{\frac{c}{a^2-ic^2}-\frac{c}{a^2-c^2}-\frac{c}{i(a^2-c^2)}+\frac{c}{ia^2-c^2}\right\}.$$

This reduces to

$$X-iY = -\frac{q^2\cdot(1-i)}{\pi\varepsilon_0}\,\frac{a^2c^3}{(a^4+c^4)(a^2-c^2)}.$$

The total force, or resultant, is the modulus of the complex number, i.e. $|X-iY| = \sqrt{(X^2+Y^2)}$, and so the force is

$$|X-iY| = \frac{q^2\sqrt{2}a^2c^3}{\pi\varepsilon_0(a^4+c^4)(c^2-a^2)}.$$

Example 3. Using the potential

$$w = \frac{I}{2\pi\sigma}\log\frac{z+ic}{z-ic},$$

or otherwise, show that the resistance per unit length between two equal infinite straight parallel cylindrical conductors of radius a placed with their axes parallel at a distance $2b$ apart in an infinite conducting medium of conductivity σ is $(1/\pi\sigma)\cosh^{-1}(b/a)$.

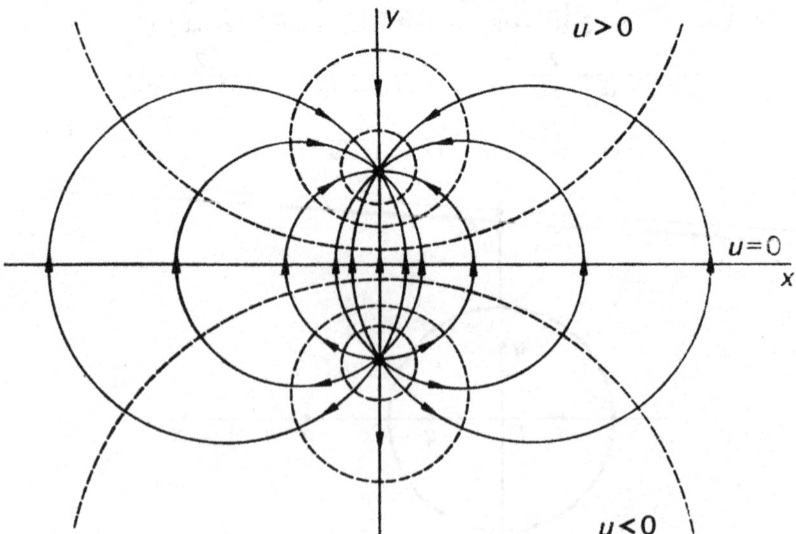

FIG. 8.23

The real and imaginary parts of w are

$$u = \frac{I}{2\pi\sigma} \ln \left| \frac{z+ic}{z-ic} \right|, \qquad v = \frac{I}{2\pi\sigma} \arg \left(\frac{z+ic}{z-ic} \right).$$

The equipotential surfaces (shown dotted in Fig. 8.23) are the coaxial circles having $\pm ic$ as limiting points. The lines of current flow are the orthogonal circles having $\pm ic$ as common points. The potential function corresponds to point electrodes of strengths $+I$, $-I$ at the points $-ic$, $+ic$ respectively.

However, we could use any of the other equipotential surfaces as electrodes as long as they have the appropriate potential. We find the relation between potential, position and radius as follows.

From the original potential

$$\frac{z+ic}{z-ic} = \exp(2\pi\sigma w/I).$$

$$\therefore \quad \frac{z+ic}{z-ic} \cdot \frac{z^*-ic}{z^*+ic} = \exp\{2\pi\sigma(w+w^*)/I\} = \exp(4\pi\sigma u/I).$$

But

$$\frac{(z+ic)(z^*-ic)}{(z-ic)(z^*+ic)} = \frac{x^2+y^2+c^2-2cy}{x^2+y^2+c^2+2cy}.$$

$$\therefore \quad (x^2+y^2+c^2)(1-e^{4\pi\sigma u/I}) = 2cy(1+e^{4\pi\sigma u/I}),$$

i.e.

$$x^2+y^2-2cy \coth(2\pi\sigma u/I)+c^2 = 0.$$

This represents a circle with centre at $ic \coth(2\pi\sigma u/I)$ and radius $c \operatorname{cosech}(2\pi\sigma u/I)$. Hence, for the circle as an electrode,

$$b = c \coth(2\pi\sigma u/I), \qquad a = c \operatorname{cosech}(2\pi\sigma u/a),$$

$$\therefore \quad \frac{b}{a} = \cosh\left(\frac{2\pi\sigma u}{I}\right).$$

If u_1 is the potential of the emitting electrode at ic and u_2 is that of the absorbing electrode at $-ic$,

$$u_1 = \frac{I}{2\pi\sigma} \cosh^{-1}\left(\frac{b}{a}\right), \qquad u_2 = -\frac{I}{2\pi\sigma} \cosh^{-1}\left(\frac{b}{a}\right).$$

Hence the resistance is

$$R = \frac{u_1-u_2}{I} = \frac{1}{\pi\sigma} \cosh^{-1}\left(\frac{b}{a}\right).$$

---

4. The use of conformal transformations

Further use can be made of the principles of the theory of functions of a complex variable to transform the solution of one problem given by one potential $w(z)$ into the solution of a different problem by means of a conformal transformation.

In a conformal transformation one complex plane, the z-plane, is mapped on another, the ζ-plane, when z $(= x+iy)$ is given as an analytic function of ζ $(= \xi+i\eta)$. If a field is represented in the z-plane by a potential $w(z)$, the transformation $z = f(\zeta)$ gives a corresponding field in the ζ-plane, the potential of the latter field being $w\{f(\zeta)\}$. The usual properties of a conformal transformation then apply to the relation between the two fields. Thus the boundaries of one transform to the boundaries of the other; the singularities (charges, dipoles, etc.) transform to corresponding singularities of the other; the field-lines are rotated and lengths magnified by the transformation to give one field from the other; the charges on corresponding sections of conductors are equal. The use of the complex potential can be regarded as a special case of a conformal transformation when the field in the z-plane is the uniform field having lines parallel to the real axis as field lines and lines parallel to the imaginary axis as equipotentials; the potential of this field in the z-plane is a constant multiple of the simple function z. Then the conformal transformation $z = w(\zeta)$ transforms the "rectilinear" field of the z-plane into the field with potential w in the ζ-plane. (The real part of $w(\zeta)$, i.e. x, is the potential of the field in the ζ-plane and the field-lines correspond to $y =$ constant.) The use of conformal transformations is illustrated in the following examples.

Example 1. We use a conformal transformation to find the potential of the field of a line-dipole placed between a pair of earthed planes inclined at an angle π/n, Fig. 8.24 (i).

First we consider the transformation $\zeta = z^n$. On the positive real axis $z = x$ $(x > 0)$, so that $\zeta = x^n$ $(x > 0)$. The positive real axis in the z-plane transforms to the positive real axis of the ζ-plane. On the line arg $z = \pi/n$, $z = re^{i\pi/n}$ $(r > 0)$.

$$\therefore \quad \zeta = r^n e^{i\pi}.$$

Hence the negative real axis of the ζ-plane coresponds to the line arg $z = \pi/n$. A point inside the "wedge" is transformed to a point in the upper half of the ζ-plane, Fig. 8.24 (ii).

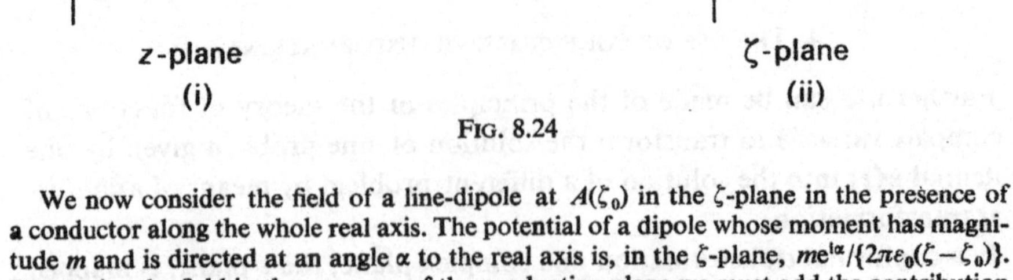

FIG. 8.24

We now consider the field of a line-dipole at $A(\zeta_0)$ in the ζ-plane in the presence of a conductor along the whole real axis. The potential of a dipole whose moment has magnitude m and is directed at an angle α to the real axis is, in the ζ-plane, $me^{i\alpha}/\{2\pi\varepsilon_0(\zeta-\zeta_0)\}$. But to give the field in the presence of the conducting plane we must add the contribution from the image dipole, viz. $me^{-i\alpha}/\{2\pi\varepsilon_0(\zeta-\zeta_0^*)\}$. Hence, in the ζ-plane,

$$w(\zeta) = \frac{m}{2\pi\varepsilon_0}\left\{\frac{e^{i\alpha}}{\zeta-\zeta_0}+\frac{e^{-i\alpha}}{\zeta-\zeta_0^*}\right\},$$

The transformed field in the z-plane, where $\zeta = z^n$, is

$$w(z) = \frac{m}{2\pi\varepsilon_0}\left\{\frac{e^{i\alpha}}{z^n - z_0^n} + \frac{e^{-i\alpha}}{z^n - (z_0^*)^n}\right\}.$$

We will also look more closely at the field near the point z_0 corresponding to A in the z-plane. By Taylor's theorem, when z is near z_0,

$$z^n = z_0^n + (z - z_0)nz_0^{n-1} + O\{(z-z_0)^2\}.$$

$$\therefore \frac{1}{z^n - z_0^n} = \frac{1}{(z-z_0)nz_0^{n-1}} + O(1).$$

$$\therefore w(z) = \frac{1}{2\pi\varepsilon_0}\left(\frac{me^{i\alpha}}{nz_0^{n-1}}\right)\cdot\frac{1}{z-z_0} + O(1).$$

The first term is the predominant term and represents the potential of a (line-)dipole in the z-plane at z_0 of (complex) moment $me^{i\alpha}/(nz_0^{n-1})$.

(This is a special case of the general result that under the transformation $\zeta = f(z)$ a dipole in one plane corresponds to a dipole in the other; the strengths are related by the magnification $|df/dz|$ and the direction is rotated through the angle arg (df/dz), the characteristic features of a conformal transformation. In detail, the dipole $me^{i\alpha}$ in the z-plane becomes the dipole $me^{i\alpha} f'(z_0)$ in the ζ-plane.)

Example 2. A thin sheet of material of conductivity σ and thickness t covers the region

$$|x| \leqslant a, \quad y \geqslant 0, \quad (x-a)^2 + y^2 \geqslant \delta^2, \quad (x+a)^2 + y^2 \geqslant \delta^2,$$

where δ/a is small. Two electrodes make perfect contact with the sheet over the two curved parts of the boundary, and a steady current flows from one electrode to the other. By using the transformation

$$Z = \sin(\pi z/2a),$$

or otherwise, prove that the resistance of the sheet is, approximately,

$$\frac{2}{\pi\sigma t}\ln\left(\frac{4a}{\pi\delta}\right).$$

First we determine the region in the Z-plane which corresponds to the conducting area in the z-plane, Fig. 8.25 (i).

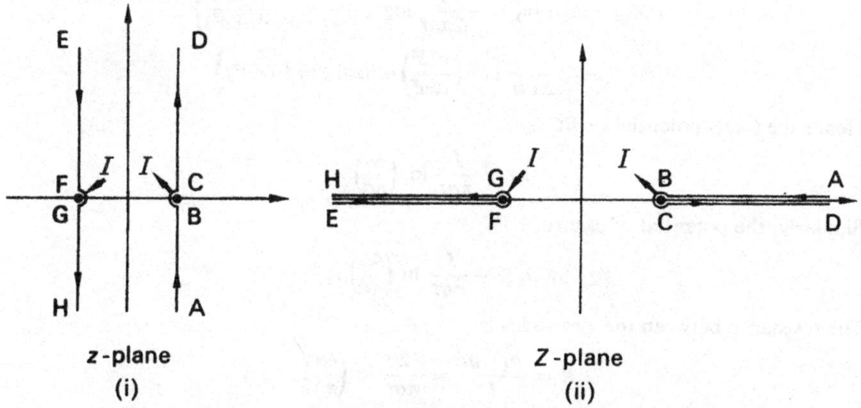

FIG. 8.25

On the boundary AB, CD we put $z = a+iy$ with $-\infty < y < -\delta$, $\delta < y < \infty$. Then

$$Z = \sin\left(\frac{\pi}{2}+\frac{i\pi y}{2a}\right) = \cosh\frac{\pi y}{2a}. \tag{1}$$

On the arc BC we put $z = a+\delta e^{i\theta}$ with $\pi/2 \leqslant \theta \leqslant 3\pi/2$. Then

$$Z = \sin\left(\frac{\pi}{2}+\frac{\pi\delta e^{i\theta}}{2a}\right) = \cos\left(\frac{\pi\delta e^{i\theta}}{2a}\right) = 1-\frac{\pi^2\delta^2 e^{2i\theta}}{8a^2}+O(\delta^4). \tag{2}$$

We deduce that the boundary $ABCD$ transforms to the "keyhole" also labelled $ABCD$, in the Z-plane, Fig. 8.25 (ii). [The radius of the circle in the Z-plane is $\pi^2\delta^2/(8a^2)$.] Since $z = 0$ corresponds to $Z = 0$ we conclude that the conducting area in the z-plane transforms to the whole of the Z-plane, outside the keyholes. Since AB, CD, EF, GH are lines of flow in the z-plane and the arcs BC, FG are electrodes emitting currents I, $-I$ respectively, then the corresponding circles in the Z-plane are also electrodes of the same strengths. The lines AB, CD, EF, GH do not affect the flow pattern in the Z-plane, which is therefore that arising from two electrodes at $Z = \pm 1$, and for which the potential is

$$w(Z) = -\frac{I}{2\pi\sigma t}\log\left(\frac{Z-1}{Z+1}\right). \tag{1}$$

(The factor t occurs in the denominator because the metal has thickness t instead of unity.) Hence the complex potential for the flow in the z-plane is

$$w(Z) = -\frac{I}{2\pi\sigma t}\log\left(\frac{\sin(\pi z/2a)-1}{\sin(\pi z/2a)+1}\right). \tag{2}$$

To find the potential on the right-hand electrode we substitute

$$z = a-\delta e^{i\theta} \text{ in (2)} \quad \left[\text{or } Z = 1-\frac{\pi^2\delta^2 e^{2i\theta}}{8a^2} \text{ in (1)}\right];$$

the real part of the resulting expression is the potential of the electrode BC. We substitute $z = -a+\delta e^{i\theta}$ with

$$-\pi/2 \leqslant \theta \leqslant \pi/2 \text{ in (2)} \quad \left[\text{or } Z = -1+\frac{\pi^2\delta^2 e^{2i\theta}}{8a^2} \text{ in (1)}\right]$$

to obtain the potential of the electrode FG. Hence, for BC,

$$w(Z_1) = u_1+iv_1 = -\frac{I}{2\pi\sigma t}\log\left\{\frac{(-\pi^2\delta^2 e^{2i\theta}/8a^2)}{2-(\pi^2\delta^2 e^{2i\theta}/8a^2)}\right\}$$

$$= -\frac{I}{2\pi\sigma t}\left\{\ln\left(\frac{\pi^2\delta^2}{16a^2}\right)+i(\pi+2\theta)+O(\delta^2)\right\}.$$

Hence the (real) potential of BC is

$$u_1 = \frac{I}{\pi\sigma t}\ln\left(\frac{4a}{\pi\delta}\right).$$

Similarly, the potential of electrode FG is

$$u_2 = -\frac{I}{\pi\sigma t}\ln\left(\frac{4a}{\pi\delta}\right).$$

The resistance between the electrodes is

$$R = \frac{u_1-u_2}{I} = \frac{2}{\pi\sigma t}\ln\left(\frac{4a}{\pi\delta}\right).$$

Exercises 8.4

1. An infinite hollow circular cylinder of dielectric, of internal and external radii a and b, $b > a$, and dielectric constant K, is set with its axis perpendicular to a uniform electric field of strength E_0. Show that the field inside the cavity is uniform and of magnitude

$$4KE_0/\{(K+1)^2-(K-1)^2\,(a/b)^2\}.$$

2. A line charge P, of strength q per unit length, is placed parallel to the axis O of an uncharged infinite cylindrical conductor of radius a, the perpendicular distance OP being c. Show that at a point Q on the cylinder, such that OQ and OP are perpendicular, the force per unit area is

$$\frac{q^2a^2}{2\pi^2\varepsilon_0(c^2+a^2)^2}.$$

3. The axis of an infinite conducting cylinder, with radius a and charge e per unit length, is parallel to, and distant b ($b > a$) from, an earthed conducting plane. Prove that the cylinder is attracted towards the plane with a force of magnitude $e^2/\{4\pi\varepsilon_0(b^2-a^2)^{1/2}\}$ per unit length, and verify that this force is the resultant of the mechanical forces acting on the elements of charge on unit length of the cylinder.

4. Show that the potential

$$\phi+i\psi = E\cos\alpha\,(z^2-a^2)^{1/2}+iEz\sin\alpha$$

(where E, α, a are real and positive) determines the field produced when a thin conducting strip of breadth $2a$ is placed in a uniform electric field E with its plane at an angle α to the direction of the field.

5. A piece of uniform plane conducting material of thickness t and conductivity σ is bounded by a circle of radius a and centre O. Two small perfectly conducting circular electrodes A, B, each of radius c, are inserted in the sheet such that the distances OA, OB are each equal to b ($b < a$), and such that OA is normal to OB. Current enters at A and leaves at B. Show that the equivalent resistance of the system is approximately

$$\frac{1}{2\pi\sigma t}\ln\left[\frac{2b^2(a^4+b^4)}{c^2(a^2-b^2)^2}\right]$$

provided c is small compared with b.

6. A thin sheet of conductivity σ and thickness t is bounded by arcs AB, CD of concentric circles of radii c and a respectively, where $c < a$, and by straight lines BC, AD. Both arcs subtend an angle θ at the centre, and BC, AD are radial lines which separate the conductor from insulating material. Show that the complex potential function $w = \log z$ can be used to find the current flow when current is fed in uniformly across AB and there is a constant potential difference between AB and CD.

Show that the effective resistance of the conductor is

$$\frac{\ln(a/c)}{\sigma\theta t}.$$

7. Two infinite circular cylindrical conductors, each of radius $2a$, are embedded with their axes parallel and at distance $5a$ apart in a medium of specific conductivity σ. A steady potential difference is maintained between the two conductors, so that a current flows through the intervening medium. Show, by the method of images or otherwise, that the lines of flow are arcs of circles, and that the resistance per unit length is $(\pi\sigma)^{-1}\ln 2$.

8. From the definition of tanh z in terms of the exponential function, show that tanh $(z+i\pi/2)=\coth z$, and show also that tanh $z=z$, approximately, when $|z|$ is small. The complex potential of an electrostatic field in the plane of $z = x+iy$ is

$$w = U+iV = E\left\{z+\frac{\pi a^2}{h}\coth\frac{\pi z}{h}\right\},$$

where E, h and a are real constants. Show that the lines $y = \pm\frac{1}{2}h$ are equipotentials $V =$ constant. Show also that, if $\pi a/h$ is small, V is approximately constant (zero) on the circle $|z| = a$.

9. Explain the use of the conjugate functions ϕ, ψ, where $\phi+i\psi = f(x+iy)$, to solve certain electrostatic problems and show that the gradient vector operator ∇ is

$$\left(\frac{1}{h}\frac{\partial}{\partial\phi}, \frac{1}{h}\frac{\partial}{\partial\psi}\right), \quad \text{where} \quad h = \left|\frac{d(x+iy)}{d(\phi+i\psi)}\right|.$$

Show that the function

$$\phi+i\psi = A\cosh^{-1}\left(\frac{x+iy}{c}\right),$$

where A and c are real constants, gives the electrostatic potential ϕ due to a freely charged elliptic conducting cylinder of infinite length. Show further that if the charge per unit length of the cylinder is q, the surface density at a point P of the surface is $q/(2\pi f)$, where f is the length of the semi-diameter conjugate to the diameter through P of the normal cross-section of the cylinder.

8.5 Gravitation

1. Fundamental principles

The basis of the theory of Newtonian gravitational potential is the inverse square law of attraction, viz. that two particles of mass m_1, m_2 separated by distance r *attract* one another with a force Gm_1m_2/r^2, where G is a constant, the constant of gravitation. The theory and results follow very closely those of electrostatics (with G put equal to $(4\pi\varepsilon_0)^{-1}$ and using SI units) except that

(1) the gravitational field at displacement $\boldsymbol{r}$ from a particle of mass m is

$$\boldsymbol{F} = -Gm\boldsymbol{r}/r^3$$

so that

(2) this field is conservative, the corresponding gravitational potential being given by

$$U = -Gm/r, \quad \boldsymbol{F} = -\textbf{grad}\ U. \tag{8.15}$$

Here we take the potential to vanish at infinity and $-U(P)$ is the work required to take a unit particle from the field point P to infinity by any path.

The inverse square law and the principle of superposition lead, as in electrostatics, to Gauss's theorem, which in this case takes the form

$$\oint\!\!\oint_{\Sigma} \boldsymbol{F} \cdot \mathbf{d}\boldsymbol{S} = -4\pi GM, \tag{8.16}$$

M being the mass enclosed by Σ. The space and surface forms of Gauss's theorem are, respectively,

$$\operatorname{div} \boldsymbol{F} = -4\pi G\varrho, \quad \nabla^2 U = 4\pi G\varrho, \tag{8.17}$$

$$\left(\frac{\partial U}{\partial n}\right)_+ - \left(\frac{\partial U}{\partial n}\right)_- = 4\pi G\sigma, \tag{8.18}$$

ϱ and σ being volume and surface densities respectively. [Although a surface layer of matter cannot exist in practice, nevertheless the idea is of value in solving certain problems. See example 6, p. 324.] Also U is continuous at a surface layer. The potential at the origin due to volume distribution ϱ, surface distribution σ and particles M_l is given by

$$U = -\frac{G}{4\pi} \iiint \frac{\varrho \, d\tau}{r} - \frac{G}{4\pi} \sum \oint\!\!\oint_{S_i} \frac{\sigma \, dS}{r} - \frac{G}{4\pi} \sum_l \frac{M_l}{r_l}, \tag{8.19}$$

[cf. the derivation of equation (2.15) of Vol. 1, p. 32]. Further, given ϱ, σ, and M_l, the potential function U is unique. The theorems of example 4, p. 267 and Exercises 8.1, nos. 1–5 hold in a suitably modified form for the gravitational field.

2. Gravitational energy

Since the work done by the forces of the field when a unit particle is taken from P to infinity is $U(P)$, the potential energy of a particle of mass m, placed at P, is $mU(P)$. We follow the technique used on p. 94 for establishing the energy of the electrostatic field and suppose now that we assemble a specified volume distribution of matter piecemeal from infinite dispersion, the final volume density and potential being ϱ, U respectively. Since the relation between ϱ and U is linear, when the density everywhere is $x\varrho$ the potential is xU and so the potential energy gained in bringing up mass $\varrho \, \delta x$ everywhere is $\iiint_{\text{all space}} (\varrho \, \delta x) xU \, d\tau$. It follows by integration that the potential energy of a gravitating system is

$$W = \tfrac{1}{2} \iiint_{\text{all space}} \varrho U \, d\tau. \tag{8.20}$$

Here we consider a volume distribution of matter only. Using Green's theorem (for all space) with $\boldsymbol{F} = -\mathbf{grad}\, U$ and $\varrho = \nabla^2 U/(4\pi G)$ and noting that, for a distribution of finite extent, the standard boundary conditions hold at infinity, we find

$$W = -\frac{1}{8\pi G} \iiint\limits_{\text{all space}} \boldsymbol{F}^2 \, d\tau, \tag{8.21}$$

suggesting that the energy of the distribution can be regarded, from the field outlook, as spread through all space with local density $-\boldsymbol{F}^2/(8\pi G)$ per unit volume.

3. The relation between fluid pressure and gravitational pressure

We define the pressure p at a point P in a fluid, which by definition is unable to sustain a shearing force, by $p = \lim_{A \to 0} T/A$ where T is the (normal) thrust across a small plane area A surrounding P. It is easy to show that the pressure at a point is independent of the direction of the area and that the equation of motion of the fluid is

$$\frac{1}{\varrho}\,\mathbf{grad}\, p = \boldsymbol{F} - \boldsymbol{f}, \tag{8.22}$$

where $\boldsymbol{f}$ is the acceleration of the fluid and $\boldsymbol{F}$ is the force per unit mass. Here we shall only be concerned with problems of equilibrium or steady state motions and ignore effects of viscosity so that we can assume that $\boldsymbol{F}$ is due solely to a gravitational field with potential U.

When the fluid is in equilibrium, $\boldsymbol{f} = \boldsymbol{0}$, eqn. (8.22) takes the form

$$\mathbf{grad}\, p = -\varrho\, \mathbf{grad}\, U. \tag{8.22}$$

An immediate consequence is that $\mathbf{grad}\, p$ is everywhere parallel to $\mathbf{grad}\, U$ and so the surface p = constant must also be a surface U = constant. Further, taking **curl** of eqn. (8.22) gives

$$(\mathbf{grad}\, \varrho) \times (\mathbf{grad}\, U) = \boldsymbol{0}, \tag{8.23}$$

so that $\mathbf{grad}\, \varrho$ and $\mathbf{grad}\, U$ are everywhere parallel. It follows that the surfaces ϱ = constant are also surfaces on which U = constant. In fact, level surfaces of p, ϱ and U coincide and any one of these variables can be regarded as a function of one of the other (two) variables.

Example 1. A uniform liquid sphere is of mass M and radius a. Calculate the gravitational field and potential everywhere. Show that the gravitational energy of the sphere is $-3GM^2/(5a)$. Find also the hydrostatic pressure at the centre of the sphere.

Since the distribution is spherically symmetric, Gauss's theorem gives $\mathbf{F} = \{F(r)\ 0\ 0\}$, using polar coordinates (r, θ, ψ) with origin at the centre of the sphere, and

$$4\pi r^2 F = -4\pi GM, \quad \text{i.e.} \quad F = -GM/r^2 \quad \text{for} \quad r \geqslant a,$$

$$4\pi r^2 F = -4\pi G \int_0^r 4\pi \varrho r^2\, dr, \quad \text{i.e.} \quad F = -4\pi G\varrho r/3 \quad \text{for} \quad r \leqslant a,$$

where ϱ is the density. In fact,

$$\mathbf{F} = -GM\mathbf{r}/r^3 \quad \text{for} \quad r \geqslant a,$$

$$\mathbf{F} = -GM\mathbf{r}/a^3 \quad \text{for} \quad r \leqslant a,$$

so that at external points the sphere attracts as if all its mass were concentrated at the centre.

Using $F = -dU/dr$, $U_\infty = 0$ and the fact that U is continuous at $r = a$ we find

$$U = -\frac{GM}{r} \quad \text{for} \quad r \geqslant a, \qquad U = -GM(3a^2 - r^2)/(2a^3) \quad \text{for} \quad r \leqslant a.$$

The gravitational energy is

$$\frac{1}{2}\iiint_{\text{all space}} \varrho U\, d\tau = -\frac{1}{2}\int_0^a \frac{3GM^2(3a^2 - r^2)}{8\pi a^6} 4\pi r^2\, dr = -\frac{3GM^2}{5a}.$$

[Note that the density vanishes for $r > a$ and so we evaluate the integral of (8.20) over the volume of the sphere only.]

The pressure equation

$$\textbf{grad}\ p = -\varrho\ \textbf{grad}\ U$$

has only one non-vanishing resolute, namely the radial one, which is

$$\frac{dp}{dr} = -\varrho\frac{dU}{dr}.$$

Integration gives, within the sphere,

$$p = \varrho(U_a - U),$$

where U_a is the potential at the free surface $r = a$ at which $p = 0$.

The central pressure is therefore

$$\frac{\varrho GM}{2a} = \frac{3GM^2}{8\pi a^4} = \frac{2\pi\varrho^2 Ga^2}{3}.$$

Similar methods can be used for solution of problems involving spheres with radial symmetry, i.e. for which $\varrho = \varrho(r)$, as illustrated by the next example.

Example 2. Prove that Poisson's equation for the gravitational potential U in a spherically symmetrical distribution of matter having density ϱ at a distance r from the centre may be written

$$\frac{1}{r}\frac{d^2}{dr^2}(rU) = 4\pi G\varrho.$$

A given spherical distribution of total mass M is given by

$$\varrho = \varrho_0 \frac{\sin(\pi r/a)}{\pi r/a} \quad \text{for} \quad 0 \leqslant r \leqslant a,$$

$$\varrho = 0 \quad \text{for} \quad r > a.$$

Show that $M = 4\varrho_0 a^3/\pi$ and prove that

$$U = -\frac{GM}{a}\left(1+\frac{\varrho}{\varrho_0}\right) \quad \text{for} \quad r \leqslant a.$$

Calculate in terms of G, M and a the self-energy of this distribution as compared with a state of infinite diffusion.

Laplace's equation (8.17) reduces in this case to

$$\frac{1}{r^2}\frac{\mathrm{d}}{\mathrm{d}r}\left(r^2\frac{\mathrm{d}U}{\mathrm{d}r}\right) = 4\pi G\varrho,$$

i.e.

$$\frac{1}{r}\left(r\frac{\mathrm{d}^2U}{\mathrm{d}r^2}+\frac{2\mathrm{d}U}{\mathrm{d}r}\right) = \frac{1}{r}\frac{\mathrm{d}^2(rU)}{\mathrm{d}r^2} = 4\pi G\varrho, \tag{1}$$

as required.

In the given case the total mass of the distribution is

$$M = \int_0^a 4\pi r^2\varrho\,\mathrm{d}r = 4a\varrho_0\int_0^a r\sin\left(\frac{\pi r}{a}\right)\mathrm{d}r$$

$$= 4a\varrho_0\left[-\frac{ar}{\pi}\cos\left(\frac{\pi r}{a}\right)+\frac{a^2}{\pi^2}\sin\left(\frac{\pi r}{a}\right)\right]_0^a = \frac{4\varrho_0 a^3}{\pi}. \tag{2}$$

As in example 1 above, for $r \geqslant a$ the system attracts as if all its mass were concentrated at the centre and so $U_{r\geqslant a} = -GM/r$. Also eqn. (1) becomes

$$\frac{\mathrm{d}^2}{\mathrm{d}r^2}(rU) = 4G\varrho_0 a\sin\left(\frac{\pi r}{a}\right),$$

which integrates to give

$$U = A+\frac{B}{r}-\frac{4G\varrho_0 a^3}{\pi^2 r}\sin\left(\frac{\pi r}{a}\right),$$

where A, B are constants. Since the potential is finite at $r = 0$ we put $B = 0$. Also continuity of U at $r = a$ gives $A = -GM/a$, and so

$$U = -\frac{GM}{a}-\frac{4G\varrho_0 a^3}{\pi^2 r}\sin\left(\frac{\pi r}{a}\right) = -\frac{GM}{a}\left(1+\frac{\varrho}{\varrho_0}\right)$$

on using (2).

The gravitational self-energy of the distribution is

$$W = \frac{1}{2}\iiint_{\text{all space}}\varrho U\,\mathrm{d}\tau = \frac{1}{2}\int_0^a 4\pi r^2\varrho U\,\mathrm{d}r = -\frac{2\pi GM}{a}\int_0^a r^2\left(1+\frac{\varrho}{\varrho_0}\right)\mathrm{d}r$$

$$= -\frac{2GM\varrho_0 a^2}{\pi^2}\int_0^\pi (\xi\sin\xi+\sin^2\xi)\,\mathrm{d}\xi$$

after making the substitution $r = a\xi/\pi$. Therefore

$$W = -\{(2GM\varrho_0 a^2)/\pi^2\}\left[-\xi\cos\xi+\sin\xi+\tfrac{1}{2}\xi-\tfrac{1}{4}\sin 2\xi\right]_0^\pi = -\frac{3GM\varrho_0 a^2}{\pi} = -\frac{3GM^2}{4a}.$$

Example 3. A globular cluster consists of a large number of stars of equal masses originally uniformly distributed within a sphere and at rest relative to their centre of mass O. The stars move under their mutual attraction and the effect of collisions and close encounters is to be neglected. If a sphere of variable radius r and centre O contains a constant mass M of the distribution, show that

$$\frac{d^2r}{dt^2} = -\frac{GM}{r^2}$$

and that the radius of the cluster is halved in a time

$$\left(\frac{\pi+2}{8}\right)\left(\frac{3}{2\pi G\varrho_0}\right)^{1/2},$$

where ϱ_0 is the initial density of the smoothed-out distribution of stars.

Since the system originally had radial symmetry and the stars were at rest, the initial accelerations, etc., are purely radial, and the cluster will preserve radial symmetry. Using Gauss's theorem, as in examples 1, 2 above, it follows that the acceleration due to gravity at the surface of the sphere of radius r is GM/r^2 inwards and so the radial equation of motion is

$$\frac{d^2r}{dt^2} = -\frac{GM}{r^2}. \tag{1}$$

Then if $R(t)$ is the radius of the cluster at time t and M_0 is its total mass,

$$\frac{d^2R}{dt^2} = -\frac{GM_0}{R^2}. \tag{2}$$

Multiplying (2) by the integrating factor dR/dt and integrating, we find

$$\frac{dR}{dt} = -\sqrt{\left\{\frac{2GM(R_0-R)}{R_0R}\right\}},$$

where $R = R_0$ at time $t = 0$. Then

$$\int_{R_0}^{R} \sqrt{\left\{\frac{R}{R_0-R}\right\}} dR = -\sqrt{\left(\frac{2GM_0}{R_0}\right)} \int_0^t dt.$$

Using the substitution $R = R_0 \sin^2\theta$ we find

$$t = \sqrt{\left(\frac{R_0^3}{2GM_0}\right)}\left[\frac{\pi}{2} - \sin^{-1}\sqrt{\left(\frac{R}{R_0}\right)} + \sqrt{\left\{\frac{R}{R_0}\left(1-\frac{R}{R_0}\right)\right\}}\right].$$

Therefore the radius of the cluster is halved in time

$$T = \sqrt{\left(\frac{R_0^3}{2GM_0}\right)}\left(\frac{\pi}{2}-\frac{\pi}{4}+\frac{1}{2}\right) = \frac{\pi+2}{4}\sqrt{\left(\frac{R_0^3}{2GM_0}\right)},$$

which is equivalent to the given result since $M_0 = 4\pi R_0^3\varrho_0/3$.

Example 4. MacCullagh's formula.

Consider a finite distribution of gravitating material with centre of mass O, Fig. 8.26. Then the potential at a point P is

$$U_p = -G\iiint \frac{\varrho_Q \, d\tau_Q}{PQ} = -G\iiint \frac{\varrho_Q \, d\tau_Q}{\sqrt{(R^2+r^2-2Rr\cos\theta)}},$$

where $\varrho_Q \, d\tau_Q$ is the mass of the element $d\tau_Q$ at Q and the integral is taken over the ume of the material. If OP is large compared with the greatest linear dimension, l, of the

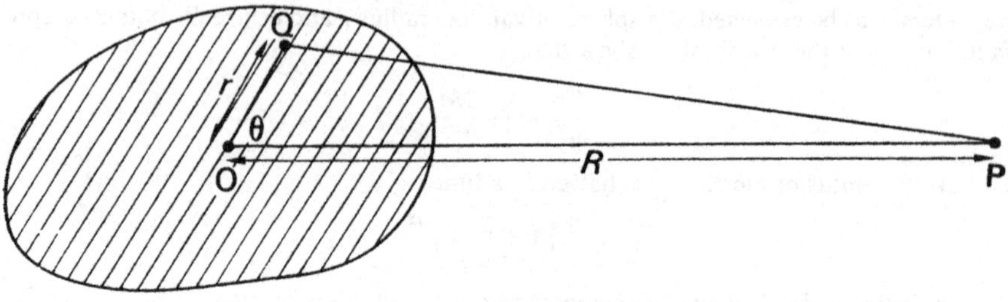

FIG. 8.26

distribution, i.e. $R \gg r$, the integrand can be expanded in terms of Legendre polynomials and we find

$$U_p = -G \iiint \frac{1}{R} \left\{ \sum_0^\infty \left(\frac{r}{R}\right)^n P_n(\cos\theta) \right\} \varrho_Q \, d\tau_Q \tag{1}$$

i.e.
$$U_p = -G \iiint \frac{1}{R} \left\{ 1 + \frac{r\cos\theta}{R} + \frac{r^2(3\cos^2\theta - 1)}{2R^2} + O\left(\frac{r^3}{R^3}\right) \right\} \varrho_Q \, d\tau_Q$$

$$= -G \left\{ \frac{M}{R} + \frac{1}{R^2} \iiint r\cos\theta \, \varrho_Q \, d\tau_Q + \frac{1}{2R^3} \iiint (2 - 3\sin^2\theta) r^2 \varrho_Q \, d\tau_Q + O\left(\frac{Gl^6}{R^4}\right) \right\}.$$

The first integral vanishes by definition of the centre of mass O. Also, if A, B, C are the moments of inertia of the distribution about any three rectangular coordinate axes $Ox_1x_2x_3$ at O and I is the moment of inertia about OP,

$$\iiint 2r^2 \varrho_Q \, d\tau_Q = A + B + C, \qquad \iiint r^2 \sin^2\theta = I.$$

$$\therefore \quad U_p = -\frac{GM}{R} - \frac{G(A+B+C-3I)}{2R^3} + O\left(\frac{Gl^6}{R^4}\right). \tag{2}$$

This is *MacCullagh's formula.*

It is usually convenient to take $Ox_1x_2x_3$ as principal axes of inertia at O. In this case, if the coordinates of P are (x_1, x_2, x_3) so that OP has direction cosines $(x_1/R, x_2/R, x_3/R)$,

$$I_{OP} = \frac{Ax_1^2 + Bx_2^2 + Cx_3^2}{R^2},$$

$$U_p = -\frac{GM}{R} - G\frac{(A+B+C)(x_1^2+x_2^2+x_3^2) - 3(Ax_1^2+Bx_2^2+Cx_3^2)}{2R^5} + O\left(\frac{Gl^6}{R^4}\right),$$

i.e.
$$U_p = -\frac{GM}{R} - \frac{G(B+C-2A)x_1^2 + (C+A-2B)x_2^2 + (A+B-2C)x_3^2}{2R^5} + O\left(\frac{Gl^6}{R^4}\right).$$

The components of the gravitational field at P can be found from the relation $F_1 = -\partial U/\partial x_1$, etc.

Note that this technique can also be applied to a distribution of electric charges. Equation (1) when applied to charges indicates that the field at great distances is the same as if all the charge were concentrated at O and so at great distances the field is almost purely radial. However, if the total charge is zero, as must be the case for an isolated neutral system, then O does not exist and the field at great distances must be that of a dipole or multipole (see § 2.10).

Example 5. The mutual potential energy of two distributions.

Let ϱ_1, ϱ_2 and U_1, U_2 be the densities and potentials respectively of two non-overlapping systems. Then

$$\iiint\limits_{\text{all space}} \varrho_1 U_2 \, d\tau = \frac{1}{4\pi G} \iiint\limits_{\text{all space}} U_2 \nabla^2 U_1 \, d\tau$$

$$= -\frac{1}{4\pi G} \iiint\limits_{\text{all space}} \mathbf{grad}\, U_2 \cdot \mathbf{grad}\, U_1 \, d\tau + \oiint U_2 \, \mathbf{grad}\, U_1 \, d\mathbf{S}.$$

The surface integral (taken over the sphere at infinity) vanishes in the usual way. By symmetry we can interchange the suffixes 1, 2 and so

$$\iiint\limits_{\text{first distribution}} \varrho_1 U_2 \, d\tau = \iiint\limits_{\text{second distribution}} \varrho_2 U_1 \, d\tau,$$

which is *our reciprocal theorem*. [Note that we take the first of these integrals over the volume of the first distribution since $\varrho_1 = 0$ outside this distribution, but equally well the integrals can be evaluated over all space.]

The combined potential energy of the two systems is

$$W = \tfrac{1}{2} \iiint (\varrho_1 + \varrho_2)(U_1 + U_2) \, d\tau$$
$$= \tfrac{1}{2} \iiint \varrho_1 U_1 \, d\tau + \tfrac{1}{2} \iiint \varrho_2 U_2 \, d\tau + \tfrac{1}{2} \iiint (\varrho_1 U_2 + \varrho_2 U_1) \, d\tau = W_1 + W_2 + W_{12},$$

where W_1, W_2 are the self-energies of the separate systems and W_{12} is the mutual potential energy of the two systems. [All these integrals are taken over all space.] Note that, by virtue of the reciprocal theorem,

$$W_{12} = \iiint \varrho_2 U_1 \, d\tau.$$

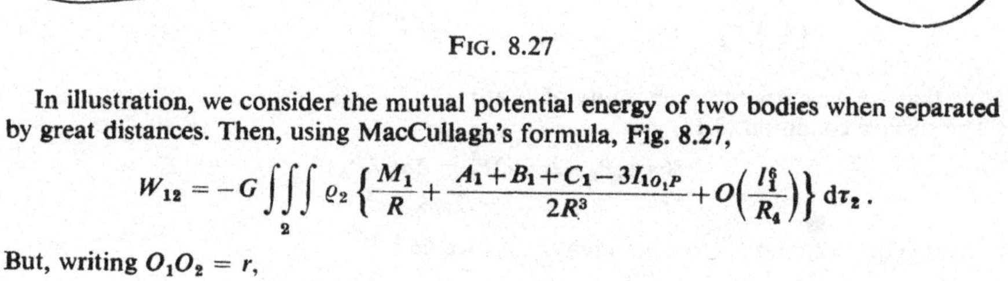

FIG. 8.27

In illustration, we consider the mutual potential energy of two bodies when separated by great distances. Then, using MacCullagh's formula, Fig. 8.27,

$$W_{12} = -G \iiint\limits_{2} \varrho_2 \left\{ \frac{M_1}{R} + \frac{A_1 + B_1 + C_1 - 3I_{1O_1P}}{2R^3} + O\left(\frac{l_1^4}{R_4}\right) \right\} d\tau_2 .$$

But, writing $O_1O_2 = r$,

$$\frac{1}{r^3} = \frac{1}{R^3} + O\left(\frac{1}{R^4}\right), \quad \frac{I_{1OP}}{R^3} = \frac{I_{1O_1O_2}}{r^3} + O\left(\frac{1}{R^4}\right).$$

Also $-\iiint \frac{\varrho_2}{R} \, d\tau_2$ is the potential at O_1 due to the distribution 2. Using MacCullagh's theorem again (for the distribution 2) we find, neglecting $O(1/R^4)$,

$$W_{12} = -G \left\{ \frac{M_1 M_2}{r} + \frac{(A_2 + B_2 + C_2 - 3I_2)}{2r^3} + \frac{(A_1 + B_1 + C_1 - 3I_1)}{2r^3} \right\}.$$

This formula enables us to calculate the force and couple exerted by two remote rigid gravitating bodies on each other.

By analysis similar to that of example 5, p. 96, we can show that the mutual potential energy of two gravitating spheres, of masses M_1, M_2, and each with spherical symmetry, is $-GM_1M_2/r$ where r is the distance between their centres. It follows that the spheres attract one another as if all their masses were concentrated at their centres.

Example 6. A solid nearly spherical core, with surface

$$r = a\{1+\varepsilon P_2(\cos\theta)\},$$

is of uniform density σ. It is surrounded by a layer of liquid of uniform density ϱ and volume $\frac{4}{3}\pi(b^3-a^3)$. Assuming that the outer surface of the liquid is

$$r = b\{1+\eta P_2(\cos\theta)\},$$

determine the gravitational potential at an external point.

Prove that
$$\eta = \frac{3a^5\varepsilon(\sigma-\varrho)}{b^2[\varrho(2b^3-5a^3)+5\sigma a^3]},$$

neglecting squares of ε.

We find, correct to the first order in ε, η, the potential at an external point (r, θ, ψ) by superposing the potentials (U_1, U_2, U_3, U_4) of

(1) a uniform sphere of density σ and radius a,
(2) a uniform shell of density ϱ with radii a, b,
(3) a layer of surface density $(\sigma-\varrho)a\varepsilon P_2(\cos\theta)$ on $r = a$,
(4) a layer of surface density $\varrho b\eta P_2(\cos\theta)$ on $r = b$.

The sum of the potentials due to (1), (2) is the same as the sum of those of a sphere of density ϱ and radius b, and a sphere of density $\sigma-\varrho$ and radius a, i.e.

$$U_1+U_2 = \frac{-4\pi G\varrho b^3}{3r}-\frac{4\pi G(\sigma-\varrho)a^3}{3r}. \tag{1}$$

[See example 1, p. 319.]

To find the potential of a surface distribution $\lambda P_2 (\lambda \ll 1)$ of matter on a sphere of radius c, we follow the technique of example 6, p. 291, and try

$$U = A\left(\frac{r}{c}\right)^2 P_2 \quad \text{for} \quad r < c, \qquad U = A\left(\frac{c}{r}\right)^3 P_2 \quad \text{for} \quad r > c. \tag{2}$$

[Note that we have used the continuity of U at $r = c$.]

The surface condition (8.18) gives

$$-3(A/c)P_2-2(A/c)P_2 = 4\pi G\lambda P_2.$$
$$\therefore\ A = -4\pi G\lambda c/5. \tag{3}$$

Applying eqns. (2) and (3) to determine U_3, U_4 we find

$$U_3 = \frac{-4\pi G(\sigma-\varrho)a^5\varepsilon P_2}{5r^3}, \qquad r > a,$$

$$U_4 = \frac{-4\pi G\varrho b^5\eta P_2}{5r^3}, \qquad r > b.$$

Therefore, neglecting second-order terms in ε, η,

$$U = -4\pi G\left\{\frac{\varrho b^3}{3r}+\frac{(\sigma-\varrho)a^3}{3r}+\frac{(\sigma-\varrho)a^5\varepsilon P_2}{5r^3}+\frac{\varrho b^5\eta P_2}{5r^3}\right\}. \tag{4}$$

Writing $r = b(1+\eta P_2)$ in (4) and neglecting second-order terms,

$$U_{\text{surface}} = \frac{4\pi G}{3}\left[\varrho b^2+(\sigma-\varrho)a^2-\left\{\varrho\eta b^2+\frac{(\sigma-\varrho)a^3\eta}{b}-3\frac{(\sigma-\varrho)a^5\varepsilon}{5b^3}-\frac{3\varrho b^2\eta}{5}\right\}P_2\right]. \quad (5)$$

At the free surface the pressure p vanishes and so V_{surface} is constant. Therefore the term involving P_2 on the right-hand side of (5) must vanish. It follows that η must take the stated value.

Example 7. A uniform gravitating liquid sphere of radius a and uniform density ϱ is made to rotate with small angular velocity ω. Find the form of the free surface and the pressure within the sphere.

The equation of motion referred to rectangular cartesian axes through the centre and with Oz along the axis of revolution has resolutes

$$\frac{1}{\varrho}\frac{\partial p}{\partial x} = -\frac{\partial U}{\partial x}-\omega^2 x, \quad \frac{1}{\varrho}\frac{\partial p}{\partial y} = -\frac{\partial U}{\partial y}-\omega^2 y, \quad \frac{1}{\varrho}\frac{\partial p}{\partial z} = -\frac{\partial U}{\partial z},$$

which integrate to give

$$\frac{p}{\varrho} = -U+\frac{1}{2}\omega^2(x^2+y^2)+\text{constant},$$

or using spherical polar coordinates,

$$\frac{p}{\varrho} = -U+\frac{1}{3}\omega^2 r^2\{1-P_2(\cos\theta)\}+\text{constant} \quad (1)$$

since $x^2+y^2 = r^2\sin^2\theta = \frac{1}{3}r^2(3-3\cos^2\theta) = \frac{2}{3}r^2\{1-P_2(\cos\theta)\}$.

Because $p = 0$ at the free surface, and ω is small, the form of (1) suggests that we try

$$r = a(1+\varepsilon P_2) \quad (2)$$

as the form of the free surface where ε is small.

[Note that this assumption agrees with the fact that the volume of the sphere is $4\pi a^3/3$ because the volume within the surface (2) is

$$\int_0^{2\pi} d\psi \int_0^{\pi} d\theta \int_0^{a\{1+\varepsilon P_2(\cos\theta)\}} r^2 \sin\theta\, dr = \frac{2\pi a^3}{3}\int_0^{\pi}\{1+\varepsilon P_2(\cos\theta)\}^3 \sin\theta\, d\theta$$

$$= \frac{2\pi a^3}{3}\int_{-1}^{1}\{1+\varepsilon P_2(\mu)^3\}\, d\mu = \frac{2\pi a^3}{3}\int_{-1}^{1}\{1+3\varepsilon P_2(\mu)+O(\varepsilon^2)\}\, d\mu = 4\pi a^3/3+O(\varepsilon^2).]$$

Then, using the same technique to determine the gravitational potential U as in example 6 above, i.e. adding to the potential of a sphere of radius r the potential of a surface layer of density $\varrho a\varepsilon P_2(\cos\theta)$ on the sphere $r = a$, we find

$$U = -\frac{4\pi G\varrho a^3}{3r}-\frac{4\pi G\varrho a^5\varepsilon P_2(\cos\theta)}{5r^3} \quad \text{for} \quad r \geqslant a(1+\varepsilon P_2), \quad (3a)$$

$$U = -\tfrac{2}{3}\pi G\varrho(3a^2-r^2)-\tfrac{4}{5}\pi G\varepsilon r^2 P_2 \quad \text{for} \quad r \leqslant a(1+\varepsilon P_2). \quad (3b)$$

Substitution from (3b) in (1) gives

$$\frac{p}{\varrho} = \frac{2}{3}\pi G\varrho(3a^2-r^2)+\frac{4}{5}\pi G\varepsilon r^2 P_2+\frac{1}{3}\omega^2 r^2(1-P_2)+C, \quad (4)$$

where C is constant. Since $r = a(1+\varepsilon P_2)$ is a surface of constant (zero) pressure

$$0 = \tfrac{2}{3}\pi G\varrho(2a^2-2a^2\varepsilon P_2)+\tfrac{4}{5}\pi G\varrho\varepsilon a^2 P_2+\tfrac{1}{3}\omega^2 a^2(1-P_2)+O(\varepsilon^2)+C.$$

Hence

$$C+\tfrac{4}{3}\pi G\varrho a^2+\tfrac{1}{3}\omega^2 a^2 = 0, \qquad (\tfrac{4}{3}-\tfrac{4}{5})\pi G\varrho\varepsilon a^2-\tfrac{1}{3}\omega^2 a^2 = 0,$$

i.e

$$\varepsilon = \frac{-5\omega^2}{8\pi G\varrho}, \qquad C = \frac{-4\pi G\varrho a^2}{3}-\frac{\omega^2 a^2}{3}.$$

Hence the free surface is the oblate spheroid

$$r = a\left(1-\frac{5\omega^2 P_2}{8\pi G\varrho}\right)$$

and the pressure within the sphere is

$$\frac{2\pi G\varrho^2}{3}(a^2-r^2)-\frac{\omega^2\varrho}{6}\{2a^2+r^2+2r^2P_2(\cos\theta)\}.$$

Exercises 8.5

1. A solid sphere of mass M and radius a with density proportional to the distance from the centre is built up from a state of infinite dispersion. Show that the loss of potential energy is $4GM^2/(7a)$.

2. A uniform sphere, of density ϱ and radius a, has a spherical cavity of radius b. If the centre of the cavity is at a distance c from the centre of the sphere, and $a > b+c$, show that the gravitational attraction is the same at all points inside the cavity.

3. The density ϱ of a sphere, of mass M and radius a, at a distance r from the centre of the sphere is given by $\varrho = \varrho_0(a-r)/a$ where ϱ_0 is a constant. Show that the gravitational attraction at a point P, distant x ($<a$) from the centre of the sphere, is $GMx(4a-3x)/a^4$. Show also that the potential at P is $GM(2a^3-2ax^2+x^3)/a^4$.

   Hence, or otherwise, show that the loss of gravitational potential energy in assembling the sphere from a state of diffusion at an infinite distance is $26GM^2/35a$.

4. A uniform solid sphere of mass M and radius a is surrounded by a uniform liquid of density ϱ and of uniform depth, whose free surface is a concentric sphere of radius b. If the pressure over this surface vanishes, prove that the pressure at a point on the inner sphere is

$$\frac{GM\varrho(b-a)}{ab}+\frac{2}{3}\pi G\varrho^2\frac{(b-a)^2(2a+b)}{b}.$$

5. A solid homogeneous sphere, of radius a and density ϱ, is divided by a plane through its centre into two hemispheres, which are placed with their plane faces coincident. By considering one of the hemispheres as made up of uniform hemispherical shells of radius r and surface density $\varrho\, dr$, show that the force of attraction of the *whole* sphere upon each such shell is $\frac{3}{4}G\pi^2\varrho^2 r^3\, dr$, and deduce that the force required to overcome the gravitational attraction pulling the two hemispheres together is $\frac{1}{3}G\pi^2\varrho^2 a^4$.

6. The position vectors of four particles m_r are $x_r = f(t)a_r$, ($r = 1, 2, 3, 4$), $f(0) = 1$, the origin being their centre of mass at time $t = 0$, when the particles are at the vertices of a regular tetrahedron of side l. Show that the motion of the particles can be explained as due to their mutual gravitation under the inverse square law, if

$$f''(t)\{f(t)\}^2 = -GM/l^3,$$

   where $M = \Sigma m_r$. Show also that if the particles are released from rest at time $t = 0$, they collide after a time $(\pi^2 l^3/8GM)^{1/2}$.

7. Find the potential of a uniform plane ring of mass M, outer radius a and inner radius b, at a point on its axis.

Hence prove that, at any point P whose distance r from the centre O of the ring lies between a and b, the potential is

$$\frac{-GM}{a^2-b^2}\left[a-\frac{b^2}{2r}-rP_1(\mu)+\frac{r^2}{2a}P_2(\mu)\right.$$
$$\left.-\sum_{n=2}^{\infty}(-1)^n\frac{1\cdot 3\ldots 2n-3}{2\cdot 4\ldots 2n}\left\{\frac{r^{2n}}{a^{2n-1}}P_{2n}(\mu)-\frac{b^{2n}}{r^{2n-1}}P_{2n-2}(\mu)\right\}\right],$$

where μ is the cosine of the angle between OP and the normal to the plane of the ring.

8. A liquid spheroid of uniform density ϱ and volume $\frac{4}{3}\pi a^3$ lies at the centre of a flat annulus of uniform surface density σ and inner and outer radii b and c respectively, where $a \ll b$ and $\sigma(c-b) \ll \varrho bc$. Prove that the form taken by the surface of the spheroid is approximately

$$r = a\{1+\varepsilon P_2(\cos\theta)\}, \quad \text{where} \quad \varepsilon = -15\sigma(c-b)/8\varrho cb$$

and θ is measured from the axis of the annulus.

9. A heavy spherical body of mass M and radius r_0 is surrounded by an atmosphere of gas in equilibrium. The pressure p and density ϱ satisfy the relation

$$\frac{p}{p_0}=\left(\frac{\varrho}{\varrho_0}\right)^{\gamma} \quad (\gamma > 1),$$

where p_0, ϱ_0 are their values at the surface of the body. Neglecting the self-gravitation of the gas, show that its density falls to zero at distance

$$r_0\left[1-\frac{\gamma}{\gamma-1}\frac{r_0 p_0}{GM\varrho_0}\right]^{-1}$$

from the centre of the body, provided that the expression in square brackets is positive.

10. If Ox, Oy, Oz are the principal axes of inertia at the mass centre of a gravitating body and A, B, C are the corresponding moments of inertia, show that the couple exerted on the body by a point-mass m situated at x, y, z is

$$\frac{3Gm}{r^5}\{(C-B)yz,\ (A-C)zx,\ (B-A)xy\},$$

in which r is the distance of m from O, and terms involving higher powers of the ratio of the linear dimensions of the solid to r are neglected.

If the body is a homogeneous oblate spheroid of semi-axes, a, a, c and density ϱ, show that the couple that tends to decrease the angle β between r and the plane $z = 0$ is

$$4\pi G\varrho m a^2 c(a^2-c^2)\sin\theta\cos\theta/5r^3.$$

Miscellaneous Exercises VIII

1. A distribution of electricity consists of volume densities β throughout certain regions of space and surface densities σ on certain surfaces, where both ϱ and σ are functions of position. This distribution gives rise to a potential V at any point of space. An independent distribution given by ϱ' and σ' gives rise to a potential V'. Prove Green's reciprocal theorem in the form

$$\iiint(\varrho V'-\varrho' V)\,d\tau+\iint(\sigma V'-\sigma' V)\,dS = 0,$$

where the volume and surface integrals are taken throughout all volumes and surfaces where ϱ, ϱ', σ, σ' are not all zero.

Find the total charge induced on the earthed conducting surface $r = a\{1+\lambda P_2(\cos\theta)\}$ by a charge Q placed on an infinitesimally small sphere at the point (r, θ, ψ), where $r > a$ and λ^2 may be neglected.

2. Prove that, on the assumption that the energy of an electrostatic field is distributed with density $\frac{1}{2}\boldsymbol{E}\cdot\boldsymbol{D}$ per unit volume, the energy contained in a region bounded by a closed surface S' is

$$\tfrac{1}{2}\iiint \varrho V\,d\tau + \tfrac{1}{2}\Sigma\iint \sigma V\,dS - \tfrac{1}{2}\iint VD_n\,dS',$$

where the first two integrals are taken over all volume and surface charges inside S', and where D_n is the outward normal component of $\boldsymbol{D}$.

A uniform insulating sphere of dielectric constant K and radius a carries on its surface a charge of density $\lambda P_n(\cos\theta)$. Prove that, on the above assumption, the interior of the sphere contributes an amount

$$\frac{4\pi\lambda^2 a^3 Kn}{\varepsilon_0(2n+1)(Kn+n+1)^2}$$

to the electrostatic energy.

3. An earthed conducting sphere of radius a is placed with its centre at a point on the axis of a circular coil of radius b, at a distance c from the centre of the coil; the coil carries a charge Q uniformly distributed. Prove that, if a is small, the force of attraction between the sphere and the coil is

$$\frac{Q^2 ac}{4\pi\varepsilon_0 f^4}\left[1+\frac{a^2}{f^2}\left(\frac{3c^2}{f^2}-1\right)+O\left(\frac{a^3}{f^3}\right)\right],$$

where $f^2 = b^2+c^2$.

4. A point charge is at the centre of a spherical shell of internal and external radii a, b of uniform material of dielectric constant K. Show by an immediate application of first principles or otherwise that the field outside the shell is independent of K.

The point charge is now replaced by a dipole of moment $\boldsymbol{M}$. Show that the external field is the same as if the shell were absent and the moment of the dipole were $\boldsymbol{M}'$, where

$$\boldsymbol{M}' = \lambda\boldsymbol{M}$$

and

$$\frac{1}{\lambda} = 1+\frac{2}{9}\frac{(K-1)^2}{K}\left(1-\frac{a^3}{b^3}\right).$$

5. A point charge Q is placed inside a large rectangular earthed conducting box at equal distances a from three faces and near the corner formed by these faces. Neglecting effects due to the distant faces, find an image system for the field inside the box.

Show that the charge density induced at any one of the three points of the box nearest to the charge is

$$-\frac{Q}{\pi a^2}\left(\frac{14}{27}-\frac{\sqrt{5}}{25}\right).$$

6. A conducting spherical shell of mass M and radius a floats on oil of dielectric constant K. When it is uncharged, a fraction $f\,(<\frac{1}{2})$ of the volume of the sphere is submerged. Show that the charge Q which will cause the sphere to be half-submerged is given by

$$Q^2 = \frac{4\pi\varepsilon_0(1-2f)(K+1)^2 Mga^2}{(K-1)f},$$

where g is the acceleration due to gravity.

7. Two charges, each of amount Q, are placed at points A, B that are at distance $2c$ apart. A uniform sphere of radius a, where $a < c$, and dielectric constant K, has centre at the mid-point of AB. Show that a repulsion

$$\frac{Q^2}{16\pi\varepsilon_0 c^2}\left[1-16(K-1)\sum_{n=1}^{\infty}\frac{n(2n+1)}{1+2n(1+K)}\left(\frac{a}{c}\right)^{4n+1}\right]$$

acts on each charge.

8. Two insulated uncharged conducting spheres of radius a are placed in a uniform electric field of intensity F perpendicular to their line of centres. If their centres are at a distance c apart, where c is large compared with a, show that there is a repulsive force between the spheres given approximately by

$$\frac{3F^2a^6}{4\pi\varepsilon_0 c^4}\left\{1-\frac{2a^3}{c^3}-\frac{8a^5}{c^5}+\frac{3a^6}{c^6}\right\}.$$

9. A conductor is bounded by $r = a(1+\lambda P_n)$, where λ^2 may be neglected. Prove that the surface density on the conductor is approximately

$$\frac{\varepsilon_0 V_0}{a}\{1+(n-1)\lambda P_n\},$$

where V_0 is the potential of the conductor.

10. If the surface density of electricity on a circular disc of radius a is given by

$$\sigma = \frac{Q}{2\pi a(a^2-r^2)^{1/2}},$$

where r is the distance from the centre, show that the potential on the axis of symmetry is

$$\frac{Q}{4\pi\varepsilon_0 a}\tan^{-1}(a/z), \quad \text{if} \quad z > 0.$$

Deduce the potential at any point (r, θ, ϕ), where $0 \leqslant r \leqslant a$ and $0 \leqslant \theta \leqslant \frac{1}{2}\pi$.

Hence verify that the disc is at a uniform potential and that its capacity is $8\varepsilon_0 a$.

11. An electrostatic field is set up by a charge Q uniformly distributed round a circular ring which has spherical polar coordinates $r = c$, $\theta = \alpha$. Find the potential of the field at any point for which $r < c$.

A sphere made of material of dielectric constant K is now placed with its centre at the origin of coordinates. If the radius a of the sphere is less than c, show that the potential of the modified field in the region $a < r < c$ is

$$\frac{Q}{4\pi\varepsilon_0}\sum_{0}^{\infty}\frac{1}{c^{n+1}}\left\{r^n-\frac{n(K-1)}{Kn+n+1}\frac{a^{2n+1}}{r^{n+1}}\right\}P_n(\cos\alpha)\,P_n(\cos\theta).$$

12. Show that the magnetic scalar potential Ω due to a magnetic dipole of moment $M\hat{k}$ (where $\hat{k}$ is a unit vector directed along the polar axis) placed at a distance c from the origin of spherical polar coordinates is expressible in the form

$$\Omega = \begin{cases} \dfrac{M}{4\pi r^2}\displaystyle\sum_{1}^{\infty} n\left(\frac{c}{r}\right)^{n-1} P_n(\cos\theta), & (r > c), \\ -\dfrac{M}{4\pi c^2}\displaystyle\sum_{0}^{\infty}(n+1)\left(\frac{r}{c}\right)^{n} P_n(\cos\theta) & (r < c) \end{cases}$$

at the point (r, θ, ϕ).

A small magnet of moment M is placed at distance $c\,(> a)$ from the centre of a soft iron sphere of radius a and uniform permeability μ, with its axis directed away from the centre of the sphere. Find the force experienced by the dipole due to the magnetization of the soft iron, showing in particular that the magnitude of this force is

$$\frac{\mu_0 M^2(\mu-\mu_0)}{4\pi c^4}\sum_0^\infty \frac{n(n+1)^2(n+2)}{\mu n+\mu_0(n+1)}\left(\frac{a}{c}\right)^{2n+1}.$$

13. Apply the transformation $w = c\cosh(\pi z/2a)$ to the following problem. In the space $x > 0$ between the three infinite earthed conducting planes $x = 0$, $y = 0$ and $y = 2a$, a long wire charged with line density q is placed along the line $x = y = a$. Show that a charge $-(2q/\pi)\cot^{-1}(\sinh\frac{1}{2}\pi)$ is induced on each unit length of the strip of the plane $x = 0$ bounded by the planes $y = 0$ and $y = 2a$.

14. Prove that the mapping $w = \log z (0 \leqslant \arg z \leqslant \pi)$ maps the upper half of the complex z-plane into a strip of the w-plane.

Two small circular electrodes each of radius δ are at a distance $2a$ apart on a line parallel to and midway between the edges of an infinite strip of width π and conductivity σ with insulated boundaries. Show that the resistance between them is approximately

$$\frac{1}{\pi\sigma}\ln\frac{\sinh 2a}{\delta}.$$

15. Show that the transformation

$$k\sin w = \sin z,$$

where $z = x+iy$, $w = u+iv$, and k is real and between 0 and 1, may be used to map the semi-infinite strip in the z-plane for which $-\frac{1}{2}\pi \leqslant x \leqslant \frac{1}{2}\pi$, $y \geqslant 0$, on a part of the region of the w-plane for which $v \leqslant 0$ and u lies between 0 and 2π.

An infinite number of thin, plane, infinitely long, conducting strips, all having the same breadth $2b$, are placed parallel to one another in the same plane, with the middle line of any one strip at a distance d (greater than $2b$ and the same for all the strips) from the middle line of either of the two strips nearest to it. The two sides of each strip together carry an electric charge q per unit length, and all the strips are at the same potential. Show that the surface density at any point of a strip is

$$\frac{q}{2d}\,\frac{\cos\dfrac{\pi x}{d}}{\left(\sin^2\dfrac{\pi b}{d}-\sin^2\dfrac{\pi x}{d}\right)^{1/2}}$$

where x is the distance from the central line of the strip.

16. A fixed uniform solid sphere of radius a is composed of material of density σ. The sphere is surrounded by an incompressible ocean of density $\varrho(<\sigma)$, and total mass $4\pi\varrho(b^3-a^3)/3$. A particle of mass m is situated at a sufficiently great distance d from the centre of the solid sphere for its gravitational effect on the surface of the ocean to be small. Show that in the approximation in which third order terms in b/d are neglected the surface of the ocean is a sphere having centre at a distance

$$\frac{3mb^3}{4\pi a^3 d^2(\sigma-\varrho)}$$

from the centre of the solid surface.

Show further that in the approximation in which terms of the third order in b/d are retained the surface of the ocean is a prolate spheroid of eccentricity

$$\left[\frac{45mb^3}{4\pi d^3\{5a^3(\sigma-\varrho)+2b^3\varrho\}}\right]^{1/2}.$$

17. A circular wire of radius a carries a steady current I, and a sphere of soft iron of uniform permeability μ is concentric with the wire. The radius of the sphere is less than a. Show that the magnetic field within the iron at a point on the axis distant z from the centre is directed along the axis, and has magnitude

$$\frac{\mu_0 I}{2a}\left[\frac{3\mu_0}{2\mu_0+\mu}+\sum_{n=1}^{\infty}(-1)^n\frac{(2n+1)(4n+3)\mu_0}{\mu_0(2n+2)+\mu(2n+1)}\;\frac{1\cdot 3\cdot 5\ldots(2n-1)}{2\cdot 4\cdot 6\ldots 2n}\left(\frac{z}{a}\right)^{2n}\right].$$

18. Two magnetic dipoles each of moment M are placed at a distance $2b$ apart with their axes in the same straight line and the same direction. Show that if a soft iron sphere of permeability μ and radius $a\,(< b)$ is placed with its centre at the mid-point of the line joining the dipoles the force of attraction between them is increased by

$$\frac{8\mu_0(\mu-\mu_0)M^2}{4\pi b^4}\sum_{0}^{\infty}\frac{(n+1)^2(2n+1)(2n+3)}{2n\mu+\mu+2n\mu_0+2\mu_0}\left(\frac{a}{b}\right)^{4n+3}.$$

19. Within a spherical cavity of radius a in a dielectric (K) is placed a charged circular ring of radius $c \sin\alpha$ with its centre C distant $c\cos\alpha$ from O, the centre of the cavity. If Q is the charge on the ring, show that the potential within the cavity for $a > r > c$ is

$$\frac{Q}{4\pi\varepsilon_0}\sum_{0}^{\infty}\left\{\frac{1}{r^{n+1}}-\frac{(K-1)(n+1)r^n}{\{K+n(K+1)\}a^{2n+1}}\right\}c^n P_n(\cos\alpha)\,P_n(\cos\theta),$$

where OC is the axis $\theta = 0$.

20. Show that the Legendre polynomial $P_n(x)$ satisfies the recurrence relation

$$(n+1)P_{n+1}+nP_{n-1} = (2n+1)xP_n.$$

The undisturbed potential of an electric field, containing no sources in the region $r \leqslant a$, is

$$\phi = \sum_{1}^{\infty}\alpha_n\left(\frac{r}{a}\right)^n P_n(\mu),$$

in spherical polar coordinates r, θ, ψ, where $\mu = \cos\theta$. An earthed conducting sphere of boundary $r = a$ is inserted in the field. Prove that there is a force on the sphere in the direction $\theta = 0$ of amount

$$(4\pi/\varepsilon_0)\sum_{1}^{\infty}(n+1)\alpha_n\alpha_{n+1}.$$

CHAPTER 9

ELECTROMAGNETIC INDUCTION AND MAGNETIC ENERGY

9.1 Faraday's and Neumann's Law

When an earthed conductor is placed near a charged conductor, the earthed conductor itself becomes charged by the process of influence. Faraday tried to find a similar effect between electric currents corresponding to electrostatic influence between charged conductors. He could find no effect when there was a *steady* current flowing in one of two circuits, but at the instant of switching the current on or off in one circuit he found that currents flowed in neighbouring circuits. It was necessary to have a *varying* current in one circuit (the primary circuit) in order to produce a current in a neighbouring circuit (the secondary circuit). Faraday conducted his researches and discovered this effect of *electromagnetic induction* in 1831. These results were formulated in a general mathematical statement by Neumann in 1845.

We do not here go into all the details of Faraday's researches but state the general result that

> when the magnetic flux linked with a circuit varies, an electromotive force is induced in that circuit proportional to the rate of change of the flux.

This is combined with a general principle, known as Lenz's Law, that when a system in a steady state is subjected to a change the system resists the change in a way which tends to restore the steady state.

When the positive sense of circulation of current in a circuit is given, the direction of positive magnetic flux is defined by the "right-hand corkscrew" rule as in Ampère's (or the Biot–Savart) Law. Neumann combined Faraday's result with Lenz's Law in the form

$$\mathcal{E} = -k\frac{dN}{dt}, \tag{9.1}$$

where $\mathscr{E}$ is the induced electromotive force in a circuit arising from a change in the magnetic flux N. The minus sign indicates that if the flux N is increased, the induced electromotive force tends to reduce the flux by sending a current in the negative sense. The quantity k (> 0) is a constant of proportionality which takes the value 1 when $\mathfrak{E}$ is in volts and N is in webers. The change in N may be the result either of a varying magnetic field, or of the motion of the circuit through a stationary magnetic field, or both.

In order to bring electromagnetic induction into our theory we must consider variable currents, so we now relax the condition that currents must be steady but retain the condition div $\boldsymbol{J} = 0$, i.e. the currents must flow in closed circuits. Neumann's Law gives the electromotive force, and not the current, the magnitude of the induced current depending upon the resistance of the secondary circuit; if the secondary circuit is not closed the electromotive force, presumably, still acts in the circuit but can cause no current to flow. In fact, inside the circuit there must be an electric field $\boldsymbol{E}$ such that $\oint \boldsymbol{E} \cdot \mathbf{d}\boldsymbol{s} = \mathscr{E}$, for this line integral, taken around the (closed) circuit, is the definition of electromotive force. However, since the tangential component of $\boldsymbol{E}$ is continuous across the surface of the conductor forming the circuit we expect $\oint \boldsymbol{E} \cdot \mathbf{d}\boldsymbol{s}$ to have the same value when the contour of integration lies in the (non-conducting) medium immediately outside the secondary circuit. The position of the secondary circuit is quite arbitrary so we adopt the generalized form of Neumann's Law as

$$-\frac{\mathrm{d}}{\mathrm{d}t}\iint \boldsymbol{B} \cdot \mathbf{d}\boldsymbol{S} = \oint \boldsymbol{E} \cdot \mathbf{d}\boldsymbol{s}, \tag{9.2}$$

where the line integral is taken around the perimeter (in the positive sense) of the open surface of the surface integral: this surface is arbitrary. To sum up this generalization we say that a changing magnetic field at any point in space gives rise to an electric field at that point, whether or not there is a conductor at that point in which a current can flow. Assuming that the surface and its perimeter in eqn. (9.2) is stationary we can write

$$\iint \left(\mathbf{curl}\, \boldsymbol{E} + \frac{\partial \boldsymbol{B}}{\partial t}\right) \cdot \mathbf{d}\boldsymbol{S} = 0,$$

from which we can deduce, since the surface is arbitrary, that

$$\mathbf{curl}\, \boldsymbol{E} + \frac{\partial \boldsymbol{B}}{\partial t} = \boldsymbol{0}. \tag{9.3}$$

Equation (9.3) is the differential, or field, form of Neumann's relation and is the generalization of the equation $\mathbf{curl}\, \boldsymbol{E} = \boldsymbol{0}$ which applies to static fields. The fact that **curl** $\boldsymbol{E}$ is not zero means that the electric field is no longer "conservative" if a magnetic field varies with time. The agent causing the variation of the magnetic field may therefore impart energy to the charges and the electric field, when the charges are allowed to move.

From eqn. (9.3) we can write, since div $\boldsymbol{B} = 0$, $\boldsymbol{B} = \textbf{curl}\ \boldsymbol{A}$ and

$$\textbf{curl}\ \boldsymbol{E} + \frac{\partial \boldsymbol{B}}{\partial t} = \textbf{curl}\left(\boldsymbol{E} + \frac{\partial \boldsymbol{A}}{\partial t}\right) = \boldsymbol{0}.$$

$$\therefore\ \boldsymbol{E} = -\textbf{grad}\ V - \frac{\partial \boldsymbol{A}}{\partial t}. \tag{9.4}$$

This is the generalization of the electrostatic field equations (**curl** $\boldsymbol{E} = \boldsymbol{0}$ and $\boldsymbol{E} = -\textbf{grad}\ V$) to the case of variable magnetic fields.

Example 1. An axially symmetric magnetic field has components $(-\frac{1}{2}B_0 r, 0, B_0 z)$ in cylindrical polar coordinates (r, θ, z), B_0 being constant. A conducting circular loop of radius a and resistance R has its centre at the point $(0, 0, h)$, and lies in the plane $z = h$. Write down the flux of the magnetic field through the loop. Hence find the current induced as h changes, neglecting self-induction, but making clear the direction in which the current flows. Find the force exerted by the magnetic field on the loop. If the loop has negligible mass, so that the force $\boldsymbol{P}$ causing the motion is always in equilibrium with the magnetic force, verify that the rate of working of $\boldsymbol{P}$ is equal to the rate of dissipation of energy by the resistance of the loop.

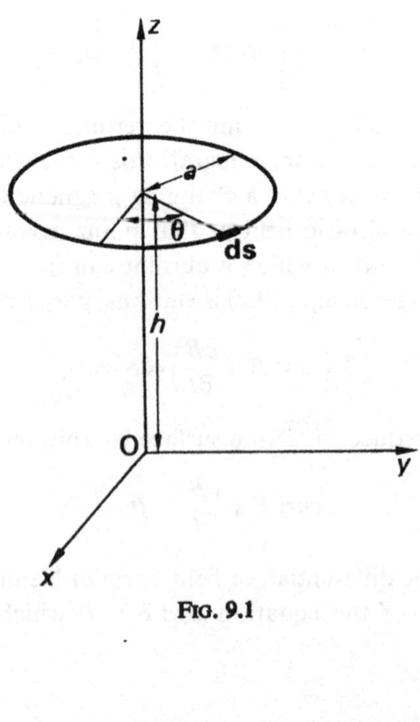

FIG. 9.1

Since we must integrate around the circuit we use cartesian axes with $x = r\cos\theta$, $y = r\sin\theta$, as appropriate. In this frame

$$\boldsymbol{B} = \{-\tfrac{1}{2}B_0 r\cos\theta \quad -\tfrac{1}{2}B_0 r\sin\theta \quad B_0 z\}.$$

The magnetic flux is $N = \pi a^2 B_0 h$.

$$\therefore\ \mathscr{E} = RI = -\pi a^2 B_0(\mathrm{d}h/\mathrm{d}t).$$

$$\therefore\ I = -\frac{\pi a^2 B_0}{R}\frac{\mathrm{d}h}{\mathrm{d}t}.$$

The element $\mathbf{d}\boldsymbol{s}$ of the circuit with position $\boldsymbol{r}$ (Fig. 9.1) is given by

$$\boldsymbol{r} = \{a\cos\theta \quad a\sin\theta \quad h\}, \qquad \mathbf{d}\boldsymbol{s} = \{-a\sin\theta \quad a\cos\theta \quad 0\}\,\mathrm{d}\theta.$$

Consequently the force on this element is

$$\begin{aligned} I\,\mathbf{d}\boldsymbol{s}\times\boldsymbol{B} &= I\{-a\sin\theta \quad a\cos\theta \quad 0\}\times\{-\tfrac{1}{2}B_0 a\cos\theta \quad -\tfrac{1}{2}B_0 a\sin\theta \quad B_0 h\}\,\mathrm{d}\theta \\ &= I\{B_0 ah\cos\theta \quad B_0 ah\sin\theta \quad \tfrac{1}{2}B_0 a^2\}\,\mathrm{d}\theta. \end{aligned}$$

On integration w.r. to θ this gives the value $-\boldsymbol{P}$ for the force exerted by the field, the force causing the motion being $+\boldsymbol{P}$, where

$$-\boldsymbol{P} = I\oint \mathbf{d}\boldsymbol{s}\times\boldsymbol{B} = \{0 \quad 0 \quad -P\} = \{0 \quad 0 \quad \pi a^2 B_0 I\}.$$

Therefore

$$P = -\pi a^2 B_0\left(-\frac{\pi a^2 B_0}{R}\frac{\mathrm{d}h}{\mathrm{d}t}\right) = \frac{\pi^2 a^4 B_0^2}{R}\left(\frac{\mathrm{d}h}{\mathrm{d}t}\right),$$

and the rate of working of this force is

$$P\frac{\mathrm{d}h}{\mathrm{d}t} = \frac{\pi^2 a^4 B_0^2}{R}\left(\frac{\mathrm{d}h}{\mathrm{d}t}\right)^2.$$

The rate of dissipation of energy by the resistance of the loop is

$$I^2 R = \frac{\pi^2 a^4 B_0^2}{R}\left(\frac{\mathrm{d}h}{\mathrm{d}t}\right)^2,$$

which shows the required equality.

Example 2. A small magnet of moment $\boldsymbol{m}$ at the origin of vectors is rotating about its centre with angular velocity $\boldsymbol{\omega}$, the magnitude of $\boldsymbol{m}$ remaining constant. Find the electric field produced at a point $\boldsymbol{r}$.

The moment of the magnet varies according to the relation

$$\frac{\mathrm{d}\boldsymbol{m}}{\mathrm{d}t} = \boldsymbol{\omega}\times\boldsymbol{m}.$$

But the vector potential of the field of the magnet is $\boldsymbol{A} = \mu_0(\boldsymbol{m}\times\boldsymbol{r})/(4\pi r^3)$.

$$\therefore\ \frac{\partial\boldsymbol{A}}{\partial t} = \frac{\mu_0}{4\pi}\frac{(\boldsymbol{\omega}\times\boldsymbol{m})\times\boldsymbol{r}}{r^3} = \frac{\mu_0}{4\pi}\frac{\boldsymbol{m}(\boldsymbol{\omega}\cdot\boldsymbol{r})-\boldsymbol{\omega}(\boldsymbol{m}\cdot\boldsymbol{r})}{r^3}.$$

It follows that, in the absence of charges, or any electrostatic field,

$$\boldsymbol{E} = -\frac{\partial\boldsymbol{A}}{\partial t} = \frac{\mu_0}{4\pi}\frac{\boldsymbol{\omega}(\boldsymbol{m}\cdot\boldsymbol{r})-\boldsymbol{m}(\boldsymbol{\omega}\cdot\boldsymbol{r})}{r^3}.$$

[This solution assumes that the changes in the magnetic field are propagated *instantaneously* from the rotating magnet. In fact, see Chapter 13, this is not correct, and so this solution is only applicable at comparatively short distances from the origin, for which the time of propagation of disturbances is negligible.]

9.2 Observers in motion: the Lorentz transformation

The previous section has dealt with only one experimental relation, Faraday's discovery, but electromagnetic induction occurs not only when the primary current grows, or decreases, but also when there is relative motion of the two circuits. A current is itself made up of charges in motion, so now we must consider more closely the resulting phenomena when observers are in relative motion. How are the measurements made by one observer who is stationary relative to a charge or circuit related to the corresponding measurements made by another observer in relative motion? We indicated in Volume I that the simple, Galilean transformation does not explain some of the experimental results adequately. We recall briefly three of these, mentioned in Volume I.

(a) Champion's experiments [§ 1.1(7a), p. 12] about the scattering of electrons by stationary electrons gave rise, as we showed, directly to a difficulty about the transformation between inertial frames. The usual assumption for this transformation indicates that the two electrons should go off at right angles. This is not observed for high energies.

(b) In a purely electromagnetic context, the experiment of Wilson and Wilson [§ 1.1(7b), p. 14], in which a moving condenser has a magnetic dielectric, raises the problem of transformation between inertial frames again. Here we have to supply the relations for the dielectric

$$\boldsymbol{B} = \mu \boldsymbol{H}, \quad \boldsymbol{D} = \varepsilon \boldsymbol{E}$$

in a frame in which the dielectric is at rest. This is not the frame in which the experimental results are found.

(c) The fact that the speed of light has certain very curious properties was shown up by Fresnel's experiment [§ 1.1(7d), p. 16] of measuring the speed of light in a moving fluid. Here again we can express the difficulty in terms of the coordinate transformation, since in a coordinate system in which the fluid is at rest the speed should be c/n, and so by making the usual transformation, eqn. (1.2), Fresnel's result should be $c/n+V$ rather than the observed

$$c/n+V(1-1/n^2).$$

Let us go back and consider in more detail the transformation between inertial frames in Newtonian mechanics. This transformation may be written in the form

$$\mathbf{r} \to \mathbf{r}' = \mathbf{r} - \mathbf{V}t, \quad t \to t' = t.$$

The second of these equations conceals rather more than one expects. It is, of course, only valid if both coordinate systems use the same units for time-measurement and the same zero. But this is not the significant part of the relation. Much more important is the fact that if two events occur at the same time in one coordinate system they occur also at the same time in another, so that the idea of simultaneity should be a well-defined concept not dependent upon the coordinate system. Einstein realized in 1904 that this apparently well-defined concept is not in fact so clearly specified as one would wish. Any physical concept such as simultaneity needs, in order to make it precise, an operational definition. That is to say, we must be able to give the set of operations for determining whether or not two given events are simultaneous (although usually we shall have no wish to go to the trouble of actually carrying out the operations). Einstein assumed that for events at one particular point it was reasonable to assume a time-reckoning as an ordering for the events. But when one has made this assumption something *further* is needed in order to allot time reckonings to distant events. Since the events are distant, some form of signalling is required, so that the distant events may be recorded locally and each fitted into its place in the time-order. Of course, any form of signal could be used and in some cases it might be convenient, for example, to use sound, but, whatever form is used, the speed with which the signals travel is sure to play an important part in the formalism. (If this is not already clear it will be exemplified below.)

Einstein recalled the curious nature of the speed of light, as exemplified in Fresnel's experiment, and so concluded that the most useful formalism would be one in which the operational method of defining the time of occurrence of distant events was with light signals. It then turns out, as will be clear shortly, that simultaneity ceases to be a well-defined concept. Events which are simultaneous in one coordinate system have to be described as non-simultaneous in another. In order to show how Einstein's convention is set up for defining time at distant events it is convenient to use a diagram in which the vertical axis represents time, and we consider, for the moment only, one space dimension, which is taken as the horizontal axis. Figure 9.2 then represents by the line OA an observer from whose point of view the diagram is drawn, so that he is at rest in it and only his time variable changes.

The line OB represents an observer who is coincident with the first observer initially (i.e. at time $t = 0$), and who is moving with a uniform speed relative to him. The dotted lines represent light signals sent from one observer to the other. The times t_1, t_3 are those of the emission of a signal from the first observer to the second and of reception of the reflected signal back again. Einstein then makes the *convention* that the time to be allotted by the first observer to the event of reflection of the signal is

$$t_2 = \tfrac{1}{2}(t_3+t_1).$$

This time will in general differ from that measured for the event by the second observer and denoted in Fig. 9.2 by t_2'.

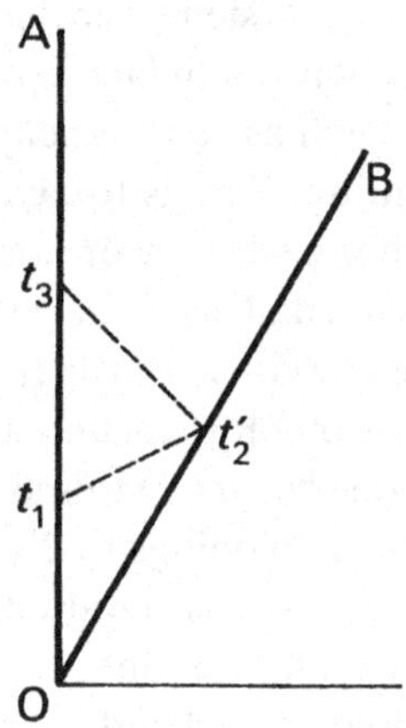

FIG. 9.2

Since, however, the motion of the second observer relative to the first is uniform, t_2' must be proportional to t_1. (Thus, for example, if t_1 were doubled, t_2' would be doubled also.) The constant of proportionality is denoted by k (a number which must depend on the value of the relative velocity of the two observers), so that $t_2' = kt_1$. It now follows from the fact that there is no preferred inertial frame at rest, but rather symmetry between the observers, that, if the corresponding diagram is drawn from the point of view of the second observer, then $t_3 = kt_2'$, with the *same* value of k.

In order to fix the dependence of k on the velocity of separation V, it is necessary to specify the distance apart of the observers. The distance of the reflection event from the first observer is

$$x = \tfrac{1}{2}c(t_3-t_1).$$

(This is, strictly speaking, a further assumption; it is equivalent to assuming that the signal velocity c is the same in both directions.) Hence the velocity of separation of the observers is

$$V = \frac{x}{t_2} = \frac{c(t_3-t_1)}{t_3+t_1} = c\frac{k^2-1}{k^2+1},$$

so that

$$k = \sqrt{\left(\frac{1+V/c}{1-V/c}\right)}.$$

Suppose now that A is transmitting pulses of light at regular intervals $T = t_1$ (which may be thought of as an idealization of transmitting a monochromatic signal of frequency $1/T$). These pulses will be observed by B at intervals

$$t_2' = kT = T\sqrt{\left(\frac{1+V/c}{1-V/c}\right)},$$

or, expanding to the first order in V/c, at intervals $T(1+V/c)$. This first-order effect is known as the *Doppler effect*. It is more familiar in everyday life in the change of pitch of the noise of an express train, a racing car or a petrol-driven model aeroplane when it passes the observer and in so doing ceases to approach and begins to recede with the same speed. However, in the case of sound there is a difference in the formulae, which is worth noting. If a source S of sound is receding with speed V from a stationary observer O (Fig. 9.3), and emits two pulses at times differing by T, the times of reception of these pulses will be

$$\frac{l}{c}, \quad T+\frac{l+VT}{c}$$

respectively, where c is now the speed of sound and l is the distance OS at time $t = 0$. These times differ by $T(1+V/c)$. If, however, the source is at rest

FIG. 9.3

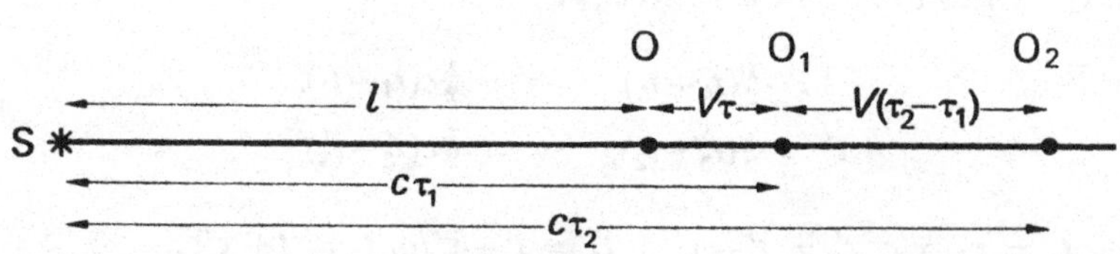

FIG. 9.4

and the observer recedes (Fig. 9.4), the times of reception will be, say, τ_1 and τ_2 where, if the distance OS is l at $t = 0$,

$$l+V\tau_1 = c\tau_1,$$

$$l+V\tau_2 = c(\tau_2-T),$$

giving

$$\tau_2-\tau_1 = T/(1-V/c).$$

The difference between these formulae (which obviously agree to the first order in V/c) shows most clearly when the source and observer are approaching each other rather than receding from one another. This causes a change of sign in V, but whereas this is of no great significance in the case of the moving observer, the case of the moving source leads to a factor multiplying T which can become very small indeed as the speed of the source approaches that of sound. This is ultimately the source of the supersonic "bang", though the full explanation of that phenomenon is more complicated.

These two different expressions for the Doppler effect for sound arise because of the existence of still air as the medium of transmission of the waves. With light, however, there is no such standard of rest and so the Doppler factor

$$\sqrt{\left(\frac{1+V/c}{1-V/c}\right)}$$

applies to both cases. If V approaches c (now the speed of light) the interval between successive pulses becomes very large. In the case of a transmitter of monochromatic radiation the frequency becomes very small, i.e. is shifted very much towards the red, and correspondingly the energy that can be transmitted by the radiation becomes very small. In the limiting case, if a source could have the speed of light, it could not be observed at all.

The Doppler shift, just noted, is one of the instances of a *first-order effect* (i.e. first order in V/c) arising from Einstein's convention. It is much more usual to have second-order effects. Such a one arises when two observers both set up time and space coordinates for an event E away from both of them (Fig. 9.5). With the notation of the figure, they allot coordinates (t, x), (t', x') to the event E, where

$$t = \tfrac{1}{2}(t_4+t_1), \qquad x = \tfrac{1}{2}c(t_4-t_1),$$

$$t' = \tfrac{1}{2}(t_3'+t_2'), \qquad x' = \tfrac{1}{2}c(t_3'-t_2').$$

Hence $t_1 = t-x/c$, $t_2' = t'-x'/c$, $t_3' = t'+x'/c$, $t_4 = t+x/c$.

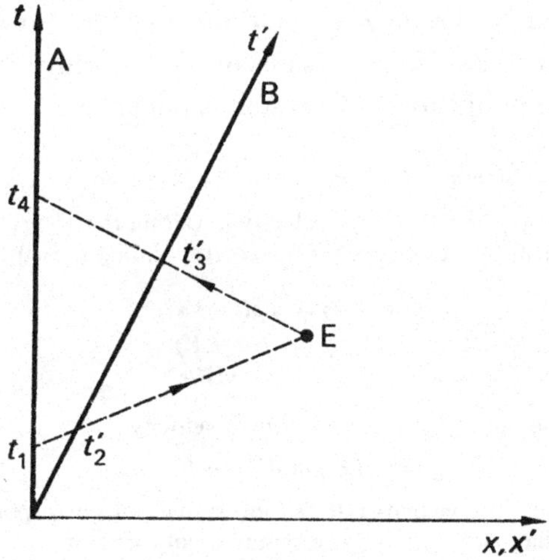

FIG. 9.5

Accordingly

$$t+\frac{x}{c} = k\left(t'+\frac{x'}{c}\right),$$

$$t'-\frac{x'}{c} = k\left(t-\frac{x}{c}\right),$$

which gives

$$t' = \frac{1}{2}\left(k+\frac{1}{k}\right)\left(t-\frac{Vx}{c^2}\right),$$

$$x' = \frac{1}{2}\left(k+\frac{1}{k}\right)(x-Vt),$$

where note has been taken of the value found above for V. Here

$$\frac{1}{2}\left(k+\frac{1}{k}\right) = \frac{1}{\sqrt{(1-V^2/c^2)}} = \beta \text{ (say)}$$

is the factor on the right-hand side, and this differs from unity only by second order terms. Indeed, the transformation equations

$$x' = \beta(x-Vt), \quad t' = \beta(t-Vx/c^2), \quad \beta = (1-V^2/c^2)^{-1/2} \tag{9.5}$$

become, to the first order in V/c,

$$x' = x-Vt, \quad t' = t,$$

exactly as in Newtonian mechanics. The accurate transformation equations (9.5) are called the *Lorentz transformation*; they appear to have been found independently first by Lorentz in 1904 and then by Einstein in 1905. The theoretical importance of Lorentz's discovery was only realized by Poincaré shortly before the publication of Einstein's paper.

Example 1. The addition of velocities

If a point P is moving with uniform velocity u' (along the x'-axis) relative to O', then $x' = u't'$. From eqns. (9.5) we obtain, for the corresponding measurements by O,

$$\beta(x-Vt) = u'\beta(t-Vx/c^2).$$

$$\therefore \quad x = \frac{t(u'+V)}{1+u'V/c^2}.$$

Hence O observes the point P to have a uniform velocity

$$u = (u'+V)/(1+u'V/c^2). \tag{1}$$

Correct to the first order this velocity is the simple sum $u'+V$ corresponding to the Galilean transformation. Similarly we can find the symmetrical formula

$$u' = \frac{u-V}{1-uV/c^2} \tag{2}$$

from the relations (9.5) when they are solved for x, t, viz.

$$x = \beta(x'+Vt'), \quad t = \beta(t'+Vx'/c^2), \quad \beta = (1-V^2/c^2)^{-1/2},$$

in which the sign of V is reversed. These results (1) and (2) are the accurate transformation formulae for addition of velocities. Since the Lorentz transformations are linear in x, t, the result also holds for the instantaneous values of non-uniform velocities, viz. if O' observes the point at x' to have velocity $u' = dx'/dt'$ at time t', then O observes the point to be at the corresponding distance x at time t moving with velocity $u = dx/dt$ where x, t, x', t' are related by eqn. (9.5) and u, u' are related as above, in (1) and (2).

Another way of deriving these results is to consider the transmission and reception of signals between three observers (Fig. 9.6a). If the appropriate k-factors between O and

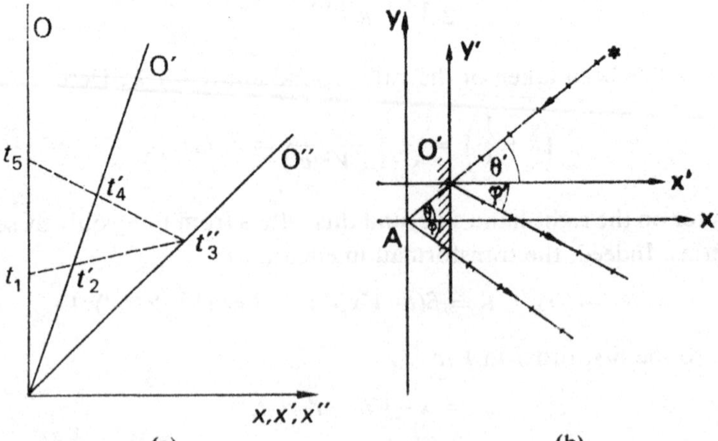

FIG. 9.6. The two frames are shown separate, although they coincide in position at the instant of reflection

O' and between O' and O'' are k_1, k_2, then

$$t_2' = k_1 t_1, \quad t_3'' = k_2 t_2'$$

so that

$$t_3'' = (k_1 k_2) t_1,$$

giving a k-factor equal to $k_1 k_2$ between O and O''. If, as before,

$$k_1 = \sqrt{\left(\frac{1+V/c}{1-V/c}\right)}$$

and if

$$k_2 = \sqrt{\left(\frac{1+u/c}{1-u/c}\right)},$$

then it is easy to see that

$$k_1 k_2 = \sqrt{\left(\frac{1+u'/c}{1-u'/c}\right)}$$

where

$$u' = \frac{u+V}{1+uV/c^2}.$$

Example 2. The Fitzgerald contraction

The measurement of the length of an object consists of two events, namely, the coincidences of the two ends of the object with the graduations of the measuring scale. When the object is moving relative to a given inertial frame, these events must be simultaneous in that inertial frame if the difference in the two space coordinates is to give what would be called the length of the object in that frame. When the object is stationary the events need not be simultaneous. In this case the observer (with his measuring scale) and the object will have the same velocity, and the length measured by this observer is called the "proper length".

We consider an object (along the x-axis) which is stationary in the $xyzt$ frame of observer O. The space separation between the two events of measurement is therefore the proper length, i.e. $dx = l_0$.

The differential form of eqn. (9.5) gives

$$dx' = \beta(dx - V\,dt), \quad dt' = \beta(dt - V\,dx/c^2).$$

When O' measures the object the two events are simultaneous, so that

$$dt' = 0, \quad dt = V\,dx/c^2.$$

Therefore

$$dx' = \beta(1 - V^2/c^2)\,dx,$$

$$l' = l_0/\beta = l_0(1 - V^2/c^2)^{1/2}.$$

Hence the length ascribed by O' to the object is less than the proper length.

Similarly if the object moved with velocity V, i.e. was stationary in the $x'y'z't'$ frame, then the length measured by O' would be l_0, the proper length, and that by O would be $l = l_0(1 - V^2/c^2)^{1/2}$. The *Fitzgerald contraction* is the shortening of the length of the object when it is in motion while being measured.

Example 3. The time dilation

If two events occur at the same position but at different times in a frame, the time interval between them is said to be the "proper" time interval, and is the value given by the observer (O say) associated with the frame, i.e.

$$dx = 0, \quad dt = t_0.$$

When these two events are observed by O', who is in motion relative to O, he ascribes separations in space and time between the events given by

$$dx' = \beta(dx - V\,dt), \quad dt' = \beta(dt - V\,dx/c^2).$$
$$\therefore \; dt' = \beta t_0.$$

Hence the time interval measured by O' is greater than that measured by O. According to O', the clock used to measure the proper time moves between the events and is running more slowly than his own clock. This is the result known as time dilation.

Example 4. Prove that the equations

$$x' = \beta x - vt, \quad y' = y, \quad t' = \beta t - vx/c^2$$

agree with those for a Lorentz transformation with velocity v along the x-axis, correct to order $(v/c)^2$.

Hence prove that the resultant of such a transformation, followed by one with velocity u along the new y-axis, followed by one with velocity $-v$ along the new x-axis, followed by one with velocity $-u$ along the new y-axis is a rotation of the axes through an angle uv/c^2.

(This effect, the *Thomas precession*, was at one time thought to be important in the theory of the spinning electron.)

The first part is obvious, and the successive transformations can be written (to the second order)

$$x_1 = \beta_1 x - vt, \quad y_1 = y, \quad t_1 = \beta_1 t - vx/c^2.$$

The transformation for velocity u along the y_1-axis is

$$x_2 = x_1, \quad y_2 = \beta_2 y_1 - ut_1, \quad t_2 = \beta_2 t_1 - uy_1/c^2.$$
$$\therefore \; x_2 = \beta_1 x - vt, \quad y_2 = \beta_2 y - ut + uvx/c^2, \quad t_2 = \beta_1\beta_2 t - vx/c^2 - uy/c^2.$$

The transformation for velocity $-v$ along the x_2-axis is

$$x_3 = \beta_1 x_2 + vt_2, \quad y_3 = y_2, \quad t_3 = \beta_1 t_2 + vx_2/c^2.$$
$$\therefore \; x_3 = \beta_1^2 x - v^2x/c^2 - uvy/c^2, \quad y_3 = y_2, \quad t_3 = \beta_1^2\beta_2 t - uy/c^2 - v^2t/c^2.$$

The final transformation with velocity $-u$ along the y_3-axis is

$$x_4 = x_3, \quad y_4 = \beta_2 y_3 + ut_3, \quad t_4 = \beta_2 t_3 + uy_3/c^2.$$
$$\therefore \; x_4 = x - (uv/c^2)y, \quad y_4 = \beta_2^2 y - ut + uvx/c^2 + ut - u^2y/c^2 = y + (uv/c^2)x,$$
$$t_4 = \beta_1^2\beta_2^2 t - uy/c^2 - v^2t/c^2 + uy/c^2 - u^2t/c^2$$
$$= (1 + u^2/c^2 + v^2/c^2)t - u^2t/c^2 - v^2t/c^2 = t.$$

The resultant transformation, correct to the second order, is, therefore,

$$x_4 = x - (uv/c^2)y, \quad y_4 = y + (uv/c^2)x, \quad t_4 = t,$$

giving the stated result.

Example 5. Obtain Einstein's formulae for the Doppler effect and the aberration effect.

A plane mirror moves in the direction of its normal with uniform velocity V relative to an observer A attached to a given inertial frame. A ray of light incident at any angle θ

to the normal is reflected at an angle ϕ, both angles being measured by A. Prove that

$$\frac{\sin\theta}{\cos\theta+V/c}=\frac{\sin\phi}{\cos\phi-V/c}.$$

Calculate the wavelength of the reflected ray (according to A) in terms of θ, ϕ and the wavelength of the incident ray (according to A).

Aberration is the name given to the apparent change in direction of light from a source (usually a star) because of the relative motion of source and observer.

We assume that the source emits a pulse, or signal, and that the event of reception of the pulse at the origin of coordinates $Axyz$ is $\{0\ 0\ 0;\ 0\}$. We suppose that a second frame of reference $O'x'y'z'$ has velocity V along the x-axis, so that the two frames are related by the Lorentz transformation

$$x'=\beta(x-Vt),\quad y'=y,\quad z'=z,\quad t'=\beta(t-Vx/c^2).$$

We choose V so that the source has velocity $\{V\ 0\ 0\}$ relative to the frame $Axyz$ at the instant of emission, i.e. S is (instantaneously) stationary in $O'x'y'z'$ when the signal is emitted.

The passage of the pulse along the incident ray between S and the origin (see Fig. 9.6b on p. 342) is given by:

in frame $Axyz$, $\{-ct\cos\theta\quad -ct\sin\theta\quad 0;\ t\}$, $-r/c\leqslant t\leqslant 0$;
in frame $O'x'y'z'$, $\{-ct'\cos\theta'\quad -ct'\sin\theta'\quad 0;\ t'\}$, $-r'/c\leqslant t'\leqslant 0$.

Here r, r' are the values assigned to the distance of the event of emission in the two frames The Lorentz transformations give

$$-ct'\cos\theta'=\beta(-ct\cos\theta-Vt),\quad -ct'\sin\theta'=-ct\sin\theta,\quad t'=\beta(t+Vt\cos\theta/c).$$

$$\therefore\quad \tan\theta'=\frac{\sin\theta}{\beta(\cos\theta+V/c)}. \tag{1}$$

Also, by considering the event of emission,

$$-\frac{r'}{c}=\beta\left(-\frac{r}{c}-\frac{Vr\cos\theta}{c^2}\right),$$

$$\therefore\quad r'=\beta\{1+(V/c)\cos\theta\}r. \tag{2}$$

We can also deduce the results

$$\cos\theta'=\frac{\cos\theta+V/c}{1+(V/c)\cos\theta},\quad \sin\theta'=\frac{\sin\theta}{\beta\{1+(V/c)\cos\theta\}}. \tag{3}$$

The results (1) and (3) give the aberration.

In a similar way the reflected ray is given by:

frame $Axyz$, $\{ct\cos\phi\quad -ct\sin\phi\quad 0;\ t\}$ $t\geqslant 0$;
frame $O'x'y'z'$, $\{ct'\cos\phi'\quad -ct'\sin\phi'\quad 0;\ t'\}$ $t'\geqslant 0$.

From these we obtain

$$ct'\cos\phi'=\beta\{ct\cos\phi-Vt\},\quad ct'\sin\phi'=ct\sin\phi,\quad t'=\beta\{t-(V/c)t\cos\phi\},$$

$$\therefore\quad \tan\phi'=\frac{\sin\phi}{\beta(\cos\phi-V/c)}. \tag{4}$$

In the frame $O'x'y'z'$, $\theta'=\phi'$,

$$\therefore\quad \frac{\sin\theta}{\cos\theta+V/c}=\frac{\sin\phi}{\cos\phi-V/c}.$$

The distances covered by the pulse along the ray between reflection and any other event are given by

$$r = ct, \quad r' = ct' \quad \text{where} \quad r' = \beta\{1-(V/c)\cos\phi\}r. \tag{5}$$

We investigate the frequency of the light as measured in the two frames by considering two events—emission of pulses—separated by a short time interval dt. These pulses are received at A separated by a time interval

$$d\tau = dt + dr/c,$$

where dr is the increase in distance from A between the emissions. Since, at the time of emission, S is moving parallel to the x-axis,

$$dx = V\,dt = dr\cos\theta - r\sin\theta\,d\theta, \qquad 0 = dr\sin\theta + r\cos\theta\,d\theta.$$

$$\therefore\quad dr = V\cos\theta\,dt.$$

$$\therefore\quad d\tau = \{1+(V/c)\cos\theta\}\,dt.$$

The estimate dt' in the frame $O'x'y'z'$ of the time interval between the two emissions is given by

$$dx' = \beta(dx - V\,dt), \quad dy' = dy, \quad dz' = dz, \quad dt' = \beta(dt - V\,dx/c^2).$$

$$\therefore\quad dx' = dy' = dz' = 0, \qquad dt' = \beta(1-V^2/c^2)\,dt = dt/\beta.$$

(This is the usual formula for time dilation.) Because S is at rest in $O'x'y'z'$ at the instant of emission, this time interval dt' is also the time interval between the reception of the two pulses at the origin O'.

$$\therefore\quad d\tau' = dt'.$$

$$\therefore\quad \frac{d\tau}{d\tau'} = \beta\{1+(V/c)\cos\theta\}. \tag{6}$$

Hence the frequencies of oscillations ν, ν' observed in the two frames are related by

$$\frac{d\tau}{d\tau'} = \frac{\nu'}{\nu} = \beta\{1+(V/c)\cos\theta\}.$$

When the source is receding along the line-of-sight $\theta = 0$ and

$$\frac{\nu'}{\nu} = \beta\left(1+\frac{V}{c}\right) = \sqrt{\left(\frac{1+V/c}{1-V/c}\right)},$$

the usual formula for the Doppler effect (see p. 339).

We can also find the relation between wavelengths if we imagine the source S one wavelength away from the origin emitting a pulse to give the reflection event at $\{0\,0\,0;\,0\}$ and a third event of arrival of the pulse one wavelength along the reflected ray. In these circumstances eqn. (2) gives for the incident wave

$$\lambda_1' = \beta\{1+(V/c)\cos\theta\}\lambda_1,$$

and eqn. (5) for the reflected ray,

$$\lambda_2' = \beta\{1-(V/c)\cos\phi\}\lambda_2.$$

In the frame $O'x'y'z'$,

$$\lambda_1' = \lambda_2'.$$

$$\therefore\quad \lambda_2 = \frac{1+(V/c)\cos\theta}{1-(V/c)\cos\phi}\lambda_1.$$

(This result is important in relativistic thermodynamics.)

Example 6. Show from first principles how, according to special relativity, the apparent *direction* of light depends on the relative motion of the source, although its *speed* does not.

Radiation emitted by an electron (undergoing acceleration in the magnetic field of a synchrotron) is effectively isotropic, from the point of view of an observer moving with the electron. Show that when the electron moves past another observer in its path at nearly the speed of light the radiation will appear to this observer (who is at rest in the laboratory) to be confined to an extremely narrow cone with its axis in the direction of motion of the electron.

Obtain an approximate formula for the ratio of apparent brightness of a light-source receding radially with velocity V, where $V \ll c$, to that of an equivalent stationary source at the same distance.

The results obtained in example 5 enable us to find the answers to this problem.

Formula (1) of that example gives the relation between the observed directions θ, θ' of a ray when S is moving *away* from the receiver, and (4) gives the relation between ϕ, ϕ' after reflection. This latter is equivalent to the directions observed when S moves *towards* the receiver.

From (4), when the source is moving towards the receiver, that half of the radiation for which $0 \leqslant \phi' \leqslant \pi/2$ is emitted in the forward direction, and is observed by A to lie within the limits of ϕ given by

$$\phi' = 0, \qquad \phi = 0,$$
$$\phi' = \pi/2, \qquad \cos\phi = V/c, \qquad \text{i.e.} \qquad \phi = \alpha.$$

When $V/c \approx 1$, α is small and A observes the radiation concentrated in a narrow cone about the direction of motion. A similar discussion based on (1) shows that the radiation emitted backwards, $\pi/2 \leqslant \theta' \leqslant \pi$, corresponds to $\pi - \alpha \leqslant \theta \leqslant \pi$. Hence the backward emission is also concentrated in a cone.

The ratio of the apparent brightness of two sources is taken to be the ratio of the energies received from the sources in a certain short interval. In order to calculate these energies we assume from quantum theory that the energy is radiated by a source as *photons*, the energy of a photon of frequency ν being $h\nu$ (*Planck's Law*). Hence the brightness will be $nh\nu$, where n is the number of photons received during the interval in question.

In example (5) we calculated the time intervals $d\tau$, $d\tau'$ between the reception of two pulses as observed by A and O' respectively. If we take these pulses to be the consecutive emissions of photons, then we can write

$$\frac{n'}{n} = \frac{d\tau}{d\tau'} = \beta\left(1+\frac{V}{c}\right).$$

(We put $\cos\theta = 1$, since the source is receding along the line of sight.)

We also calculate the ratio of the observed frequencies as before,

$$\frac{\nu'}{\nu} = \beta\left(1+\frac{V}{c}\right).$$

Then

$$\frac{b}{b'} = \frac{n\nu}{n'\nu'} = \frac{1}{\beta^2(1+V/c)^2} = \frac{1-V/c}{1+V/c}$$
$$\approx 1-\frac{2V}{c} \quad \text{when} \quad (V \ll c).$$

Here b stands for the brightness observed by the observer A, and b' for that observed by O', who is moving with the source. As $V \to c$, $b \to 0$ and the brightness decreases as the velocity of recession increases.

So far all the results found have reduced to the Newtonian form when we neglect second-order terms; but this will not always be the case. There will be some cases when the first-order terms provide a measurable difference (such a case is provided by Fresnel's experiment, as will be seen below), although in nearly all cases the first-order formulae provide some guide as to what result may be expected. There is, however, one important difference between the first-order transformation and the accurate one. In the full transformation t and x are involved in a completely symmetric way; that is, both t' and x' depend on both t and x. In the first-order equations, however, t plays a somewhat ambiguous role. Although x' does depend on both x and t, t' is dependent on t alone; indeed, t is an invariant of the transformation. Thus although the accurate transformations have an apparently more complicated form than the first-order ones, they are really conceptually simpler.

The Fresnel experiment can now be explained very simply by making use of example 1, p. 342. In a coordinate system in which the water is at rest, the speed of light is c/n. Putting

$$u' = \frac{\mathrm{d}x'}{\mathrm{d}t'} = \frac{c}{n}$$

in the transformation equations for velocity, with the sign of V changed, gives

$$u = \frac{\mathrm{d}x}{\mathrm{d}t} = \frac{(c/n)+V}{1+(V/cn)}.$$

Expanding the denominator correct to the first order gives

$$\frac{\mathrm{d}x}{\mathrm{d}t} \approx \left(\frac{c}{n}+V\right)\left(1-\frac{V}{cn}\right)$$

$$\approx \frac{c}{n}+V\left(1-\frac{1}{n^2}\right),$$

in agreement with Fizeau's result. This is an example of even the first-order expression being different from the Newtonian, a fact which explains why the experimental result could be found with sufficient accuracy so long ago.

In order to explain Champion's experiments a slight further elaboration is required since these experiments are essentially two-dimensional and we have only derived the transformation for one space dimension. The corresponding first-order transformation to the one we have derived is merely the Newtonian one:

$$x' = x-Vt, \quad y' = y, \quad z' = z, \quad t' = t.$$

It is quite evident that in this case the y- and z-coordinates are involved in a symmetrical fashion, that is, there is *axial symmetry* about the direction of relative separation of the coordinate systems. We shall assume that this axial symmetry is preserved even under the Lorentz transformations—that is to say, the transformation of y and z must be unchanged under rotations in the yz-plane and so we must be able to put it into the form

$$y' = \lambda y, \quad z' = \lambda z.$$

Here λ may be a function of the velocity of separation, but since the inverse of the transformation corresponding to velocity V is that corresponding to velocity $-V$ (a fact which it is instructive for the reader to check in the case of the Lorentz transformation, see example 1, p. 342) it follows that

$$\lambda(-V) = 1/\{\lambda(V)\};$$

in other words, the coefficient λ must depend upon the direction of separation of the coordinate systems if it depends on the velocity at all, whereas (since space is isotropic) one would expect no such dependence. Accordingly λ must simply be a constant, and by a suitable choice of units in the two coordinate systems the transformations of coordinates at right angles to the direction of separation become simply

$$y' = y, \quad z' = z.$$

In the experiment of Champion we have to investigate the transformation of velocities in various directions. Suppose that the impact shown in Fig. 9.7 is one in a coordinate system in which the particles enter symmetrically; we take this as the $xyzt$ coordinate system. Applying the transformation (9.5), viz.

$$x' = \beta(x-Vt), \quad y' = y, \quad z' = z, \quad t' = \beta(t-Vx/c^2),$$

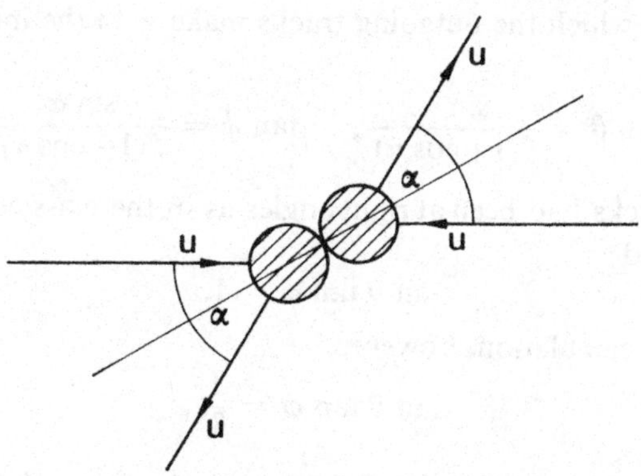

FIG. 9.7

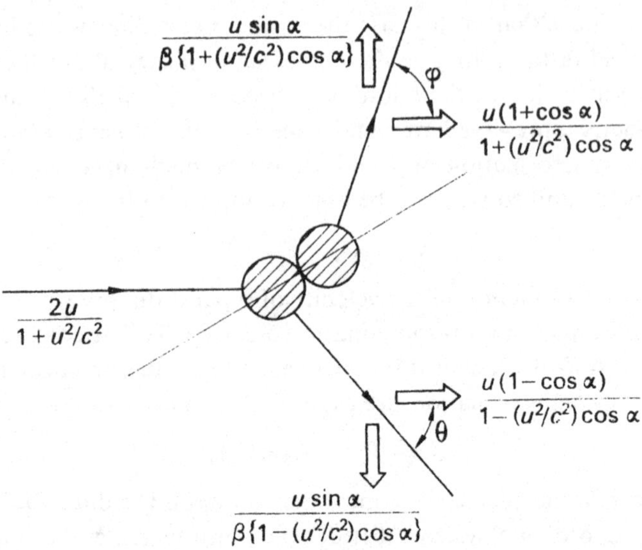

FIG. 9.8

it follows at once by differentiation that in any other coordinate system

$$u'_x = \frac{dx'}{dt'} = \frac{u_x - V}{1 - Vu_x/c^2}, \quad u'_y = \frac{u_y}{\beta(1 - Vu_x/c^2)}, \quad u'_z = \frac{u_z}{\beta(1 - Vu_x/c^2)}. \quad (9.6)$$

Since the right-hand particle has to be reduced to rest by this transformation we have

$$V = -u.$$

Accordingly Fig. 9.8 shows the components of velocity in the coordinate system in which one of the electrons is at rest initially (the "laboratory frame"). The incoming velocity is no longer $2u$ because of the way that velocities add together, as explained in the discussion of Fresnel's experiment. The two angles which the outgoing tracks make with the incoming one are now given by

$$\tan\theta = \frac{\sin\alpha}{\beta(1+\cos\alpha)}, \qquad \tan\phi = \frac{\sin\alpha}{\beta(1-\cos\alpha)}.$$

If these two tracks had been at right angles as in the classical case, then we should have had

$$\tan\theta \tan\phi = 1.$$

In the accurate calculation, however,

$$\tan\theta \tan\phi = \frac{1}{\beta^2},$$

a result which agrees with the classical one to the first order.

Example. The explanation just given of Champion's experiments involves a quantity β which is expressed in terms of the incident velocity in the centre-of-mass coordinate system. Express the product in terms of the velocity in the laboratory coordinate system.

$$\tan\theta\tan\phi = \frac{1}{\beta^2}.$$

Here
$$\beta = \beta(u) = \frac{1}{\sqrt{(1-u^2/c^2)}}.$$

We need to express this in terms of

$$v = \frac{2u}{1+u^2/c^2}.$$

Clearly

$$\sqrt{\left(1-\frac{v^2}{c^2}\right)} = \frac{1-u^2/c^2}{1+u^2/c^2}.$$

So putting

$$\beta(v) = \frac{1}{\sqrt{(1-v^2/c^2)}},$$

$$\frac{u^2}{c^2} = \frac{\beta(v)-1}{\beta(v)+1}$$

and

$$\beta(u) = \frac{1}{\sqrt{\left(1-\dfrac{\beta(v)-1}{\beta(v)+1}\right)}} = \sqrt{\left\{\frac{1}{2}\,[\beta(v)+1]\right\}},$$

it follows that
$$\tan\theta\tan\phi = \frac{2}{1+\beta(v)}.$$

Notice that this is approximately

$$\tan\theta\tan\phi = \frac{1}{1+\dfrac{1}{4}\dfrac{v^2}{c^2}}$$

$$\approx \frac{1}{1+\dfrac{E}{2mc^2}},$$

where m is the mass of an electron and E is the incident energy of the electrons (that is, the Newtonian expression $\frac{1}{2}mv^2$, which is correct if E is not too large). Experiments with electrons give results in agreement with this prediction.

9.3 The Lorentz transformation of field quantities

The eqn. (9.3) expressing Faraday's result must be true, i.e. have the same form, in any appropriate coordinate system. We will therefore begin by obtaining the first-order equations giving the relation between components $\boldsymbol{E}$ and $\boldsymbol{B}$ as measured by observers O and O', who are in relative motion with velocity $\boldsymbol{V}$. The details of the first-order transformation are more straightforward and serve as a guide for obtaining the accurate transformation.

The Newtonian coordinate transformation, i.e. the first-order approximation to the Lorentz transformation, is

$$\boldsymbol{r}' = \boldsymbol{r} - \boldsymbol{V}t, \quad t' = t,$$

so that we may express eqn. (9.3) in terms of differentiations in the new coordinate system by making the substitution

$$\nabla \equiv \nabla', \quad \frac{\partial}{\partial t} \equiv \frac{\partial}{\partial t'} - \boldsymbol{V}\cdot\nabla'. \tag{9.7}$$

[It may surprise the reader to notice that the time-differentiation is not the same in the two coordinate systems. This is because in partial differentiation with respect to time the other (space-)coordinates are being held fixed, and therefore the coordinates being held fixed differ in the two coordinate systems.] When the derivatives are replaced by their values in the new coordinate system we obtain

$$\mathbf{curl}'\,\boldsymbol{E} + \frac{\partial \boldsymbol{B}}{\partial t'} - \boldsymbol{V}\cdot\nabla'\boldsymbol{B} = \boldsymbol{0}.$$

This equation is apparently of a different form from the original one, but when we recall that, because

$$\operatorname{div} \boldsymbol{B} = 0,$$

it follows that

$$\mathbf{curl}\,(\boldsymbol{V}\times\boldsymbol{B}) = -(\boldsymbol{V}\cdot\nabla)\boldsymbol{B}.$$

The new equation can be written

$$\mathbf{curl}'\,(\boldsymbol{E} + \boldsymbol{V}\times\boldsymbol{B}) + \frac{\partial \boldsymbol{B}}{\partial t'} = \boldsymbol{0}, \tag{9.8}$$

which will be of the same form as the old one so long as the first-order transformation of the electric and magnetic vectors is

$$\begin{aligned} \boldsymbol{E} \to \boldsymbol{E}' &= \boldsymbol{E} + \boldsymbol{V}\times\boldsymbol{B}, \\ \boldsymbol{B} \to \boldsymbol{B}' &= \boldsymbol{B}. \end{aligned} \tag{9.9}$$

It is to be emphasized that this calculation is correct only to the first order, and just as there is the peculiar asymmetry between $\boldsymbol{r}$ and t in the first-order coordinate transformation, so there is an asymmetry here between $\boldsymbol{E}$ and $\boldsymbol{B}$. The vector $\boldsymbol{E}'$ depends upon both $\boldsymbol{E}$ and $\boldsymbol{B}$ in the old coordinate system, whereas the new $\boldsymbol{B}'$ depends only upon the old $\boldsymbol{B}$.

It is now time to work out the transformation accurately and for this purpose it is convenient to suppose that the velocity of separation is, as usual, V along the x-axis. The transformation of the coordinates is then

$$x' = \beta(x-Vt), \qquad y' = y, \qquad z' = z, \qquad t' = \beta(t-Vx/c^2),$$

so that the derivatives satisfy

$$\frac{\partial}{\partial x} \equiv \beta\left(\frac{\partial}{\partial x'}-\frac{V}{c^2}\frac{\partial}{\partial t'}\right), \quad \frac{\partial}{\partial y} \equiv \frac{\partial}{\partial y'}, \quad \frac{\partial}{\partial z} \equiv \frac{\partial}{\partial z'}, \quad \frac{\partial}{\partial t} \equiv \beta\left(\frac{\partial}{\partial t'}-V\frac{\partial}{\partial x'}\right). \tag{9.10}$$

Writing out the three components of eqn. (9.3) gives

$$\begin{aligned} &\frac{\partial E_3}{\partial y'}-\frac{\partial E_2}{\partial z'}+\beta\frac{\partial B_1}{\partial t'}-\beta V\frac{\partial B_1}{\partial x'} = 0, \\ &\frac{\partial E_1}{\partial z'}-\beta\frac{\partial E_3}{\partial x'}+\frac{\beta V}{c^2}\frac{\partial E_3}{\partial t'}+\beta\frac{\partial B_2}{\partial t'}-\beta V\frac{\partial B_2}{\partial x'} = 0, \\ &\beta\frac{\partial E_2}{\partial x'}-\frac{\beta V}{c^2}\frac{\partial E_2}{\partial t'}-\frac{\partial E_1}{\partial y'}+\beta\frac{\partial B_3}{\partial t'}-\beta V\frac{\partial B_3}{\partial x'} = 0. \end{aligned} \tag{9.11}$$

Taking the last two of these equations, it will be seen that they are of the original form so long as

$$B_2' = \lambda\beta\left(B_2-\frac{VE_3}{c^2}\right),$$

$$B_3' = \lambda\beta\left(B_3-\frac{VE_2}{c^2}\right),$$

together with transformations of the electric vector very similar to the first-order transformations

$$E_1' = \lambda E_1, \quad E_2' = \lambda\beta(E_2-VB_3), \quad E_3' = \lambda\beta(E_3+VB_2).$$

In all of these transformations the constant λ which enters will in general depend upon the velocity V. Going back to the first equation, we can re-arrange it so as to show the transformation of B_1. To do this we must use the other equation

$$\operatorname{div} \boldsymbol{B} = 0$$

which we have found already. [It may be asked whether we can be sure, now that we have relaxed constraints so that **curl** $\boldsymbol{E}$ no longer vanishes, that div $\boldsymbol{B}$ will still vanish. However, Faraday's relation (9.3) shows that, in any case,

$$\frac{\partial}{\partial t}(\operatorname{div} \boldsymbol{B}) = 0,$$

so that div $\boldsymbol{B}$ must be constant in time at each point of space. The equation

$$\operatorname{div} \boldsymbol{B} = 0$$

is not therefore an entirely new one, but simply an *initial condition*.] If, then, we multiply the first of Faraday's relations by $\lambda\beta$, it can be written

$$\frac{\partial E_3'}{\partial y'}-\frac{\partial E_2'}{\partial z'}-\lambda\beta V\left(\frac{\partial B_2}{\partial y'}+\frac{\partial B_3}{\partial z'}\right)+\lambda\beta^2\frac{\partial B_1}{\partial t'}-\lambda\beta^2 V\frac{\partial B_1}{\partial x'}=0.$$

Since, however, $\partial/\partial y = \partial/\partial y'$, $\partial/\partial z = \partial/\partial z'$, the middle term becomes

$$-\lambda\beta V\left(\frac{\partial B_2}{\partial y'}+\frac{\partial B_3}{\partial z'}\right)=\lambda\beta V\frac{\partial B_1}{\partial x}=\lambda\beta^2 V\left(\frac{\partial B_1}{\partial x'}-\frac{V}{c^2}\frac{\partial B_1}{\partial t'}\right).$$

In all, then,

$$\frac{\partial E_3'}{\partial y'}-\frac{\partial E_2'}{\partial z'}+\lambda\beta^2\left(1-\frac{V^2}{c^2}\right)\frac{\partial B_1}{\partial t'}=0.$$

The remaining component of the vector must transform according to the relation

$$B_1' = \lambda B_1.$$

Consider first the case in which there is no electric field in the initial coordinate system. From the axial symmetry in this case it follows that, since the direction of V becomes unimportant, $\lambda(V) = \lambda(-V)$. However, by looking at the transformation of the first components, we also have (since the reciprocal transformation corresponds to $-V$),

$$\lambda(-V)=\frac{1}{\lambda(V)}.$$

Since the first-order transformation for the magnetic vector is simply the identity transformation it follows that, on taking the square root,

$$\lambda(V)=1.$$

Therefore, the transformations are

$$\begin{aligned} E_1' &= E_1, & B_1' &= B_1,\\ E_2' &= \beta(E_2-VB_3), & B_2' &= \beta(B_2+VE_3/c^2),\\ E_3' &= \beta(E_3+VB_2), & B_3' &= \beta(B_3-VE_2/c^2). \end{aligned} \qquad (9.12,13)$$

The formulae that we found for transforming the magnetic vector have important consequences, particularly because they determine the relationship between the two constants ε_0, μ_0 introduced in the definitions of the electric and magnetic quantities.

To show that this is so we consider a line of (stationary) charges along the x-axis of the $xyzt$ frame. If each charge is of strength q and there are n such charges (densely) packed along the axis in unit distance, the field produced is that of an infinite line charge of line-density nq. Therefore

$$\boldsymbol{E} = \frac{nq}{2\pi\varepsilon_0}\left\{0 \quad \frac{y}{y^2+z^2} \quad \frac{z}{y^2+z^2}\right\}.$$

Because of the Fitzgerald contraction the observer O', with frame $x'y'z't'$, will observe the n charges distributed on a length β, instead of unit length, i.e. a line-density βnq. This charge is moving along the x'-axis with speed $-V$ and so (assuming that q still represents the magnitude of each charge in the $x'y'z't'$ frame, since this magnitude is simply the number of elementary particles in the charge) it constitutes a current of strength $I' = -\beta nqV$.

The transformation equations (9.12), (9.13) give for the fields in the $x'y'z't'$ frame

$$\boldsymbol{E}' = \frac{nq}{2\pi\varepsilon_0}\left\{0 \quad \frac{\beta y}{y^2+z^2} \quad \frac{\beta z}{y^2+z^2}\right\}$$

$$= \frac{\beta nq}{2\pi\varepsilon_0}\left\{0 \quad \frac{y'}{y'^2+z'^2} \quad \frac{z'}{y'^2+z'^2}\right\}, \tag{9.14}$$

where we have used the fact that $y' = y$, $z' = z$. Also

$$\boldsymbol{B}' = \frac{\beta nqV}{2\pi\varepsilon_0 c^2}\left\{0 \quad \frac{z}{y^2+z^2} \quad \frac{-y}{y^2+z^2}\right\}$$

$$= \frac{I'}{2\pi\varepsilon_0 c^2}\left\{0 \quad \frac{-z'}{y'^2+z'^2} \quad \frac{y'}{y'^2+z'^2}\right\}. \tag{9.15}$$

But (9.14) is the electrostatic field produced by a charge of line-density βnq; the magnetic field produced by a line-current of strength I' is

$$\frac{\mu_0 I'}{2\pi}\left\{0 \quad \frac{-z'}{y'^2+z'^2} \quad \frac{y'}{y'^2+z'^2}\right\}. \tag{9.16}$$

A comparison of this with (9.15) shows that $\mu_0 = 1/(\varepsilon_0 c^2)$,

i.e.
$$c = \frac{1}{(\varepsilon_0\mu_0)^{1/2}}. \tag{9.17}$$

The fact that the magnetic and electric constants are connected in this way with the speed of light is often thought to be a consequence of the whole Maxwell theory. Our argument shows that this is not so. In fact, the units must be related in this way already, and the fact that, as will be seen later, the Maxwell theory predicts that electromagnetic waves move with this same speed is an additional result, and so counts as evidence in support of the hypothesis that light is a form of electromagnetic radiation.

9.4 The Lorentz force on a moving charge

The formula we discuss in this section, although known by the name of Lorentz, was actually found by Heaviside as early as 1889. It is easier to give the first-order approximation to the calculation and so we begin with this first-order approximation.

Suppose that in a frame of reference (with observer attached) having origin O there is a non-uniform but constant (in time) magnetic field $\mathbf{B}$ and a point charge Q situated at some point. The observer O will not observe any acceleration of the point charge for he would see no cause for a force to act on it.

Suppose that a second frame of reference (with observer) $x'y'z't'$ moves relative to O with constant velocity $\mathbf{V}$. Then in the estimation of O' the charge Q has a velocity $-\mathbf{V}$ and the magnetic field at any point in the $x'y'z't'$ frame varies with the time. For a point in the frame of O' the magnetic field at a later instant, $t+\delta t$, will take the value which prevailed at time t at a point in the $xyzt$ frame distant $\mathbf{V}\delta t$. Hence, in the estimation of O',

$$\frac{\partial \mathbf{B}'}{\partial t} = (\mathbf{V}\cdot\nabla)\mathbf{B}' = -\,\mathbf{curl}\,(\mathbf{V}\times\mathbf{B}')$$

since $\mathbf{V}$ is constant. According to Faraday's law of electromagnetic induction, because $\mathbf{B}'$ varies there is an electric field $\mathbf{E}'$ given by (9.3). Therefore,

$$\mathbf{E}' = \mathbf{V}\times\mathbf{B}'.$$

(We assume that $\mathbf{E}'$ must vanish if $\mathbf{V}$ is zero, for then the two frames are identical.) According to O', because of the field $\mathbf{E}'$, there is a force $Q\mathbf{E}'$ on the charge; but he observes no acceleration (since he is in *uniform* motion relative to O, who also observes no acceleration). Therefore, a charge in motion through a magnetic field must experience a force which, in the case we are considering, must cancel out the force $Q\mathbf{E}'$. Hence a charge in motion through a magnetic field with velocity $-\mathbf{V}$ experiences a total force

$$\mathbf{F} = Q\{\mathbf{E}'+(-\mathbf{V})\times\mathbf{B}'\}. \tag{9.18}$$

This is exactly the expression we expect from eqn. (9.9) giving the first-order transformation for the fields $\boldsymbol{E}'$, $\boldsymbol{B}'$. Here, of course, we are making the assumption that the force on the particle exerted by the fields depends only on its position and velocity and not, for instance, on its acceleration. We make this same assumption in the accurate calculation given below.

However, before we can obtain the accurate expression for the force on a charge we must know the relation between components of force in frames related by a Lorentz transformation. The simple form of Newton's law of motion for a particle,

$$m\frac{\mathrm{d}^2\boldsymbol{r}}{\mathrm{d}t^2} = \boldsymbol{F},$$

is not invariant under a Lorentz transformation. To obtain an invariant form we first notice that a quantity s (the proper time) is an invariant. If two events have space and time separations $\mathrm{d}x$, $\mathrm{d}t$ in one frame and $\mathrm{d}x'$, $\mathrm{d}t'$ in another frame, related by the transformations

$$x' = \beta(x-Vt), \quad y' = y, \quad z' = z, \quad t' = \beta(t-Vx/c^2),$$

then

$$\begin{aligned} \mathrm{d}s^2 &= \mathrm{d}t'^2-(\mathrm{d}\boldsymbol{r}'^2)/c^2 \\ &= \beta^2(\mathrm{d}t-V\,\mathrm{d}x/c^2)^2-\beta^2(\mathrm{d}x-V\,\mathrm{d}t)^2/c^2-(\mathrm{d}y^2+\mathrm{d}z^2)/c^2 \\ &= \beta^2(1-V^2/c^2)(\mathrm{d}t^2-\mathrm{d}x^2/c^2)-(\mathrm{d}y^2+\mathrm{d}z^2)/c^2 \\ &= \mathrm{d}t^2-(\mathrm{d}\boldsymbol{r}^2)/c^2, \end{aligned} \tag{9.19}$$

and the invariance is established. If the two events take place at the same position in the $x'y'z't'$ frame, $\mathrm{d}\boldsymbol{r}' = \boldsymbol{0}$ and

$$\mathrm{d}s^2 = \mathrm{d}t'^2.$$

This time interval is the proper time (see example 3, p. 343) since it is measured by an observer at rest relative to the events. According to this observer, O', Newton's law of motion can be expressed in either of the equivalent forms

$$\frac{\mathrm{d}}{\mathrm{d}t'}\left(m_0\frac{\mathrm{d}\boldsymbol{r}'}{\mathrm{d}t'}\right) = \boldsymbol{F}' \quad \text{or} \quad \frac{\mathrm{d}}{\mathrm{d}s}\left(m_0\frac{\mathrm{d}\boldsymbol{r}'}{\mathrm{d}s}\right) = \boldsymbol{F}'. \tag{9.20}$$

If we take the second form of this equation together with (an identity for O')

$$\frac{\mathrm{d}}{\mathrm{d}s}\left(m_0\frac{\mathrm{d}t'}{\mathrm{d}s}\right) = 0, \tag{9.21}$$

we have four equations in which the numerators of the left-hand sides transform according to the Lorentz transformation and the denominators are invariant. We now postulate that Newton's equation of motion is exactly valid in the $x'y'z't'$ frame *when the particle is instantaneously at rest in that frame* and is given in any other frame by the Lorentz transformations. We use m_0 to denote the mass—called the "proper" mass or "rest-mass"—of the particle.

Now we suppose that the frame $xyzt$ is such that O', and his frame together with the particle at the instant in question, has a velocity V along the x-axis. Then the coordinates and time transform according to (9.5) and so the eqns. (9.20), (9.21) become in this frame

$$\frac{d}{ds}\left(m_0\frac{dx}{ds}\right) = F_1, \quad \frac{d}{ds}\left(m_0\frac{dy}{ds}\right) = F_2,$$
$$\frac{d}{ds}\left(m_0\frac{dz}{ds}\right) = F_3, \quad \frac{d}{ds}\left(m_0\frac{dt}{ds}\right) = F_0, \tag{9.22}$$

where $\{F_1 F_2 F_3;\ F_0\}$ are obtained from $\{F_1' F_2' F_3';\ 0\}$ by a Lorentz transformation, viz.

$$F_1 = \beta(F_1'+V\cdot 0), \quad F_2 = F_2', \quad F_3 = F_3', \quad F_0 = \beta(0+VF_1'/c^2).$$

From eqn. (9.19) we see that

$$\left(\frac{ds}{dt}\right)^2 = 1-\frac{V^2}{c^2} \quad \text{or} \quad \frac{dt}{ds} = \beta \tag{9.23}$$

and eqns. (9.22) become

$$\frac{d}{ds}\left(m_0\beta\frac{dx}{dt}\right) = \beta F_1', \quad \frac{d}{ds}\left(m_0\beta\frac{dy}{dt}\right) = F_2', \quad \frac{d}{ds}\left(m_0\beta\frac{dz}{dt}\right) = F_3' \tag{9.24}$$

together with the fourth equation

$$\frac{d}{ds}(m_0\beta) = \frac{VF_1'}{c^2}. \tag{9.25}$$

We use (9.23) again to turn the first differentiation into one w.r. to t and obtain for (9.24)

$$\frac{d}{dt}\left(m\frac{dx}{dt}\right) = F_1', \quad \frac{d}{dt}\left(m\frac{dy}{dt}\right) = \frac{F_2'}{\beta}, \quad \frac{d}{dt}\left(m\frac{dz}{dt}\right) = \frac{F_3'}{\beta} \tag{9.26}$$

together with

$$\frac{d}{dt}(m) = \frac{VF_1}{c^2}. \tag{9.27}$$

Provided that we interpret $m = m_0\beta$ as the mass of the particle, eqns. (9.26) can be regarded as Newton's equations of motion of a particle in the *xyzt* frame, under the action of a force $\{F_1\ F_2\ F_3\}$ where

$$F_1 = F_1', \quad F_2 = F_2'/\beta, \quad F_3 = F_3'/\beta. \tag{9.28}$$

[Since, according to O, the particle instantaneously has the velocity $\{V\ 0\ 0\}$, eqn. (9.27) implies that $\mathrm{d}(mc^2)/\mathrm{d}t$ is the rate of increase of energy of the particle.] Equations (9.28) give the transformation law for force between the two frames, provided we regard the mass of the particle as varying with its velocity according to

$$m = m_0\beta = \frac{m_0}{(1-V^2/c^2)^{1/2}}. \tag{9.29}$$

Such an increase of mass with velocity had already been found by Kaufmann in 1901.

Suppose now that in the frame $x'y'z't'$, there is an electromagnetic field $\boldsymbol{E}'$, $\boldsymbol{B}'$ estimated by O'; then the force acting on a small charge q would be given by O' to be

$$\boldsymbol{F}' = q\boldsymbol{E}'; \quad \text{or} \quad F_1' = qE_1', \quad F_2' = qE_2', \quad F_3' = qE_3'.$$

The corresponding observations made by O would be a force with components

$$F_1 = qE_1', \quad F_2 = qE_2'/\beta, \quad F_3 = qE_3'/\beta$$

acting on the particle when it is in motion through an electromagnetic field $\boldsymbol{E}$, $\boldsymbol{B}$ related to $\boldsymbol{E}'$, $\boldsymbol{B}'$ by eqns. (9.12), (9.13). Hence the force on q, according to O, is

$$F_1 = qE_1, \quad F_2 = q(E_2 - VB_3), \quad F_3 = q(E_3 + VB_2)$$

or, in vector terms,

$$\boldsymbol{F} = q(\boldsymbol{E} + \boldsymbol{V} \times \boldsymbol{B}). \tag{9.30}$$

This is the accurate formula, for at no stage have we made use of first-order expressions, and is called the Lorentz force on a moving charge. (*In this case the accurate formula is identical with the first-order result.*)

Example 1. Show that a transformation of coordinates from x, y, z, t to x', y', z', t' which transforms

$$c^2\,\mathrm{d}t^2 - \mathrm{d}x^2 - \mathrm{d}y^2 - \mathrm{d}z^2 \quad \text{into} \quad c^2\,\mathrm{d}t'^2 - \mathrm{d}x'^2 - \mathrm{d}y'^2 - \mathrm{d}z'^2$$

is necessarily a linear transformation. Find such a linear transformation (not the identity) in which $y = y'$ and $z = z'$.

It will be convenient to simplify the algebra by using x_0, x_1, x_2, x_3 for t, x, y, z, and to write

$$c^2\,dt^2 - dx^2 - dy^2 - dz^2 = \sum_{i,j=0}^{3} a_{ij} x_i x_j.$$

Then the condition is

$$\sum_{i,j,p,q} a_{ij} \frac{\partial x_i'}{\partial x_p} \frac{\partial x_j'}{\partial x_q}\, dx_p\, dx_q = \sum a_{pq}\, dx_p\, dx_q,$$

so that

$$\sum_{i,j} a_{ij} \frac{\partial x_i'}{\partial x_p} \frac{\partial x_j'}{\partial x_q} = a_{pq}.$$

Differentiating gives

$$\sum_{i,j} a_{ij} \left(\frac{\partial^2 x_i'}{\partial x_p\, \partial x_r} \frac{\partial x_j'}{\partial x_q} + \frac{\partial x_i'}{\partial x_p} \frac{\partial^2 x_j'}{\partial x_q\, \partial x_r} \right) = 0,$$

and we could write this equation in an abbreviated form as

$$\{q\} + \{p\} = 0.$$

Similarly

$$\{r\} + \{q\} = 0,$$

and

$$\{p\} + \{r\} = 0,$$

so that

$$\{p\} = \{q\} = \{r\} = 0,$$

that is,

$$\sum_{i,j} a_{ij} \frac{\partial x_i'}{\partial x_p} \frac{\partial^2 x_j'}{\partial x_q\, \partial x_r} = 0.$$

Multiplying by $\partial x_p / \partial x_k'$, and summing over p,

$$\sum_{j} a_{kj} \frac{\partial^2 x_j'}{\partial x_q\, \partial x_r} = 0$$

and, as a consequence,

$$\frac{\partial^2 x_j'}{\partial x_q\, \partial x_r} = 0,$$

which implies that the transformation is linear.

Of course, such a linear transformation is

$$x' = \beta(x - Vt), \quad y' = y, \quad z' = z, \quad t' = \beta(t - Vx/c^2).$$

Example 2. Calculate the mechanical force on an electron of rest-mass m and charge $-e$ moving with velocity $\boldsymbol{V}$ in a magnetic field $\boldsymbol{B}$, and hence establish the theory of Larmor precession.

The calculation of the force on the electron is given on p. 359; in this example we introduce the characteristic effect of a constant magnetic field on a charged particle. The equation of motion of the particle is given by

$$m\boldsymbol{f} = -e(\boldsymbol{E} + \boldsymbol{v} \times \boldsymbol{B}) + \boldsymbol{F},$$

where $\boldsymbol{f}$ is the acceleration and $\boldsymbol{F}$ stands for any forces acting on the particle, other than those arising from the electric and magnetic fields.

If the frame of reference is made to rotate with angular velocity $\boldsymbol{\omega}$, the expressions for $\boldsymbol{f}$, $\boldsymbol{v}$ are

$$\boldsymbol{f} = \frac{\partial^2 \boldsymbol{r}}{\partial t^2} + 2\boldsymbol{\omega} \times \frac{\partial \boldsymbol{r}}{\partial t} + \boldsymbol{\omega} \times (\boldsymbol{\omega} \times \boldsymbol{r}),$$

$$\boldsymbol{v} = \frac{\partial \boldsymbol{r}}{\partial t} + (\boldsymbol{\omega} \times \boldsymbol{r}).$$

Hence the equation of motion of the electron is

$$m\left\{\frac{\partial^2 \boldsymbol{r}}{\partial t^2}+2\boldsymbol{\omega}\times\frac{\partial \boldsymbol{r}}{\partial t}+\boldsymbol{\omega}\times(\boldsymbol{\omega}\times\boldsymbol{r})\right\} = -e\left\{\boldsymbol{E}-\boldsymbol{B}\times\left(\frac{\partial \boldsymbol{r}}{\partial t}+\boldsymbol{\omega}\times\boldsymbol{r}\right)\right\}+\boldsymbol{F}.$$

If we choose

$$\boldsymbol{\omega} = e\boldsymbol{B}/(2m),$$

this equation becomes

$$m\left\{\frac{\partial^2 \boldsymbol{r}}{\partial t^2}-\boldsymbol{\omega}\times(\boldsymbol{\omega}\times\boldsymbol{r})\right\} = -e\boldsymbol{E}+\boldsymbol{F}.$$

Since $\boldsymbol{B}$ is small we may neglect the second-order term $\boldsymbol{\omega}\times(\boldsymbol{\omega}\times\boldsymbol{r})$ and obtain

$$m\frac{\partial^2 \boldsymbol{r}}{\partial t^2} = -e\boldsymbol{E}+\boldsymbol{F}. \tag{1}$$

This is the equation of motion of a particle under the influence of the electric field and other forces only; the effect of the (constant) magnetic field is to add the rotation $e\boldsymbol{B}/(2m)$. This rotation is called the *Larmor rotation* (or *precession*) and gives the characteristic feature of the effect of a magnetic field on the motion of a particle. If the motion of the particle given by (1) is a straight line parallel to $\boldsymbol{B}$ the path of the particle is a spiral. In other circumstances the path of the particle is a cycloid. If a system is composed of particles all having the same value for e/m the rotation applies equally to every particle and the whole system rotates with the Larmor rotation.

Example 3. If the *relativistic acceleration* f of a particle P moving along a straight line is defined as the Newtonian acceleration measured in that frame of reference in which P is instantaneously at rest, prove that, in any frame of reference with respect to which P has velocity v,

$$f = \beta^3\frac{dv}{dt} = \frac{d}{dt}(\beta v),$$

where $\beta = (1-v^2/c^2)^{-1/2}$.

If a particle moves from rest with constant relativistic acceleration g, prove that after time t it has travelled a distance

$$\frac{c^2}{g}\{(1+g^2t^2/c^2)^{1/2}-1\}.$$

We consider first the motion of a particle relative to a frame which has a uniform velocity V. Then the velocity is given by (9.6), viz.

$$u' = \frac{u-V}{1-uV/c^2}.$$

Hence for small intervals,

$$du' = du\left\{\frac{1}{1-uV/c^2}+\frac{V(u-V)}{c^2(1-uV/c^2)^2}\right\} = \frac{du(1-V^2/c^2)}{(1-uV/c^2)^2}.$$

But, since $dt' = \beta(dt-V\,dx/c^2)$,

$$a' = \frac{du'}{dt'} = \frac{du(1-V^2/c^2)}{(1-uV/c^2)^2}\cdot\frac{1}{\beta(dt-V\,dx/c^2)} = \frac{du}{dt}\,\frac{1-V^2/c^2}{\beta(1-uV/c^2)^3}.$$

We now take the special case of this result in which $u = V = v$, so that the particle is instantaneously at rest in the frame $Ox't'$. Then

$$f = \frac{du'}{dt'} = \frac{dv}{dt}\cdot\frac{1-v^2/c^2}{(1-v^2/c^2)^3}\,\frac{1}{\beta} = \frac{dv}{dt}\cdot\frac{1}{(1-v^2/c^2)^{3/2}} = \beta^3\frac{dv}{dt}.$$

Also

$$\frac{d}{dt}(\beta v) = \beta\frac{dv}{dt}+v\frac{d\beta}{dt} = \beta\frac{dv}{dt}+\frac{v}{2(1-v^2/c^2)^{3/2}}\frac{2v}{c^2}\frac{dv}{dt},$$

$$= \beta\frac{dv}{dt}\left\{1+\frac{v^2/c^2}{1-v^2/c^2}\right\} = \frac{\beta}{1-v^2/c^2}\frac{dv}{dt} = \beta^3\frac{dv}{dt}.$$

The second result is obtained from the fact that, when the relativistic acceleration is constant

$$\frac{d}{dt}(\beta v) = g, \quad v = \frac{gt}{\beta} = gt(1-v^2/c^2)^{1/2},$$

since $v = 0$ when $t = 0$.

$$\therefore \quad v^2 = g^2t^2 - g^2t^2v^2/c^2,$$

$$\therefore \quad v^2(1+g^2t^2/c^2) = g^2t^2,$$

$$\therefore \quad v = \frac{dx}{dt} = \frac{gt}{(1+g^2t^2/c^2)^{1/2}}.$$

Integration of this, with the condition that $x = 0$ when $t = 0$, leads to the quoted expression for x.

Example 4. Obtain the formula

$$m' = \frac{m}{\sqrt{(1-v^2/c^2)}}$$

for the relative mass of a particle of proper mass m moving with relative speed v.

A particle of proper mass m moves about a stationary particle of proper mass M under the law of gravitation GMm/r^2, where r denotes the distance of m according to an observer associated with M. Assuming that the only modification of the classical equations of motion is the introduction of the Special Relativity formulae for linear momentum and kinetic energy, obtain the differential equation

$$\frac{d^2u}{d\theta^2}+(1-\alpha^2)u = \text{constant},$$

where $u = 1/r$, (r, θ) are polar coordinates, $\alpha = GMm/ch$, and h denotes the angular momentum.

Deduce that, if α is small, the apsidal line advances through an approximate angle $\pi\alpha^2$ per revolution.

(The proper mass and relative mass was discussed on p. 358.)

Since Newton's law

$$\frac{d}{dt}\left(m\frac{dr}{dt}\right) = F \tag{1}$$

is valid for low velocities, one assumes it in the rest-frame, in which it also has the form

$$\frac{d}{ds}\left(m\frac{dr}{ds}\right) = F, \tag{2}$$

where

$$ds^2 = dt^2 - \frac{1}{c^2}dr^2,$$

i.e.

$$\left(\frac{ds}{dt}\right)^2 = 1-v^2/c^2, \quad \beta = dt/ds. \tag{3}$$

Now ds is unchanged under the Lorentz transformation, so that this equation has the same form in any inertial frame (so long as $\boldsymbol{F}$ is correctly transformed). But in such a frame (2) can be written

$$\frac{dt}{ds}\frac{d}{dt}\left(m\frac{dt}{ds}\frac{d\boldsymbol{r}}{dt}\right) = \boldsymbol{F}, \tag{4}$$

corresponding to a measured mass

$$m' = m\frac{dt}{ds} = \frac{m}{\sqrt{(1-v^2/c^2)}}. \tag{5}$$

The modification described here results in the angular momentum equation being

$$m'r^2\frac{d\theta}{dt} = mr^2\frac{d\theta}{ds} = h,$$

where m' is the relative mass.

The energy equation is

$$m'c^2 - GMm/r = E,$$

and can be written

$$c^2\beta = E' + GM/r$$

(where $E' = E/m$), and since, from above,

$$1 = \left(\frac{dt}{ds}\right)^2 - \frac{1}{c^2}\left\{\left(\frac{dr}{ds}\right)^2 + r^2\left(\frac{d\theta}{ds}\right)^2\right\}, \tag{6}$$

we can follow the usual Newtonian technique of introducing $u = 1/r$, and putting

$$m\frac{d\theta}{ds} = hu^2,$$

$$m\frac{dr}{ds} = m\frac{d\theta}{ds}\frac{d}{d\theta}\left(\frac{1}{u}\right) = -h\frac{du}{d\theta}.$$

Now the energy equation gives an expression for $\beta = dt/ds$, so that from (6) we obtain

$$m^2 = \frac{m^2}{c^4}(GMu+E')^2 - \frac{h^2}{c^2}\left\{u^2 + \left(\frac{du}{d\theta}\right)^2\right\}.$$

Differentiating w.r. to θ and cancelling $du/d\theta$ gives

$$h^2\left(u + \frac{d^2u}{d\theta^2}\right) = \frac{GMm^2}{c^2}(GMu+E').$$

$$\therefore \quad \frac{d^2u}{d\theta^2} + (1-\alpha^2)u = \text{constant} = K,$$

where $\alpha = \dfrac{GMm}{ch}$.

If α is small, the solution is

$$u = \frac{K}{1-\alpha^2} + A\cos\{(1-\alpha^2)^{1/2}\,\theta + \phi\}$$

which is almost an ellipse, except that successive apse-lines ($du/d\theta = 0$) are at intervals

$$\frac{\pi}{\sqrt{(1-\alpha^2)}} = \pi\left(1 + \frac{1}{2}\alpha^2\right).$$

There are two apses per revolution, giving an advance of apse-line of $\pi\alpha^2$ per revolution.

Exercises 9.4

1. A point P has velocity components (u, v, w), (u', v', w') in rectangular frames S, S' respectively, the corresponding axes being parallel and S' having velocity V relative to S in the direction Ox.

   Obtain the law of composition of velocities and show that, when P is instantaneously at rest in S', the acceleration components satisfy the equations

$$\dot{u}' = \beta^3\dot{u}, \quad \dot{v}' = \beta^2\dot{v}, \quad \dot{w}' = \beta^2\dot{w}$$

   where $\beta = (1-u^2/c^2)^{-1/2}$.

   Deduce the law of transformation

$$X' = X, \quad Y' = \beta Y, \quad Z' = \beta Z$$

   for the components of force acting on a particle at P and moving with the velocity of P.

2. Define s, the separation, or interval, between two events as observed in a frame S, and verify that s is an invariant under the Lorentz transformation.

   In an inertial frame S the events P and Q are assigned coordinates $(x_1, 0, 0, t_1)$ and $(x_2, 0, 0, t_2)$ respectively, where $x_2 > x_1$ and $t_2 > t_1$. Show that

   (i) if the interval PQ is real (timelike), then all inertial observers will agree that P preceded Q, and there is one frame in which P and Q occur at the same spatial point;

   (ii) if the interval PQ is imaginary (spacelike), there is no such agreement, and there is one frame in which the events are simultaneous.

   Determine the velocities of these two frames relative to S.

3. S and S' are two inertial frames in standard configuration, the Lorentz transformation between them being

$$x' = \gamma(v)(x-vt), \quad y' = y, \quad z' = z, \quad t' = \gamma(v)\left(t-\frac{vx}{c^2}\right),$$

   where

$$\gamma(v) = \left(1-\frac{v^2}{c^2}\right)^{-1/2}.$$

   Find the transformation between velocities.

   A particle moves along the x-axis in frame S. Show that its acceleration α with respect to its instantaneous rest-frame is equal to $\gamma^3(u)\,(du/dt)$, where u is its velocity at time t.

   If α is constant, and the particle starts from rest at the origin, deduce that the equation of motion is

$$\alpha x^2 + 2c^2x - \alpha c^2t^2 = 0.$$

   Prove that light signals emitted after time $t = c/\alpha$ at the origin will never reach the receding particle.

4. A plane polarized electromagnetic wave is propagating in free space in a direction parallel to the x-axis in an inertial frame S, its field components being given by

$$\mathbf{E} = \left\{0, \ \alpha \sin\omega\left(t-\frac{x}{c}\right), \ 0\right\},$$

$$\mathbf{B} = \left\{0, \ 0, \ \frac{a}{c}\sin\omega\left(t-\frac{x}{c}\right)\right\}.$$

Prove that the electromagnetic wave observed from an inertial frame S' moving with velocity V in a direction parallel to the x-axis has amplitude λa and frequency $\lambda\omega$, where $\lambda = \sqrt{[(1-V/c)(1+V/c)]}$.

5. A particle moves in an electric field $E\hat{\imath}$ and carries charge e; its rest mass is m. If at time $t = 0$, the particle is moving in the $\hat{\jmath}$ direction with linear momentum p_0, show that at time t with suitable choice of origin,

$$x = \frac{1}{eE}\sqrt{[\gamma_0^2+(ceEt)^2]},$$

$$y = \frac{p_0 c}{eE}\sinh^{-1}\left(\frac{ceEt}{\gamma_0}\right),$$

where $\gamma_0^2 = c^2(p_0^2+m^2c^2)$, and find the equation of the path described.

6. Show that if a particle of rest mass m and charge e is moving with velocity v in a field of electric intensity E and magnetic flux density B the force on the particle is

$$e\{E+v\times B\}.$$

If $E = cja \cos n(z-ct)$, $B = -ia \cos n(z-ct)$, where i, j, k are unit vectors along Ox, Oy, Oz respectively, and the particle is at rest at O at $t = 0$, show that the particle remains in the plane $x = 0$ and that if

$$\beta^{-2} = 1-v^2/c^2, \quad \text{then} \quad \beta\dot{z} = c(\beta-1).$$

Find the y and z co-ordinates of the particle at time t, showing that $y = d(1-\cos\theta)$, $8z = nd^2(\sin 2\theta-2\theta)$ where $\theta = n(z-ct)$ and $mn^2cd = ae$.

9.5 Non-steady currents

In Chapter 5 we discussed the conditions concerning the flow of *steady* currents in various circumstances. Here we investigate the effects of electromagnetic induction when the currents vary with time, but still flow in closed circuits. We consider in detail only the case of currents in linear circuits.

It is clear that inductance (see § 6.7), both self- and mutual-inductance, will have important effects because a variation in a current induces electromotive forces in any circuits linked magnetically, i.e. in cases for which $L_{pq} \neq 0$, with the given circuit.

When considering networks of linear circuits carrying *steady* currents we found two conditions (Kirchhoff's laws) which had to be satisfied. These two laws were the forms taken by, for the first law div $J = 0$, and for the second law $J = \sigma(E+E')$, when applied to the special case of linear networks.

The condition div $J = 0$ must still apply and so Kirchhoff's first law also applies, viz. the sum of the current strengths entering any node is zero.

The second law was obtained by integrating $\boldsymbol{J}/\sigma = \boldsymbol{E}+\boldsymbol{E}'$ around any closed loop of the network, remembering that, since $\operatorname{curl} \boldsymbol{E} = \boldsymbol{0}$, $\oint \boldsymbol{E}\cdot\mathbf{ds} \equiv 0$. The field $\boldsymbol{E}'$ represented the effect of batteries or cells, which reduced to discontinuities in the potential in most cases. The modification we must make is to integrate $\boldsymbol{J}/\sigma = \boldsymbol{E}+\boldsymbol{E}'$ around a closed loop of the network subject now to the fact that

$$\operatorname{curl} \boldsymbol{E} = -\frac{\partial \boldsymbol{B}}{\partial t}.$$

$$\therefore \quad \oint \boldsymbol{E}\cdot\mathbf{ds} = \iint \operatorname{curl} \boldsymbol{E}\cdot\mathbf{d}\boldsymbol{S} = -\iint \frac{\partial \boldsymbol{B}}{\partial t}\cdot\mathbf{d}\boldsymbol{S} = -\frac{\mathrm{d}N}{\mathrm{d}t},$$

where the double integral is taken over any surface spanning the loop (which we assume is stationary) and N is the total magnetic flux through the loop. The integration is done in the form

$$\oint \frac{\boldsymbol{J}\cdot\mathbf{ds}}{\sigma} = \oint \boldsymbol{E}\cdot\mathbf{ds} + \oint \boldsymbol{E}'\cdot\mathbf{ds} = -\frac{\mathrm{d}N}{\mathrm{d}t} + \mathcal{E},$$

where $\mathcal{E}$ stands for the electromotive force, or sum of the discontinuities in the potential which occur in the loop. In the case of a linear circuit, $\oint \boldsymbol{J}\cdot\mathbf{ds}/\sigma = \Sigma RI$, where R is the resistance of a section of the loop and I is the current flowing in that section. Hence, the basic equation corresponding to Kirchhoff's second law is

$$\mathcal{E} = \frac{\mathrm{d}N}{\mathrm{d}t} + \Sigma RI, \tag{9.31}$$

and, for a given system, there is one such equation for each closed loop, and a complete system must include all circuits, whether in electrical connection or not, which have a non-zero value for the mutual induction.

Equation (9.31) states that the electromotive force $\mathcal{E}$ in a loop arising from batteries and discontinuities must be balanced by $\mathrm{d}N/\mathrm{d}t$ and ΣRI. These terms are regarded as "back e.m.f.'s" which arise from the typical elements in the circuits, inductances—$\mathrm{d}N/\mathrm{d}t$—and resistances—RI—when a current I flows through the elements. The equation for the circuit is obtained by balancing the given e.m.f.'s against the "back e.m.f.'s". We now look more closely at the terms $\mathcal{E}$ and $\mathrm{d}N/\mathrm{d}t$.

In systems of steady currents we did not need to consider condensers (or capacitors) for a steady current cannot pass through such an arrangement. However, if a condenser, from which there is no "leakage" of the field (see § 3.3) carries a charge Q on one plate it must carry a charge $-Q$ on the other plate, Fig. 9.9. Also, a current I can flow towards one plate and increase Q provided that an equal current flows away from the other plate to correspond to the change in $-Q$. Because of the presence of the charges $Q, -Q$ there is a potential difference between the points A and B given by $V_A - V_B = Q/C$, where C is the capacitance. For most purposes, since there is no

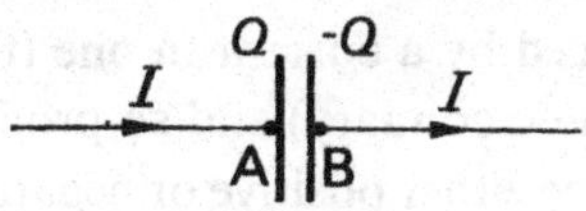

FIG. 9.9

leakage of electric field, the capacitor simply produces a discontinuity Q/C in the potential at a point ($A \equiv B$) in the circuit which is carrying a current I. Because $I = \mathrm{d}Q/\mathrm{d}t$, such an arrangement cannot occur in connection with *steady* currents where both potentials and currents have to be steady. In eqn. (9.31) we must include in $\mathcal{E}$ the discontinuities such as Q/C which occur where capacitors are put in any circuit.

The term $\mathrm{d}N/\mathrm{d}t$ in eqn. (9.31) refers to the magnetic flux N through the loop considered, N arising from *all* sources, i.e. from the current flowing in the loop itself and from the currents in any neighbouring circuits. The expression for this flux is given in eqn. (9.32) where the flux in circuit p is given by

$$N_p = \sum_q L_{pq} I_q .$$

Hence the term $\mathrm{d}N/\mathrm{d}t$, the back e.m.f. of all the inductances, is given by

$$\frac{\mathrm{d}N_p}{\mathrm{d}t} = \Sigma L_{pq} \frac{\mathrm{d}I_q}{\mathrm{d}t} . \tag{9.32}$$

As was pointed out in § 6.7, the letter L (without suffixes) is used to denote self-inductance so that the contribution to $\mathrm{d}N/\mathrm{d}t$ from the current I flowing in the given loop is $L(\mathrm{d}I/\mathrm{d}t)$; the letter M is used to denote mutual inductance, so that the contribution to $\mathrm{d}N/\mathrm{d}t$ from a current I_1 flowing in the neighbouring circuit is $M(\mathrm{d}I_1/\mathrm{d}t)$, and

$$\frac{\mathrm{d}N}{\mathrm{d}t} = L \frac{\mathrm{d}I}{\mathrm{d}t} + M \frac{\mathrm{d}I_1}{\mathrm{d}t} . \tag{9.33}$$

The unit of inductance is the *henry*, and $N = LI$ gives the flux in *webers* when the current is in ampères.

A self-inductance L is usually made in the form of a solenoid so that the magnetic field produced by I interlaces all the turns of the solenoid and so produces a large value for N. Since the positive sense of N is defined as that of flux produced by a positive current, it follows that L must always be positive. (It should be noted that even a single linear conductor has a small value for L, which, however, can usually be neglected.) If two circuits, insulated from one another, are closely wound on the same solenoid, then the magnetic field produced by a current in one (the primary) interlaces all the turns of the other (the secondary) and so produces a large value for M. However, this value may be either positive or negative. The use of "soft-iron" cores in the above arrangements increases the values of L, M by increasing the strength of the magnetic field. A *transformer*, consisting of two (or more) coils, a primary and a secondary, wound on the same core, which is usually a loop of "soft-iron", is designed to have chosen values for the mutual inductance between the primary and secondary windings.

In example 1 following we consider a simple circuit containing inductance, resistance and capacitance, and point out an important analogy. The reader should also note the diagrammatic representation of the different elements of a circuit.

Example 1. We consider the circuit shown in Fig. 9.10 which consists of an inductance L between A and B, a capacitance C between B and D, a resistance R between D and F, i.e. all in series, and the ends A, F are connected to a source of constant e.m.f. E.

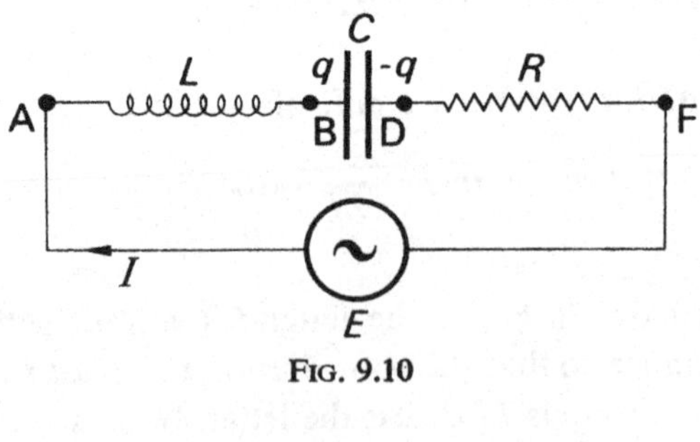

FIG. 9.10

The form taken by eqn. (9.31) when applied to this circuit is

$$L\frac{dI}{dt}+RI = E-\frac{q}{C}, \qquad I = \frac{dq}{dt},$$

where q is the charge on plate B of the condenser. This gives

$$L\frac{d^2q}{dt^2}+R\frac{dq}{dt}+\frac{1}{C}q = E. \tag{1}$$

There are three possible forms for the solution of this differential equation:

(a) $R^2 > 4(L/C)$, $q = e^{-Rt/(2L)}\{A \cosh mt + B \sinh mt\} + CE$, where $m^2 = \dfrac{R^2}{4L^2} - \dfrac{1}{LC}$;

(b) $R^2 = 4(L/C)$, $q = e^{-Rt/(2L)}\{A + Bt\} + CE$;

(c) $R^2 < 4(L/C)$, $q = e^{-Rt/(2L)}\{A \cos nt + B \sin nt\} + CE$, where $n^2 = \dfrac{1}{LC} - \dfrac{R^2}{4L^2}$.

The corresponding values for I may be obtained by differentiation. It will be noticed that case (c) corresponds to a damped oscillation of the charge to and fro around the circuit.

For a mechanical analogy we now consider a mass m attached to a spring, of natural length a and modulus of elasticity λ; the motion of the mass is subject to a resistance kv and there is a constant force F also acting on the mass, Fig. 9.11. The equation of motion of the mass in this case is

$$m\frac{d^2x}{dt^2} = F - k\frac{dx}{dt} - \frac{\lambda x}{a},$$

i.e.

$$m\frac{d^2x}{dt^2} + k\frac{dx}{dt} + \frac{\lambda}{a}x = F. \tag{2}$$

This is an equation of exactly the same form as (1) with x—the extension of the spring from its unstretched position—replacing q, m replacing L, k replacing R, λ/a replacing $1/C$ and F replacing E.

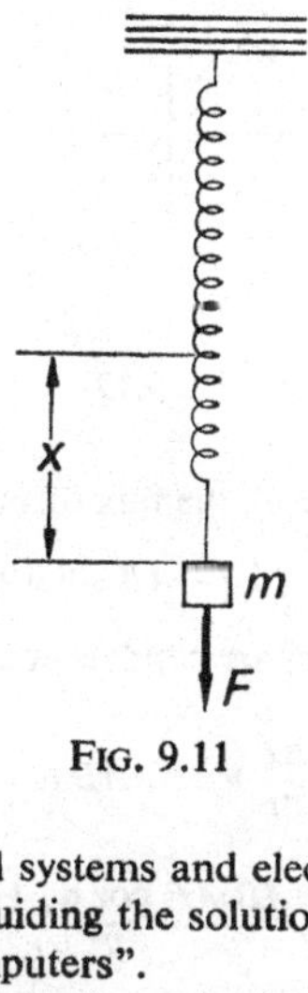

FIG. 9.11

This analogy between mechanical systems and electrical systems can be drawn for all such circuits and is very useful in guiding the solution of electrical problems in addition to being the basis of "analogue computers".

It will be seen that case (a) corresponds to the "dead beat" motion of the mechanical system in which, under the influence of the force F, the extension increases monotonically until, as $t \to \infty$, it eventually reaches the position of equilibrium in which the tension in the spring is balanced by the force F, or the charge on the condenser corresponds to the applied e.m.f. E.

Case (b) is the case of "critical" damping. Damped oscillations result in case (c), for which the resistance is small.

If the resistance is zero, eqn. (1) becomes

$$L\frac{d^2q}{dt^2} + \frac{1}{C}q = E$$

with solution

$$q = EC + A\cos nt + B\sin nt, \quad n^2 = 1/(LC),$$

indicating that undamped simple harmonic oscillations of the charge take place. In practice no electrical circuit is without resistance (any more than a mechanical system is entirely free from friction) and these (natural) oscillations die out eventually. In all the above cases the constants A and B are arbitrary and must be determined by means of initial conditions. The terms of the complementary function usually contain an exponential term $e^{-Rt/(2L)}$ or a similar term and are called *transients* because, after an appreciable lapse of time, the exponential factor has become negligibly small.

Example 2. A plane circuit, of area A, resistance R and self-inductance L, is made to rotate in a constant magnetic field B with constant angular velocity ω about an axis perpendicular to the magnetic field. Find the couple opposing the rotation at time t, given that at time $t = 0$ the plane of the circuit is parallel to the magnetic field, and show that its time-average over a very long time is

$$\frac{A^2B^2R\omega}{2(R^2+\omega^2L^2)}.$$

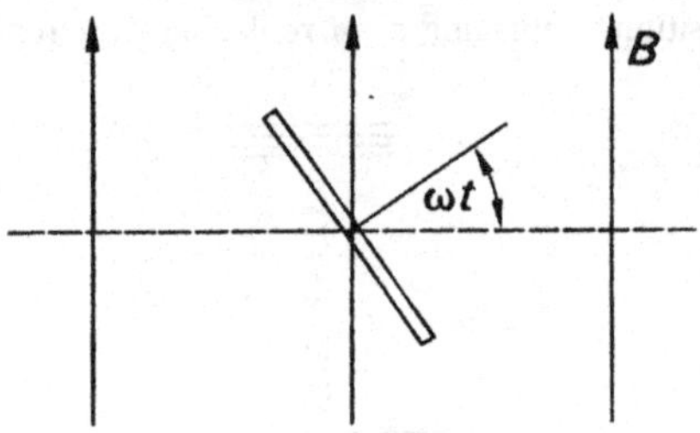

FIG. 9.12

When a current I is flowing in the coil, the flux through the circuit is given by

$$N = LI + AB\sin\omega t,$$

see Fig. 9.12. Since there is no battery or condenser in this circuit, application of (9.31) gives

$$\frac{dN}{dt}+RI = L\frac{dI}{dt}+RI+AB\,\omega\cos\omega t = 0,$$

which has solution

$$I = C_1e^{-Rt/L} - \frac{AB\omega(R\cos\omega t+L\omega\sin\omega t)}{R^2+L^2\omega^2}.$$

Since we are required to consider a time-average over a long period we assume that the transient $C_1e^{-Rt/L}$ is negligible and take

$$I = -\frac{AB\omega(R\cos\omega t+L\omega\sin\omega t)}{R^2+L^2\omega^2}.$$

When a current I flows in the coil it behaves, in the uniform field, as a magnetic dipole of moment AI directed along the axis of the coil. Hence the couple, G, acting on the coil is $AIB\cos\omega t$.

$$\therefore\quad G = -\frac{A^2B^2\omega(R\cos^2\omega t+L\omega\sin\omega t\cos\omega t)}{R^2+L^2\omega^2}.$$

The average value of G, from the term in $\cos^2 \omega t$, is

$$G = -\frac{A^2B^2R\omega}{2(R^2+L^2\omega^2)}.$$

The minus sign indicates that the couple tends to reduce ω.

Example 3. A magnetic dipole of magnitude m is placed on the axis of a fixed circular wire of resistance R and self-inductance L, distant h from the plane of the wire and with its axis along the axis of the wire. The dipole undergoes small longitudinal oscillations $\varkappa \sin pt$ about this point. Show that, at time t in the steady state, the current induced in the wire is approximately

$$\frac{3ma^2ph\varkappa}{\mu_0(a^2+h^2)^{5/2}} \frac{(Lp \sin pt+R \cos pt)}{(R^2+p^2L^2)}.$$

Show also that the maximum force on the wire is of magnitude

$$\frac{9a^4m^2ph^2\varkappa}{\mu_0^2(a^2+h^2)^5 (R^2+p^2L^2)}.$$

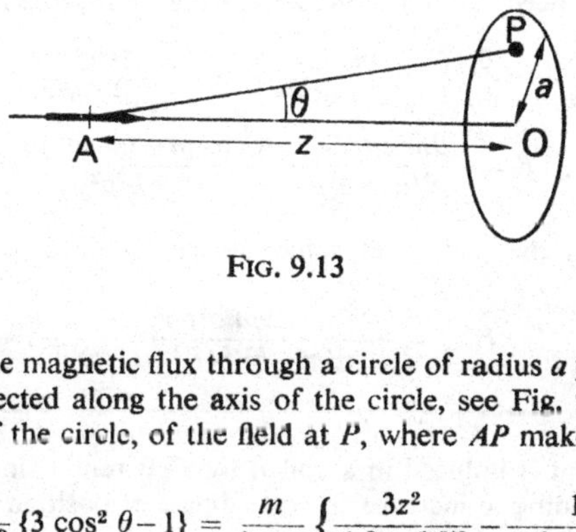

FIG. 9.13

First we calculate the magnetic flux through a circle of radius a produced by a magnet of moment m and directed along the axis of the circle, see Fig. 9.13. The component, normal to the plane of the circle, of the field at P, where AP makes an angle θ with the axis, is

$$\frac{m}{4\pi\mu_0 AP^3}\{3 \cos^2 \theta-1\} = \frac{m}{4\pi\mu_0}\left\{\frac{3z^2}{(z^2+r^2)^{5/2}} - \frac{1}{(z^2+r^2)^{3/2}}\right\},$$

where $OP = r$. Integrating over the circle, we find the flux to be

$$N_1 = \frac{m}{2\mu_0}\int_0^a \left\{\frac{3z^2}{(z^2+r^2)^{5/2}} - \frac{1}{(z^2+r^2)^{3/2}}\right\} r\,dr = \frac{ma^2}{\mu_0(z^2+a^2)^{3/2}}$$

The small oscillations of the question imply that we must put $z = h+\varkappa \sin pt$. Then

$$N_1 = \frac{ma^2}{\mu_0(a^2+h^2+2\varkappa h \sin pt)^{3/2}} = \frac{ma^2}{\mu_0(a^2+h^2)^{3/2}}\left\{1+\frac{2\varkappa h \sin pt}{a^2+h^2}\right\}^{-3/2}$$

$$= \frac{ma^2}{\mu_0(a^2+h^2)^{3/2}}\left\{1-\frac{3\varkappa h \sin pt}{a^2+h^2}\right\}$$

correct to the first order in $\varkappa$.

The equation giving the current I in the wire is

$$L\frac{dI}{dt}+\frac{dN_1}{dt}+RI = 0.$$

$$\therefore\quad L\frac{dI}{dt}+RI = -\frac{dN_1}{dt} = \frac{3ma^2hp}{\mu_0(a^2+h^2)^{5/2}}\varkappa\cos pt = b\cos pt.$$

This is an equation similar to that in example 2 and is solved in the same way, giving

$$I = \frac{b(R\cos pt+Lp\sin pt)}{R^2+L^2p^2} = \frac{3ma^2hp\varkappa}{\mu_0(a^2+h^2)^{5/2}}\cdot\frac{R\cos pt+Lp\sin pt}{R^2+L^2p^2}$$

on omitting the transient term $C_1e^{-Rt/L}$.

The work function for calculating the force on the wire is

$$W = -NI = -LI^2-N_1I.$$

To calculate the force on the wire, i.e. the force which tends to alter h, we differentiate W, keeping the current constant, and obtain

$$F = -\frac{dW}{dh} = I\frac{dN_1}{dh}.$$

Now I is of $O(\varkappa)$ so we need dN_1/dh in the zero order of approximation, i.e.

$$\frac{dN_1}{dh} = \frac{d}{dh}\left\{\frac{ma^2}{\mu_0(a^2+h^2)^{3/2}}\right\} = -\frac{3ma^2h}{\mu_0(a^2+h^2)^{5/2}}.$$

$$\therefore\quad F = -\frac{9m^2a^4ph^2\varkappa}{\mu_0^2(a^2+h^2)^5}\cdot\frac{R\cos pt+Lp\sin pt}{R^2+L^2p^2}.$$

It is easily shown that the maximum value of the factor depending on the time is $(R^2+L^2p^2)^{1/2}$.

$$\therefore\quad F_{max} = -\frac{9m^2a^4ph^2\varkappa}{\mu_0^2(a^2+h^2)^5(R^2+L^2p^2)^{1/2}}.$$

Example 4. A current is induced in a coil B by a current $I\sin pt$ in a coil A. Show that the mean force tending to increase any coordinate of position θ is

$$-\frac{I^2p^2LM}{R^2+L^2p^2}\frac{\partial M}{\partial\theta},$$

where L, M, N are the coefficients of induction of the coils, and R is the resistance of B.

We denote the current in B by i and write down the expression for the flux B as

$$N_B = Li+MI\sin pt.$$

Hence the equation for the circuit is

$$\frac{dN_B}{dt}+Ri = L\frac{di}{dt}+MIp\cos pt+Ri = 0,$$

of which the solution is

$$i = C_1e^{-Rt/L}-\frac{MIp(R\cos pt+Lp\sin pt)}{R^2+L^2p^2}.$$

In order to calculate the force on the coil we use the work-function (see p. 202) $W = N_B i$ and differentiate it, assuming the currents are constant. As in example 1, when calculating the mean value we assume that the transient is negligible so that

$$W = -Li^2 - MIi \sin pt = -Li^2 + \frac{M^2 I^2 p(R \cos pt + Lp \sin pt) \sin pt}{R^2 + L^2 p^2}.$$

Only the factor M^2 in this expression can vary with θ, a coordinate fixing the relative position of the two circuits. Hence the force tending to increase θ is (see § 9.8)

$$-\frac{\partial W}{\partial \theta} = -\frac{2MI^2 p(R \cos pt + Lp \sin pt) \sin pt}{R^2 + L^2 p^2} \frac{\partial M}{\partial \theta}.$$

The average value of this is, from the term $\sin^2 pt$,

$$-\frac{MI^2 Lp^2}{R^2 + L^2 p^2} \frac{\partial M}{\partial \theta}.$$

Exercises 9.5

1. A flat circular coil of resistance R and radius a is made from a length $2\pi an$ of thin wire. The coil is initially at rest in the plane of the magnetic meridian and is turned with constant angular acceleration through a right angle about its vertical diameter. Neglecting the self-inductance, show that the current in the coil is greatest after the angle turned through is θ, where $\cot \theta = 2\theta$.

2. An electric generator consists of a plane coil of wire which rotates with constant angular velocity ω about a fixed axis in its plane. The coil has n turns each of area A, and the axis of rotation is perpendicular to a constant uniform magnetic field $\mathbf{B}$. At time $t = 0$ the plane of the coil contains $\mathbf{B}$. The generator is connected to a circuit containing an inductance L and a resistance R. Prove that the current J satisfies

$$L\frac{dJ}{dt} + RJ = -\omega nAB \cos(\omega t),$$

and find its value after the transient current has become negligible.

Find the ultimate average rate at which energy is dissipated in the circuit. Assuming that the couple acting on the circuit is $nABJ \cos(\omega t)$, verify the conservation of energy.

3. A square circuit of side $2a$ has two sides parallel to a fixed axis u passing through the centre of the square. A current j flows in an infinite straight wire parallel to and at distance d from the axis u. Prove that the magnetic flux through the circuit is

$$\frac{\mu_0 aj}{2\pi} \ln\left(\frac{a^2 + 2ad \cos \theta + d^2}{a^2 - 2ad \cos \theta + d^2}\right),$$

where θ is the inclination of the plane of the circuit to the plane through the infinite wire and the axis u.

If the circuit be made to rotate with constant spin ω about the axis u, find the current induced in the circuit, and show that the average couple opposing the motion is approximately

$$\frac{2\mu_0^2 a^4 j^2 \omega R}{\pi^2 d^2 (R^2 + L^2 \omega^2)},$$

where R and L are the resistance and self-induction of the circuit, and terms of order higher than $(a/d)^3$ are neglected.

4. A circular coil of wire has radius b, resistance R and self-inductance L. A bar magnet of length $2a$ and moment $2am$ has its poles on the axis of the coil, and is made to oscillate so that the distance x between the centre of the coil and the centre of the magnet is $k \sin \omega t$, where k is small compared with $\sqrt{(a^2+b^2)}$, and $1/\omega$ is large compared with b/c. Prove that the induced current is

$$J = \frac{Ak^2\omega}{\sqrt{(R^2+4L^2\omega^2)}} \sin(2\omega t - \alpha),$$

where

$$A = \frac{3\mu_0 mab^2}{2(a^2+b^2)^{5/2}} \quad \text{and} \quad \alpha = \tan^{-1}\frac{2L\omega}{R}.$$

Verify that the mean rate of working of the force required to maintain the oscillation of the magnet is equal to the mean value of RJ^2.

5. A plane circuit, area S, rotates with uniform angular velocity ω about the axis of z, which lies in its plane at a distance h from the centre of gravity of the area. A magnetic dipole of strength m is fixed in the axis of x at a great distance a from the origin, pointing in the direction Ox. Prove that the current at time t is approximately

$$\frac{2\mu_0 S\omega m}{4\pi a^3(R^2+L^2\omega^2)^{1/2}} \cos(\omega t - \varepsilon) + \frac{9\mu_0 S\omega mh}{4\pi a^4(R^2+4L^2\omega^2)^{1/2}} \cos(2\omega t - \eta),$$

where η, ε are determinate constants.

9.6 Initial conditions: switching

The neutral state of any circuit is that in which there is no charge on any condenser, no current flows in the circuit and no e.m.f. acts; in the analogous mechanical arrangement the mass is stationary with the spring at its natural length under no tension and no external force acts on the particle. In the preceding examples we obtained solutions by assuming that the lapse of time has made the transients negligible; we have not considered initial conditions. In general, whenever a system differs from its neutral state we must solve the relevant differential equations subject to appropriate initial conditions. In some cases a discontinuous change occurs in the conditions governing the motion, e.g. the mass moving on the end of the spring may coalesce with a stationary particle. In such a case the subsequent motion satisfies different equations and is determined by solving the new equations and adopting the values of velocity and displacement immediately after the discontinuity as initial values for this later motion. In order to investigate how such discontinuities affect the displacement and velocity we sometimes use the following device: we assume that the change takes place continuously over a short but finite period of time τ and then integrate with respect to the time the differential equations which must apply during this period and take the limit $\tau \to 0$. (The methods of mechanics used for impulsive motion are good examples of this technique.)

The chief discontinuous operation which occurs in connection with electrical circuits is that of "switching in" or "switching out" certain components or breaking a circuit. In the latter case an infinite resistance is substituted for a finite resistance and the current strength is instantaneously reduced to zero. If there is any inductance in a circuit this breaking of the circuit induces an infinite e.m.f. in the circuit. (It is this which in practice leads to sparks occurring when a switch is opened; the air cannot withstand the voltage generated when the current strength drops so rapidly.) If the circuit which is broken is inductively coupled to a secondary circuit, an infinite e.m.f. is instantaneously generated in the secondary circuit.

In many cases the initial conditions, or the discontinuities, can be seen "on inspection", but in other cases we have to use the integration procedure. In such cases the mechanical analogue can be of considerable assistance. Thus, an impulsive change occurs in a mechanical system because a large force acts for a very short period; during this period the displacement of a mass does not change and the velocity alters discontinuously. In an electrical case a circuit may receive an "impulse" when a neighbouring circuit is broken; during the period of this impulse the charges on condensers will not be altered (analogous to displacements) but the current will change discontinuously. (Current is the analogue of velocity.) The following are examples of the use of initial conditions and integration procedures.

Example 1. We consider the discharge of a condenser which initially has charges $\pm Q_0$ on its plates when it is connected to a resistance and an inductance in series.

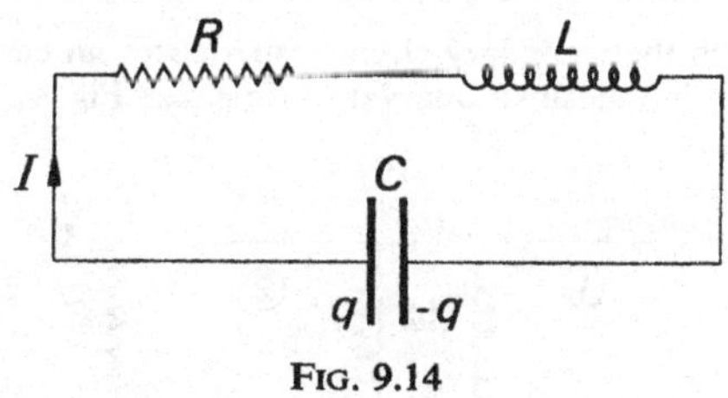

FIG. 9.14

We consider the state of affairs for a subsequent time as shown in Fig. 9.14. The differential equations are

$$RI + L\frac{dI}{dt} = \frac{q}{C}, \quad I = -\frac{dq}{dt}.$$

$$\therefore \quad L\frac{d^2q}{dt^2} + R\frac{dq}{dt} + \frac{q}{C} = 0.$$

This is the same as eqn. (1) of example 1 on p. 368 and has the three possible solutions given there with $E = 0$.

At time $t = 0$ when the condenser is connected, $q = Q_0$ and $I = 0$ and we deduce three possible types of solution with these initial conditions.

(a) $$R^2 > 4L/C, \quad q = Q_0 e^{-Rt/(2L)}\left\{\cosh mt + \frac{R}{2Lm}\sinh mt\right\},$$

where $$m^2 = \frac{R^2}{4L^2} - \frac{1}{LC}.$$

(b) $$R^2 = 4L/C, \quad q = Q_0 e^{-Rt/(2L)}\left(1 + \frac{Rt}{2L}\right).$$

(c) $$R^2 < 4L/C, \quad q = Q_0 e^{-Rt/(2L)}\left\{\cos nt + \frac{R}{2Ln}\sin nt\right\},$$

where $$n^2 = \frac{1}{LC} - \frac{R^2}{4L^2}.$$

In the extreme case when $R = 0$, we obtain from (c) that

(d) $$R = 0, \quad q = Q_0 \cos\left\{\frac{t}{\sqrt{(LC)}}\right\}$$

which is a simple harmonic oscillation of the charge from one plate to the other through the inductance.

The other extreme case, $L = 0$, in which the condenser is discharged through a non-inductive resistance, is obtained from the differential equation by putting $L = 0$. The result is

(e) $$L = 0, \quad q = Q_0 \exp\{-t/(RC)\}.$$

Example 2. A primary and a secondary circuit have a mutual inductance M, and the secondary circuit has a self-inductance L and resistance R. When the primary circuit is carrying a steady current i_0 it is suddenly broken and the primary current reduced instantaneously to zero. Find the current induced in the secondary circuit.

This is a case in which the secondary circuit experiences an electrical impulse, i.e. an infinite e.m.f. acts for an infinitesimal interval of time (see Fig. 9.15).

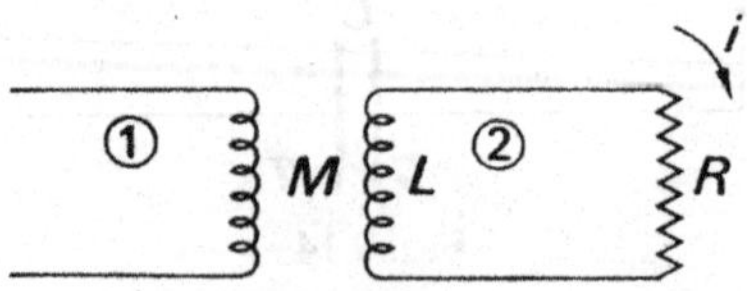

FIG. 9.15

In general the equation for the secondary circuit is, during the time that the primary current i_1 varies,

$$L\frac{di}{dt} + Ri = -M\frac{di_1}{dt}. \tag{1}$$

The term $-M\,di_1/dt$ gives the electromotive force acting in the secondary circuit arising from the change in the primary current. We assume that the "switching off" takes place in a short time τ in which period i_1 is reduced from i_0 to zero; also i increases from its neutral value zero to the initial value i_2 for the subsequent "motion". We integrate eqn.

(1) over this period τ; thus

$$L\int_0^\tau \frac{di}{dt}\,dt + R\int_0^\tau i\,dt = -M\int_0^\tau \frac{di_1}{dt}\,dt.$$

Then

$$L[i]_0^\tau + R\int_0^\tau i\,dt = -M[i_1]_0^\tau.$$

During the rapid change i remains finite, between zero and i_2, so that $\int_0^\tau i\,dt = \tau\bar{i}$, where $\bar{i}$ is a mean value.

$$\therefore\quad L(i_2-0)+R\tau\bar{i} = -M(0-i_0).$$

Hence in the limit $\tau \to 0$ we deduce that

$$Li_2 = Mi_0. \tag{2}$$

For the subsequent behaviour the primary current must be zero and, corresponding to eqn. (1), we have for the secondary circuit

$$L\frac{di}{dt}+Ri = 0.$$

The solution of this differential equation is

$$i = i_2 e^{-Rt/L} = (Mi_0/L)e^{-Rt/L}.$$

There is an alternative view of the condition (2). When the primary current is steady and the secondary current zero the magnetic flux in the secondary is $L.0+Mi_0$; when the secondary current has the value i_2 and the primary current is zero the flux in the secondary is $Li_2+M.0$. Condition (2) therefore shows that the magnetic flux in the secondary is unaltered during the impulse. We can draw an analogy between this result and two particles interacting through, say, a string. If the first particle jerks a second particle into motion by pulling a string taut suddenly and in the process is itself brought to rest, the momentum imparted to the second particle is equal to that of the first. But mass and velocity are the mechanical analogues of inductance and current, so our result of "conservation of flux" is the analogue of the law of conservation of momentum. However, an argument of the kind leading to (2) from (1) must be used to get the exact relation. Argument by analogy is not proof!

Example 3. A circuit contains a condenser of capacity C, a wire of resistance R, and a coil of self-inductance $L = 2R^2C$ and negligible resistance. The coil is in parallel with the wire and the condenser is in series with them. At time $t = 0$ the charge of the condenser is Q_0 and there is no current passing through the coil. Show that the charge of the condenser at a later time t is $Q_0e^{-kt}(\cos kt - \sin kt)$, where $k = (2RC)^{-1}$.

We include this example to illustrate the application of the generalization of Kirchhoff's second law. (The assumed currents in Fig. 9.16 satisfy the first law.) There are two independent loops in this network, viz. R, C and the conductor, and L, C and the conductor. The equations for the loops are

$$Ri_1 = -q/C,\quad L(di_2/dt) = -q/C,\quad i_1+i_2 = dq/dt,$$

i.e.

$$i_1 = -2kq,\quad di_2/dt = -2k^2q,\quad di_1/dt+di_2/dt = d^2q/dt^2.$$

$$\therefore\quad \frac{d^2q}{dt^2}+2k\frac{dq}{dt}+2k^2q = 0.$$

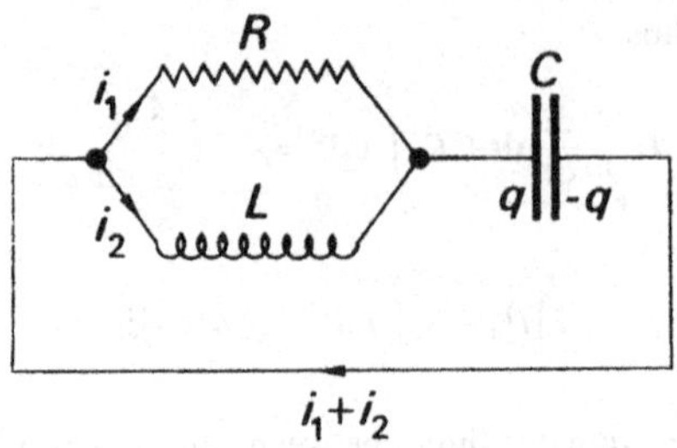

FIG. 9.16

The general solution is

$$q = e^{-kt}(A \cos kt + B \sin kt), \qquad \frac{dq}{dt} = ke^{-kt}\{(B-A) \cos kt - (B+A) \sin kt\}.$$

The initial conditions given are $q = Q_0$ and $i_2 = 0$; from the equation for the R, C loop we find the initial value of i_1 is $-Q_0/(CR) = -2kQ_0$. Hence initially $(dq/dt)_0 = -2kQ_0$.

$$\therefore \quad Q_0 = A, \qquad -2kQ_0 = k(-A+B), \qquad \text{i.e.} \qquad B = -Q_0.$$

$$\therefore \quad q = Q_0 e^{-kt}(\cos kt - \sin kt).$$

Example 4. A circuit of self-inductance L_1 and resistance R_1 contains a condenser of capacity C, and is coupled with a second circuit of negligible resistance and self-inductance L_2. The mutual induction of the two circuits is M. The condenser is given a charge Q before being switched into the first circuit at time $t = 0$. Show that, at time t, the current in the second circuit is of the form $A(e^{-\alpha t} - e^{-\beta t})$, and verify that the total quantity of electricity which traverses the second circuit is MQ/L_2.

Prove also that the heat generated in the first circuit is $\frac{1}{2}Q^2/C$.

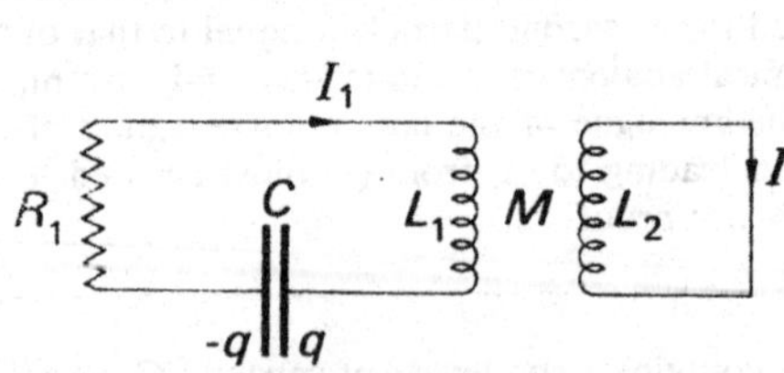

FIG. 9.17

The circuit is shown in Fig. 9.17. The application of eqn. (9.31) to each of these circuits gives

$$\frac{d}{dt}(L_1I_1 + MI_2) + R_1I_1 = -\frac{q}{C}, \tag{1}$$

$$\frac{d}{dt}(MI_1 + L_2I_2) = 0, \tag{2}$$

$$I_1 = \frac{dq}{dt}. \tag{3}$$

Since the two circuits are in the neutral state when the condenser carrying a charge Q is connected, the initial conditions are

$$q = Q, \quad I_1 = 0 = I_2$$

(cf. releasing a mechanical system from rest in a displaced position). We integrate (2) to give $MI_1+L_2I_2 = 0$, and from (1) and (3) obtain

$$L_1\frac{d^2I_1}{dt^2}+M\frac{d^2I_2}{dt^2}+R_1\frac{dI_1}{dt}+\frac{1}{C}I_1 = 0,$$

and, on eliminating I_2,

$$(L_1L_2-M^2)\frac{d^2I_1}{dt^2}+R_1L_2\frac{dI_1}{dt}+\frac{L_2}{C}I_1 = 0. \tag{1}$$

Since $I_1 = 0$ when $t = 0$, the solution of this differential equation has the form

$$I_1 = D(e^{p_1t}-e^{p_2t}),$$

where p_1, p_2 are the roots of

$$(L_1L_2-M^2)p^2+R_1L_2p+L_2/C = 0.$$

This solution is of the required form.

Because $L_1L_2-M^2 > 0$ and all the other coefficients of eqn. (1) are positive it follows that $p_1p_2 > 0$ and $p_1+p_2 < 0$. Therefore, if p_1 and p_2 are real they must both be negative, and if they are complex conjugates their real part is negative [being equal to $\frac{1}{2}(p_1+p_2)$]. Hence I_1, and also I_2, tend to zero as $t \to \infty$.

The total charge traversing the second circuit is

$$Q_2 = \int_0^\infty I_2\,dt = -\frac{M}{L_2}\int_0^\infty I_1\,dt = -\frac{M}{L_2}\int_0^\infty \frac{dq}{dt}\,dt = -\frac{M}{L_2}[q]_0^\infty = \frac{MQ}{L_2}.$$

The rate of generation of heat in the first circuit is RI_1^2 and

$$R_1I_1^2 = -I_1\left(L_1\frac{dI_1}{dt}+M\frac{dI_1}{dt}+\frac{q}{C}\right) = -I_1\left\{\left(L_1-\frac{M^2}{L_2}\right)\frac{dI_1}{dt}+\frac{q}{C}\right\}.$$

$$\therefore\quad H = \int_0^\infty R_1I_1^2\,dt = -\left(L_1-\frac{M^2}{L_2}\right)\int_0^\infty I_1\frac{dI_1}{dt}\,dt-\frac{1}{C}\int_0^\infty \frac{dq}{dt}\,q\,dt$$

$$= -\left(L_1-\frac{M^2}{L_2}\right)[I_1^2]_0^\infty-\frac{1}{2C}[q^2]_0^\infty.$$

From the initial values we see that $H = \frac{1}{2}Q^2/C$.

Since the first circuit is the only circuit containing resistance all energy dissipated as heat is dissipated in this circuit, and the total energy available for this is the original electrostatic energy of the condenser.

Example 5. Two condensers of capacity C have one plate of each earthed. Initially, the first condenser has a charge Q and the second is uncharged. The two insulated plates are connected by a wire of resistance R and inductance $\frac{1}{8}CR^2$. Show that, when the rate of dissipation of energy is a maximum, the energy still to be dissipated is $5Q^2/(4Ce^2)$.

We represent the state of affairs at time t after switching on by the diagram of Fig. 9.18. Application of eqn. (9.4) leads to

$$\frac{1}{8}CR^2\frac{dI}{dt}+RI = \frac{q_1}{C}-\frac{q_2}{C}, \tag{1}$$

$$I = -\frac{dq_1}{dt} = +\frac{dq_2}{dt}. \tag{2}$$

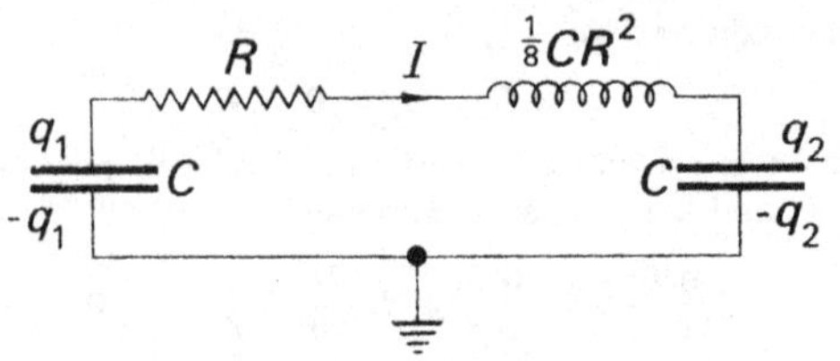

FIG. 9.18

In this case the initial conditions are $q_1 = Q$, $q_2 = 0$, $I = 0$. (Compare this with the release of a mechanical system from rest in a displaced position.) Elimination of q_1 and q_2 gives

$$C^2R^2\frac{d^2I}{dt^2}+8CR\frac{dI}{dt}+16I = 0. \tag{3}$$

Since q_1 and q_2 are given initially, eqn. (1) implies that the initial value of dI/dt is $(dI/dt)_0 = 8Q/(C^2R^2)$. The solution of (3) subject to the initial conditions on I and dI/dt is

$$I = \frac{8Q}{C^2R^2}te^{-4t/(CR)}, \qquad \frac{dI}{dt} = \frac{8Q}{C^2R^2}\left(1-\frac{4t}{CR}\right)e^{-4t/(CR)}. \tag{4}$$

The rate of production of heat is RI^2, which has a stationary value either when $I = 0$ or when $t = \frac{1}{4}CR$, the latter value giving the maximum. Hence the energy dissipated subsequently is

$$\int_{CR/4}^{\infty} RI^2\,dt = \frac{64Q^2}{C^4R^3}\int_{CR/4}^{\infty} t^2e^{-8t/(CR)}\,dt = \frac{5Q^2}{4Ce^2}.$$

(The integral is evaluated by parts.)

Example 6. Two circuits, each containing a condenser of capacity C and a coil of self-inductance L and negligible resistance, are coupled with mutual inductance M. Find the periods of free electric oscillation in the circuits.

Let the currents flowing round the circuits 1, 2 and the charges on the condensers be as shown in Fig. 9.19. Then using the methods of p. 367 the current equations are

$$L\frac{dI_1}{dt}+M\frac{dI_2}{dt} = \frac{q_1}{C}, \qquad M\frac{dI_1}{dt}+L\frac{dI_2}{dt} = \frac{q_2}{C},$$
$$\frac{dq_1}{dt} = -I_1, \qquad \frac{dq_2}{dt} = -I_2. \tag{1}$$

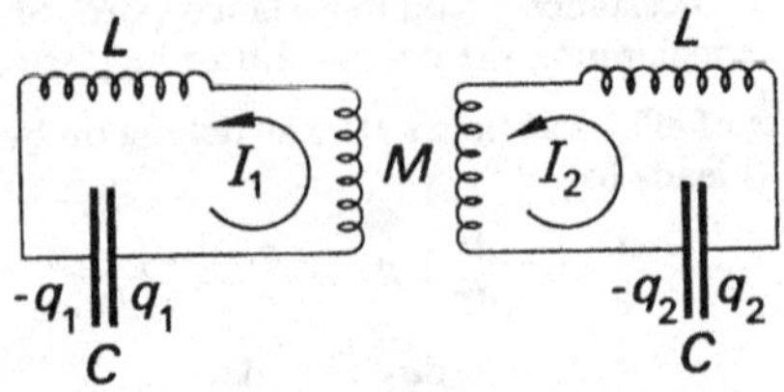

FIG. 9.19

For free oscillation with period $2\pi/\omega$, we seek a solution such that all variables depend on t through the factor $e^{i\omega t}$, i.e. as suggested by

$$I_1 = I_{10}e^{i\omega t}, \quad I_2 = I_{20}e^{i\omega t}, \quad q_1 = q_{10}e^{i\omega t}, \quad q_2 = q_{20}e^{i\omega t}.$$

Substituting in eqn. (1) and cancelling the exponential factor we find

$$\begin{aligned} Li\omega I_{10}+Mi\omega I_{20} &= \frac{q_{10}}{C}, & Mi\omega I_{10}+Li\omega I_{20} &= \frac{q_{20}}{C}, \\ i\omega q_{10} &= -I_{10}, & i\omega q_{20} &= -I_{20}. \end{aligned} \tag{2}$$

Eliminating I_{10}, etc., we find that for a non-trivial solution of eqns. (2), ω must satisfy the equation

$$\left(Li\omega-\frac{1}{i\omega C}\right)^2-(Mi\omega)^2 = 0,$$

i.e.

$$\left\{(L-M)\omega-\frac{1}{\omega C}\right\}\left\{(L+M)\omega-\frac{1}{\omega C}\right\} = 0.$$

The possible values of ω are therefore $1/\sqrt{\{(L-M)C\}}$, $1/\sqrt{\{(L+M)C\}}$ and the corresponding periods of free electric oscillation are $2\pi\sqrt{\{(L-M)C\}}$, $2\pi\sqrt{\{(L+M)C\}}$.

Note that when $M \to 0$, we obtain the period $2\pi\sqrt{(LC)}$ of a single uncoupled circuit without ohmic resistance.

The method we employed here is the exact analogue of the method for finding the frequencies of the normal oscillations of a mechanical system.

Example 7. A primary circuit of negligible resistance and total self-inductance nL consists of n equal coils and a capacitance C/n connected in series. The coils are each coupled by a mutual inductance M to one of n similar separate secondary circuits; no two coils are coupled to the same secondary circuit and there is no coupling between any two secondary circuits. The secondary circuits each consist of a capacitance C and a self-inductance L, and are also of negligible resistance. Show that a current in the primary circuit varies with time according to the sum of two terms, one representing a sinusoidal oscillation of period $2\pi\sqrt{\{C(L+M)\}}$ and the other a sinusoidal oscillation of period $2\pi\sqrt{\{C(L-M)\}}$.

A battery of electromotive force E and resistance R is now connected across the capacitance in the primary circuit, and the system is allowed to reach a steady state. The battery connections are then broken. Find the current in the primary circuit and the current in each secondary circuit at a time t after the breaking of the connections.

The current I and charge $\pm q$ in the primary circuit and the respective currents I_r and charges q_r give the state of affairs at an arbitrary time t (the battery is not connected), Fig. 9.20. The equations arising from (9.31) are:

for the primary,
$$nL\frac{dI}{dt}+M\sum_{r=1}^{n}\frac{dI_r}{dt} = \frac{q}{C/n}, \quad \frac{dq}{dt} = -I; \tag{1}$$

for each secondary,
$$L\frac{dI_r}{dt}+M\frac{dI}{dt} = \frac{q_r}{C} \tag{2}$$

and
$$\frac{dq_r}{dt} = -I_r \quad (r = 1, 2, \ldots, n). \tag{3}$$

We eliminate q, q_r and obtain

$$nL\frac{d^2I}{dt^2}+M\sum_1^n\frac{d^2I_r}{dt^2}+\frac{nI}{C} = 0, \tag{4}$$

$$L\frac{d^2I_r}{dt^2}+M\frac{d^2I}{dt^2}+\frac{I_r}{C} = 0 \quad (r = 1, 2, \ldots, n). \tag{5}$$

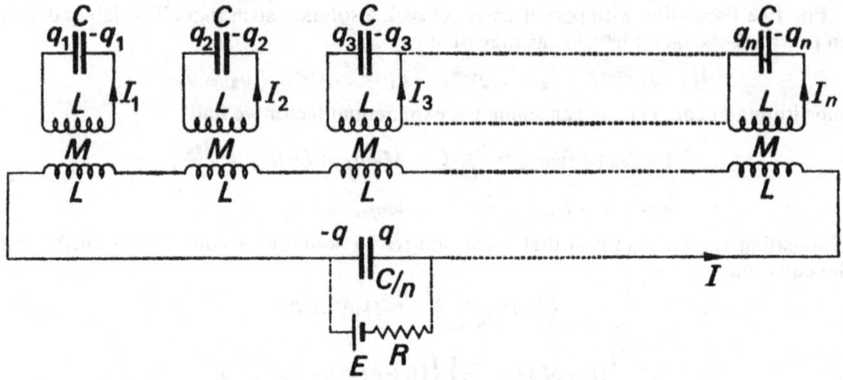

FIG. 9.20

If we add all the equations (5) together and write $J = \sum_{1}^{n} I_r$, we obtain,

$$nL\frac{d^2I}{dt^2}+M\frac{d^2J}{dt^2}+\frac{nI}{C} = n\left(L\frac{d^2}{dt^2}+\frac{1}{C}\right)I+M\frac{d^2J}{dt^2} = 0, \tag{4a}$$

$$nM\frac{d^2I}{dt^2}+L\frac{d^2J}{dt^2}+\frac{J}{C} = nM\frac{d^2I}{dt^2}+\left(L\frac{d^2}{dt^2}+\frac{1}{C}\right)I = 0. \tag{5a}$$

We eliminate J from these equations and obtain

$$\left\{\left(L\frac{d^2}{dt^2}+\frac{1}{C}\right)\left(L\frac{d^2}{dt^2}+\frac{1}{C}\right)-M^2\frac{d^4}{dt^4}\right\}I = \frac{1}{C^2}\left\{C^2(L^2-M^2)\frac{d^4}{dt^4}+2LC\frac{d^2}{dt^2}+1\right\}I = 0. \tag{6}$$

Equation (6) has a general solution

$$I = A_1 \cos(\omega_1 t+\alpha_1)+A_2 \cos(\omega_2 t+\alpha_2),$$

where ω_1^2, ω_2^2 are the roots of

$$C^2(L^2-M^2)x^2-2LCx+1 = 0,$$

i.e.

$$\omega_1^2 = \{C(L+M)\}^{-1}, \quad \omega_2^2 = \{C(L-M)\}^{-1}. \tag{7}$$

Therefore the primary current consists of two oscillations as stated.

When the battery is connected the steady state of the system corresponds to a charge nE/C on the primary condenser, and a steady current E/R in the primary circuit; all the charges and currents in the secondary circuits are zero. When the battery connection is broken the e.m.f. ceases to act, but the primary current can continue to flow, by means of the discharge of the condenser. Hence the initial conditions are:

$$q = nE/C, \quad I = E/R, \quad q_r = 0, \quad I_r = 0 \quad (r = 1, 2, \ldots, n). \tag{8}$$

Because of the symmetry of the secondary circuits and of the initial conditions applying to them we can write

$$q_1 = q_2 = \ldots = q_n = q', \quad \text{and} \quad I_1 = I_2 = \ldots = I_n = I' \quad (= J/n).$$

Equations (1) and (2) become

$$L\frac{dI}{dt}+M\frac{dI'}{dt} = \frac{q}{C}, \qquad \frac{dq}{dt} = -I, \tag{9}$$

$$L\frac{dI'}{dt}+M\frac{dI}{dt} = \frac{q'}{C}, \qquad \frac{dq'}{dt} = -I'. \tag{10}$$

Adding and subtracting eqns. (9) and (10) we obtain

$$C(L\pm M)\frac{d^2}{dt^2}(q\pm q')+(q\pm q') = 0,$$

so that

$$q+q' = A\sin\omega_1 t+B\cos\omega_1 t, \qquad q-q' = F\sin\omega_2 t+G\cos\omega_2 t,$$

$$I+I' = \omega_1(A\cos\omega_1 t-B\sin\omega_1 t), \qquad I-I' = \omega_2(F\cos\omega_2 t-G\sin\omega_2 t).$$

The initial conditions (8) now lead to

$$I+I' = E\left(\frac{1}{R}\cos\omega_1 t-\frac{\omega_1 n}{C}\sin\omega_1 t\right), \qquad I-I' = E\left(\frac{1}{R}\cos\omega_2 t-\frac{\omega_2 n}{C}\sin\omega_2 t\right),$$

from which I and I' can be obtained separately.

Example 8. A primary circuit contains an inductance L and a capacitance C in series and a secondary circuit also consists of an inductance L and a capacitance C in series; the mutual inductance between the circuits is M. There is an electromotive force $v(t)$ in the primary circuit for time $t > 0$ and the charges on the capacitors and currents in the circuits are zero at time $t = 0$. Prove that, at time $t > 0$, the charge in the capacitor in the secondary is

$$\tfrac{1}{2}C\int_0^t v(\tau)\{m\sin m(t-\tau)-n\sin n(t-\tau)\}\,d\tau,$$

where $$m^{-2} = (L+M)C, \qquad n^{-2} = (L-M)C.$$

If $v(t)$ has the constant value v_0 for $kT < t < (k+1)T$, where $k = 0, 2, 4, \ldots$, and is zero for other values of t, find the charge on the capacitor in the secondary at any time t.

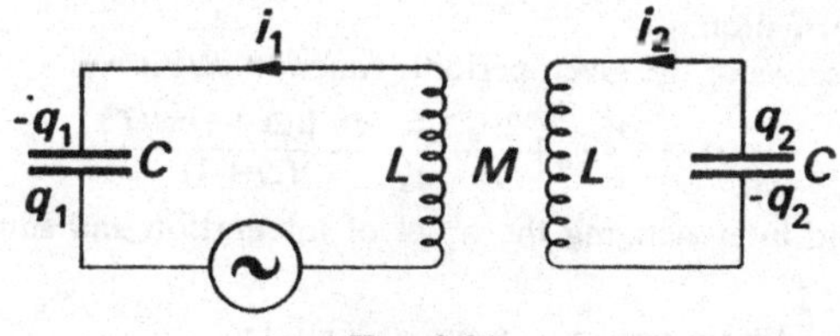

FIG. 9.21

With reference to Fig. 9.21, the equations for the circuits are

$$L\frac{di_1}{dt}+M\frac{di_2}{dt} = \frac{q_1}{C}+v(t),$$

$$M\frac{di_1}{dt}+L\frac{di_2}{dt} = \frac{q_2}{C},$$

$$\frac{dq_1}{dt} = -i_1, \qquad \frac{dq_2}{dt} = -i_2,$$

and the initial conditions are $i_1 = 0 = i_2$, $q_1 = 0 = q_2$ at $t = 0$.

We solve these equations for q_2 by use of the Laplace transformation. The transformed equations are

$$pLI_1+pMI_2 = \frac{Q_1}{C}+V(p),$$

$$pMI_1+pLI_2 = \frac{Q_2}{C},$$

$$pQ_1 = -I_1, \quad pQ_2 = -I_2,$$

where a capital letter denotes the Laplace transform of the corresponding small letter. Solving for Q_2, we find

$$Q_2 = \frac{Mp^2V(p)}{\left\{\left(p^2L+\frac{1}{C}\right)^2-p^4M^2\right\}} = \frac{Mp^2V(p)}{\left\{p^2(L+M)+\frac{1}{C}\right\}\left\{p^2(L-M)+\frac{1}{C}\right\}}$$

$$= \frac{\frac{1}{2}V(p)}{p^2(L-M)+\frac{1}{C}}-\frac{\frac{1}{2}V(p)}{p^2(L+M)+\frac{1}{C}},$$

i.e.

$$Q_2 = \frac{1}{2}C\left\{\frac{m^2V(p)}{p^2+m^2}-\frac{n^2V(p)}{p^2+n^2}\right\}, \tag{1}$$

where m, n are as defined in the question. But

$$\frac{MV(p)}{p^2+m^2} = \text{L}\{\sin mt\}\,\text{L}\{v(t)\}$$

and it follows from the convolution rule that

$$\text{L}^{-1}\left\{\frac{mV(p)}{p^2+m^2}\right\} = \int_0^t v(\tau)\sin m(t-\tau)\,d\tau. \tag{2}$$

From (1), (2),

$$q_2 = \tfrac{1}{2}C\int_0^t v(\tau)\{m\sin m(t-\tau)-n\sin n(t-\tau)\}\,d\tau. \tag{3}$$

When $v(t)$ is the stated (square pulse) function we can proceed in two ways, obtaining either (a) a series solution using Fourier series, or (b) a (closed) wave-form solution using the Laplace transformation.

(a) The Fourier series of the given periodic function $v(t)$ is

$$v(t) = \frac{1}{2}v_0+\frac{2v_0}{\pi}\sum_{r=1}^{\infty}\frac{\sin\{(2r+1)\pi t/T\}}{(2r+1)}. \tag{4}$$

Using (4) in (3) and interchanging the order of integration and summation we obtain

$$\frac{2q_2}{Cv_0} = \frac{1}{2}\int_0^t\{m\sin m(t-\tau)-n\sin n(t-\tau)\}\,d\tau$$

$$+\frac{2}{\pi}\sum_{r=1}^{\infty}\int_0^t\left[\frac{\sin\{(2r+1)\pi\tau/T\}}{(2r+1)}\{m\sin m(t-\tau)-n\sin n(t-\tau)\}\right]d\tau$$

$$= \frac{1}{2}(\cos mt-\cos nt)$$

$$+\frac{2Tm}{\pi}\sum_{r=1}^{\infty}\frac{1}{(2r+1)}\left\{\frac{(2r+1)\pi\sin mt-mT\sin(2r+1)\pi t/T}{(2r+1)^2\pi^2-m^2T^2}\right\}$$

$$-\frac{2Tn}{\pi}\sum_{r=1}^{\infty}\frac{1}{2r+1}\left\{\frac{(2r+1)\pi\sin nt-nT\sin(2r+1)\pi t/T}{(2r+1)^2\pi^2-n^2T^2}\right\}$$

as the solution.

(b) We derive the wave-form solution from eqn. (1) as follows:

$$V(p) = \mathrm{L}\{v(t)\} = \int_0^\infty v(t)\mathrm{e}^{-pT}dt = \frac{v_0}{p(1+\mathrm{e}^{-pT})},$$

$$= \frac{v_0}{p}\sum_{r=0}^{\infty}(-1)^r\,\mathrm{e}^{-rpT}.$$

$$\therefore\ \frac{2Q_2}{v_0C} = \sum_{r=0}^{\infty}(-1)^r\left\{\frac{m^2}{p(p^2+m^2)}-\frac{n^2}{p(p^2+n^2)}\right\}\mathrm{e}^{-rpT}$$

$$= \sum_{r=0}^{\infty}(-1)^r\left\{\frac{p}{p^2+m^2}-\frac{p}{p^2+n^2}\right\}\mathrm{e}^{-rpT}.$$

$$\therefore\ q_2 = \tfrac{1}{2}v_0C\sum_{r=0}^{\infty}(-1)^r\,H(t-rT)\{\cos m(t-rT)-\cos n(t-rT)\},$$

where $H(t-a)$ is the Heaviside unit function. Thus,

for $0 < t < T$, $\quad q_2 = \tfrac{1}{2}v_0C\,(\cos mt-\cos nt)$,

for $T < t < 2T$, $\quad q_2 = \tfrac{1}{2}v_0C\,\{\cos mt-\cos nt-\cos m(t-T)+\cos n(t-T)\}$.

Exercises 9.6

1. A condenser holding charge Q is discharged by a momentary current through a ballistic galvanometer, in which the current causes a suspended coil to rotate between fixed magnets. Assuming (a) that the discharge is completed while the angle θ of deflexion of the coil is still small, and (b) that the disturbing torque due to current of strength i is Ki, where K is constant, prove that the initial angular velocity of the coil is KQ/I, where I is the moment of inertia of the coil about its axis of rotation.

   If the subsequent free oscillation of θ is a slightly damped harmonic motion of period $2\pi/n$, and θ_1 and θ_2 are the first and second throws, prove that $Q = nI\theta_1^{3/2}\theta_2^{-1/2}K^{-1}$ approximately. (A *throw* is a maximum deflexion in either direction.)

2. A coil of self-inductance L and resistance R is connected across terminals A and B in parallel with a condenser of capacity C. An e.m.f. $E\cos\omega t$ is applied between A and B. Prove that the current I flowing through the source of the e.m.f. is

$$\left(I_0-\frac{ER}{R^2+\omega^2L^2}\right)\mathrm{e}^{-Rt/L}+\frac{E}{R^2+\omega^2L^2}\{R\cos\omega t+[\omega L-\omega C(R^2+\omega^2L^2)]\sin\omega t\},$$

   where I_0 is the value of I at time $t = 0$.

3. A circuit consists of a capacitor C, a wire of resistance R, a coil of self-inductance $L = 2R^2C$ and zero resistance, and a generator of voltage $E_0\sin kt$, where $k = (2RC)^{-1}$. The wire and coil are in parallel, and the generator and capacitor in series with them. Show that, whatever the initial values of the current and the charge Q on the capacitor, asymptotically Q approaches

$$\frac{E_0}{5kR}(3\sin kt-\cos kt).$$

4. The primary circuit of a transformer is of resistance R and of self-inductance L. The secondary circuit is of resistance S and self-inductance N and the mutual induc-

tance of the transformer coils is M. A battery, of constant e.m.f. E, is connected across the primary of the transformer. Show that the heat subsequently generated in the secondary circuit is

$$\frac{E^2M^2}{2R(RN+SL)}.$$

5. A circuit containing inductance and resistance in series is coupled by mutual inductance M to a secondary circuit containing inductance L, resistance R and capacity C in series. A steady current i_0 is originally flowing in the primary circuit when it is opened at time $t = 0$. Show that the current in the secondary circuit at any subsequent time is given by

$$i = Mi_0e^{-\lambda t}(n \cos nt - \lambda \sin nt)/(nL)$$

where $$\lambda = R/(2L) \quad \text{and} \quad n^2 = (LC)^{-1} - \lambda^2 > 0.$$

6. Two condensers of capacities C_1 and C_2 each have one plate earthed. Initially the first condenser is charged and the second is uncharged. The two insulated plates are connected at time $t = 0$ by a wire of resistance R and self-inductance L. Show that the current i flowing in the wire satisfies the equation

$$L\frac{d^2i}{dt^2} + R\frac{di}{dt} + \left(\frac{1}{C_1} + \frac{1}{C_2}\right)i = 0.$$

If $R^2 = 4L(1/C_1 + 1/C_2)$, show that the current in the wire reaches its maximum at time $t = 2L/R$.

7. Two condensers of capacity C and C' are connected in series with an inductance L and negligible resistance. Initially the charge on C is q and that on C' is zero. A switch between them is now closed. Show that the current through the inductance at any subsequent time t is

$$i = q\left\{\frac{C'}{LC(C+C')}\right\}^{1/2} \sin\left\{\left(\frac{C+C'}{LCC'}\right)^{1/2} t\right\}.$$

Show that if $C > C'$, the sign of the charge on the first condenser is never reversed.

8. Each of two similar circuits contains a condenser of capacity C, and each has self-induction L; their mutual induction is M and their resistances are negligible. Initially one circuit is closed, without charge or current, and the other is open and has charges $\pm Q$ on the condenser plates. If the latter circuit is closed at time $t = 0$, find the charges on both condensers at any later time.

If M/L is small, show that the variation of each charge may be treated approximately as a simple harmonic oscillation of period $2\pi\sqrt{(LC)}$ and of amplitude which varies slowly with period $2\pi\sqrt{(L^3C/M^2)}$.

9. The plates of a condenser of capacity C are joined by wires of negligible resistance to points A and B, which are connected by a coil of resistance R and self-inductance L. Find the condition that the discharge of the condenser may be oscillatory.

In this case, if A and B are also connected by a wire of resistance S whose self-inductance and mutual inductance with the first circuit are both negligible, show that the discharge will no longer be oscillatory if

$$S < \frac{L}{CR + 2\sqrt{(LC)}}.$$

10. The terminals of a generator producing a variable voltage $V(t)$ are connected through a resistance R in series with a self-inductance L. A capacity C is in parallel with R. Find the differential equation satisfied by the current J through the resistance.

Show that if $L < 4R^2C$ and

$$V(t) = \begin{cases} 0 & (t < 0), \\ V_0 & (t \geqslant 0), \end{cases}$$

where V_0 is a constant, then, for $t > 0$, J is given by

$$J(t) = \frac{V_0}{R}\left[1 - \left(\cos \omega t + \frac{\alpha}{\omega} \sin \omega t\right) \exp(-\alpha t)\right],$$

where $$\alpha = \frac{1}{2RC}, \quad \omega^2 = \frac{1}{LC} - \frac{1}{4R^2C^2}.$$

11. Four points P, M, N, O are connected by branches containing resistances (R) and self-inductances (L) to form a network, all mutual inductances in which are negligible. The branch PM contains $2R$ and $2L$ in series. Each of PN and MN contains R and L in series, the latter containing also a battery of e.m.f. E and a switch S; OP, OM, ON each contains L only. Show that if S is closed at $t = 0$, the current through the battery at time t is

$$\frac{E}{4R}\left(4 - e^{-Rt/L} - e^{-Rt/3L} - 2e^{-Rt/4L}\right).$$

12. Each of two electric circuits of negligible resistance includes a condenser and they are tuned to the same period of oscillation τ when at a large distance apart. Prove that if they are brought into such a position that their coefficient of mutual induction is M, the system possesses two periods

$$\{\tau^2 \pm 4\pi^2 M \sqrt{(C_1C_2)}\}^{1/2},$$

where C_1 and C_2 are the capacities of the two condensers.

13. The connections between four terminals $ABCD$ are of negligible resistance and ar otherwise as follows. Between A and B there is a capacity C_1 and between B an C there is an inductance L_1. Between A and D there is a capacity C_2 and between D and C there is an inductance L_2. Between A and C there is a capacity C_3. The mutual inductance between L_1 and L_2 is negligible. Show that the periods of free oscillations of the circuit are given by $2\pi/\omega_1$ and $2\pi/\omega_2$, where ω_1 and ω_2 are the positive roots of the equation

$$\frac{1}{C_2C_3} + \frac{1}{C_3C_1} + \frac{1}{C_1C_2} - \omega^2\left[L_1\left(\frac{1}{C_2} + \frac{1}{C_3}\right) + L_2\left(\frac{1}{C_3} + \frac{1}{C_1}\right)\right] + L_1L_2\omega^4 = 0.$$

9.7 Quasi-steady currents

When a coil is rotated steadily in a constant magnetic field the magnetic flux through the coil varies sinusoidally and there is a corresponding sinusoidal e.m.f. in the coil. This arrangement is basically that of most large-scale generators of electricity. There are many advantages in the use of *alternating currents* (A.C.)—as such sinusoidally varying currents and voltages are called—for the distribution of electrical power; such currents also occur in radio and broadcasting. We cannot discuss here all the ramifications of electrical engineering, radio and communications; instead we confine our attention to the simple phenomena of circuits in which there are alternating currents and voltages.

The basic equation governing such circuits is (9.31), where, now, we take the electromotive force to be $\mathfrak{E} = E_0 \begin{smallmatrix}\sin\\ \cos\end{smallmatrix}(\omega t+\delta)$. The *frequency* of the alternations is $\omega/(2\pi)$ and the constant δ is the *phase angle*. (The terms used are mostly the same as those used in connection with any simple harmonic variations or oscillations.) When an alternating e.m.f. acts in a circuit governed by eqn. (9.31) the solution of the equation, which gives the current in the circuit, or the charge on the plate of a condenser, consists of the transient and the particular integral. The transient, for any circuit containing resistance, will die out and the particular integral will be a function which varies harmonically with the time with the frequency $\omega/(2\pi)$ and which persists after the transient has died out. In terms of the mechanical analogy, the particular integral represents the *forced vibration*. We can sum up the methods we introduce below as methods of finding the particular integrals, or forced vibrations, corresponding to the presence of alternating e.m.f.'s in the differential equations of circuits obtained from (9.31).

The method employed is based on the use of complex numbers. The number $z = z_0 e^{i\omega t}$ in an Argand diagram represents a complex number z_0 made to rotate in the diagram about the origin with frequency $\omega/(2\pi)$. If $\arg z_0 = \delta$, then the real and imaginary parts of z are

$$\operatorname{Re} z = |z_0| \cos(\omega t+\delta), \quad \operatorname{Im} z = |z_0| \sin(\omega t+\delta).$$

Hence, by the use of a suitable complex number z_0, any alternating quantity can be represented by a complex number $z = z_0 e^{i\omega t}$, the phase angle being given by $\arg z_0$ and the "peak value" being given by $|z_0|$ (see Fig. 9.22).

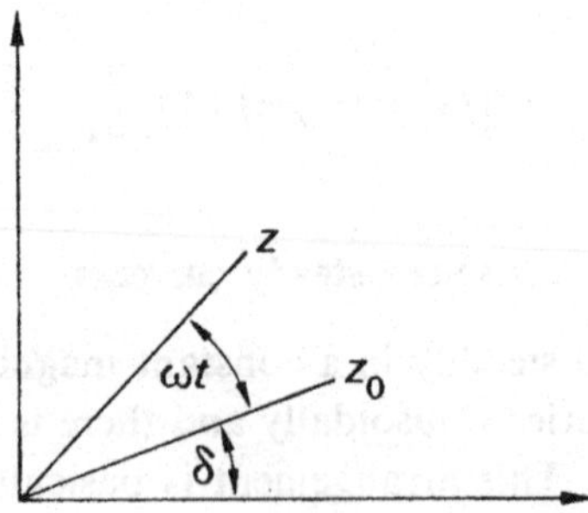

FIG. 9.22

Since the time variation of all alternating quantities is given by the factor $e^{i\omega t}$, the differential equations become algebraic equations between complex numbers and the time factor can be cancelled. This leaves one or more equations between complex numbers such as z_0 above, and these equations correspond to applications of the generalized forms of Kirchhoff's two laws.

To illustrate the method we consider the circuit of Fig. 9.10, p. 368, in which the constant e.m.f. is replaced by sinusoidal e.m.f. $\mathfrak{E} = E_0 e^{i\omega t}$. The differential equations of the circuit are

$$L\frac{dI}{dt}+RI = E_0 e^{i\omega t}-\frac{q}{C}, \quad I = \frac{dq}{dt}. \tag{9.34}$$

$$\therefore \quad L\frac{d^2I}{dt^2}+R\frac{dI}{dt}+\frac{I}{C} = i\omega E_0 e^{i\omega t}.$$

To find the forced oscillation we look for a solution $I = I_0 e^{i\omega t}$ and obtain

$$-\omega^2 L I_0 + i\omega R I_0 + (1/C) I_0 = i\omega E_0.$$

$$\therefore \quad I_0\left(i\omega L + R + \frac{1}{i\omega C}\right) = E_0. \tag{9.35}$$

We consider three special cases.

(a) *Only R is present.* In this case the terms $L(dI/dt)$, q/C are absent (i.e. $L = 0$, $C \to \infty$). Then (9.35) reduces to

$$I_0 R = E_0.$$

This is simply Ohm's law, as for a direct current, and we may regard it as expressing the equality of the "back e.m.f." across the resistance R and the applied e.m.f. E_0.

(b) *Only L is present.* In this case (9.35) becomes (with $R = 0$, $C \to \infty$)

$$I_0(i\omega L) = E_0. \tag{9.36}$$

This has the same form as Ohm's law but uses a "complex resistance" $i\omega L$ to give the "back e.m.f." across the inductance. Equation (9.36) gives

$$I_0 = E_0/(i\omega L) = E_0 e^{-i\pi/2}/(\omega L).$$

Hence the complex number I_0 is $\pi/2$ behind E_0 in phase; this means that the alternating current flowing through an inductance lags in phase by $\pi/2$ behind the alternating e.m.f. driving it through the inductance.

(c) *Only C is present.* In this case (9.36) becomes

$$\frac{I_0}{i\omega C} = E_0, \tag{9.37}$$

which again has the form of Ohm's law with a "complex resistance" $1/(i\omega C)$. The relation (9.37) also implies that when an alternating e.m.f. is applied to the plates of a condenser an alternating current flows through the condenser which leads the voltage in phase by $\pi/2$. Alternatively the "back e.m.f." across the terminals of a condenser carrying a current $I_0 e^{i\omega t}$ is $[I_0/(i\omega C)]e^{i\omega t}$. In considering quasi-steady currents we regard a condenser rather as a circuit element across which there is a "back e.m.f." when it carries a current than as a discontinuity in the potential.

We can now formulate our method using complex numbers. Instead of the term "resistance" we refer to the (complex) *impedance* Z of an element of a circuit; for a resistance $Z = R$, for an inductance $Z = i\omega L$, and for a capacitance $Z = 1/(i\omega C)$. When the (complex) impedance is multiplied by the complex current $I_0 e^{i\omega t}$, the product $ZI_0 e^{i\omega t}$ is the *reactance* of the element to the passage of the current. We now apply Kirchhoff's two laws as follows:

the *first law* requires the sum of the complex currents entering a node to be zero;

the *second law* requires that the sum of the reactances $\Sigma I_0 Z e^{i\omega t}$, corresponding to the sum ΣIR for direct currents, taken around any closed loop is equal to the applied complex e.m.f. $E_0 e^{i\omega t}$.

The solution of network problems with alternating current now follows exactly the same form as with direct currents. Since we use complex numbers each equation implies two equations between real numbers; either the real and imaginary parts are equated separately or the moduli and arguments of each side are identical. The use of the method will be clearer after a study of the following examples.

Example 1. We find expressions for the instantaneous values of the current and charge in the circuit of the example leading to (9.35) given that the applied e.m.f. is $E_0 \sin \omega t$.

When we take the e.m.f. as $E_0 e^{i\omega t}$ we must add the proviso, in this case, that the actual values are the imaginary parts of complex quantities since $E_0 \sin \omega t$ is the imaginary part of $E_0 e^{i\omega t}$.

We assume that all quantities vary with the time through the factor $e^{i\omega t}$.

$$\therefore \quad I_0 e^{i\omega t} = I = \frac{dq}{dt} = \frac{d}{dt}(q_0 e^{i\omega t}) = i\omega q_0 e^{i\omega t}.$$

$$\therefore \quad q_0 = I_0/(i\omega) = (I_0/\omega)e^{-i\pi/2}.$$

From eqn. (9.35), $I_0 = E_0/Z$,

where
$$Z = i\omega L + R + 1/(i\omega C) = R + i\{\omega L - 1/(\omega C)\} = Z_0 e^{i\alpha}$$
with
$$Z_0 = |Z| = [R^2 + \{\omega L - 1/(\omega C)\}^2]^{1/2},$$
$$\cos\alpha = R/Z_0, \qquad \sin\alpha = \{\omega L - 1/(\omega C)\}/Z_0.$$

Hence
$$I = I_0 e^{i\omega t} = (E_0/Z_0)e^{i(\omega t - \alpha)},$$

and the instantaneous value of the current is the imaginary part $(E_0/Z_0)\sin(\omega t - \alpha)$ showing that the voltage leads the current by a phase angle α.

Similarly

$$q = q_0 e^{i\omega t} = (I_0/\omega)\, e^{i(\omega t - \pi/2)} = \frac{E_0}{\omega Z_0}\, e^{i(\omega t - \alpha - \pi/2)}.$$

Hence the instantaneous value of the charge is

$$\frac{E_0}{\omega Z_0} \sin(\omega t - \alpha - \pi/2) = -\frac{E_0}{\omega Z_0} \cos(\omega t - \alpha).$$

Example 2. An electromotive force $E_0 \cos nt$ is applied to a circuit whose resistance is R and self-induction L. Prove that the periodic part of the current in the circuit is

$$\frac{E_0}{\varrho} \cos(nt - \theta),$$

where $\varrho^2 = R^2 + n^2L^2$, and $\tan\theta = nL/R$.

Prove that the proximity of a similar circuit (containing no source of electromotive force) effectively increases R to R' and decreases L to L', where

$$R' = R\left(1 + \frac{n^2M^2}{\varrho^2}\right), \quad L' = L\left(1 - \frac{n^2M^2}{\varrho^2}\right),$$

M being the coefficient of mutual induction, so that the current is now

$$\frac{E_0}{\varrho'} \cos(nt - \theta'),$$

where $\varrho'^2 = R'^2 + n^2L'^2$, and $\tan\theta' = nL'/R'$.

Prove also that the current in the second circuit is

$$\frac{nME_0}{\varrho\varrho'} \sin(nt - \theta - \theta').$$

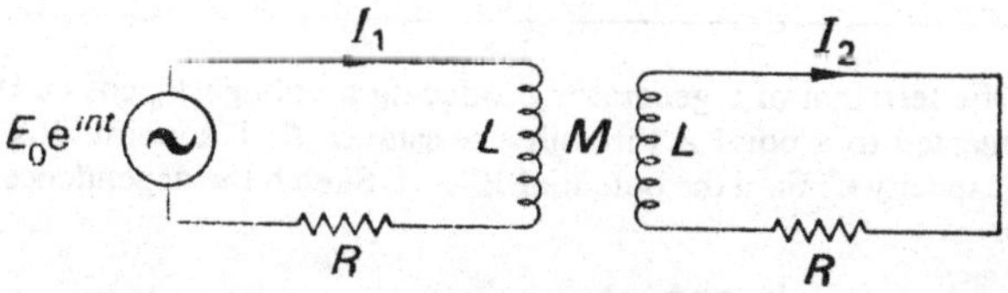

FIG. 9.23

We include this example to show how mutual inductance between two neighbouring circuits is included in the method of complex impedances, Fig. 9.23.

In the absence of the secondary circuit, the equation for the isolated primary is

$$(inL + R)I_0 = E_0, \tag{1}$$

i.e.

$$I_0 = \frac{E_0}{R + inL} = \frac{E_0}{\varrho e^{i\theta}},$$

where $\varrho^2 = R^2 + n^2L^2$, $\tan\theta = nL/R$, $0 \leqslant \theta \leqslant \pi/2$. Hence the periodic part of the current is obtained by taking real parts thus:

$$\text{Re}\,(E_0/\varrho)e^{i(nt-\theta)} = (E_0/\varrho)\cos(nt - \theta).$$

When the secondary circuit is present, the differential equations for the two circuits are

$$L\frac{dI_1}{dt}+M\frac{dI_2}{dt}+RI_1 = E_0e^{int},$$

$$M\frac{dI_1}{dt}+L\frac{dI_2}{dt}+RI_2 = 0.$$

To find the "forced oscillation" we look for solutions which vary with time through the factor e^{int}, viz. $I_1 = I_{10}e^{int}$, $I_2 = I_{20}e^{int}$. Then

$$inLI_{10}+inMI_{20}+RI_{10} = E_0, \quad (3)$$

$$inMI_{10}+inLI_{20}+RI_{20} = 0, \quad (4)$$

after cancelling the factor e^{int}. Elimination of I_{20} leads to

$$\left\{R\left(1+\frac{n^2M^2}{R^2+n^2L^2}\right)+inL\left(1-\frac{n^2M^2}{R^2+n^2L^2}\right)\right\}I_{10} = E_0,$$

i.e.

$$(R'+inL')I_{10} = E_0.$$

Comparison with the solution of (1) leads to the required results. From eqn. (4) we obtain

$$I_{20} = -\frac{inM}{R+inL}I_{10} = -\frac{inME_0}{(R+inL)(R'+inL')} = -\frac{inME_0}{\varrho\varrho' e^{i(\theta+\theta')}}.$$

The current flowing in the secondary circuit has the complex expression

$$I_2 = I_{20}e^{int} = -\frac{inME_0e^{i(nt-\theta-\theta')}}{\varrho\varrho'}.$$

The instantaneous value of the current is the real part of this, viz.,

$$\frac{nME_0\sin(nt-\theta-\theta')}{\varrho\varrho'}.$$

Example 3. One terminal of a generator producing a voltage $V_0 \cos \omega t$ is earthed and the other is connected to a point A through a resistance R. The point A is connected to earth through a capacity C. Find the potential W at A. Sketch the dependence of the amplitude of W on ω.

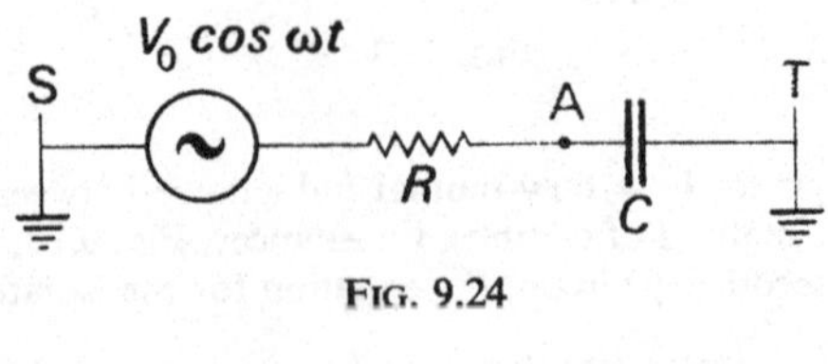

FIG. 9.24

The complex impedance of the capacitor is $1/(i\omega C)$. Therefore, if $I_0e^{i\omega t}$ is the complex current flowing from S to T, Fig. 9.24,

$$V_0 = RI_0+\frac{I_0}{i\omega C}.$$

$$\therefore\ I_0 = \frac{V_0 i\omega C}{1+i\omega CR}.$$

Corresponding to the flow of current from A to T across the capacitor, the complex voltage (complex current × complex impedance of condenser) at A is

$$W = \frac{I_0 e^{i\omega t}}{i\omega C} = \frac{V_0 e^{i\omega t}}{1+i\omega CR}.$$

$$\therefore \quad W = \frac{V_0 e^{i(\omega t-\delta)}}{\sqrt{(1+\omega^2C^2R^2)}},$$

where $\cos\delta = 1/\sqrt{(1+\omega^2C^2R^2)}$, $\sin\delta = \omega CR/\sqrt{(1+\omega^2C^2R^2)}$, i.e. $\tan\delta = \omega CR$ and $0<\delta<\pi/2$. It follows that (from real parts) the instantaneous value of the potential at A is

$$\{V_0\cos(\omega t-\delta)\}/\sqrt{(1+\omega^2C^2R^2)}.$$

The relevant graph of the amplitude, $V_0/\sqrt{(1+\omega^2C^2R^2)}$, of W against ω is shown in Fig. 9.25.

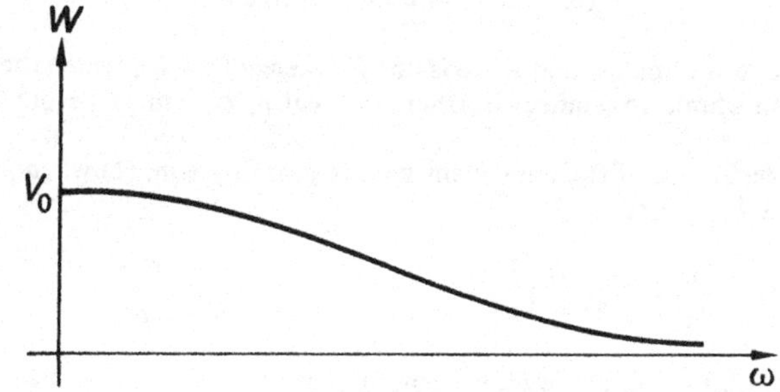

FIG. 9.25

Example 4. A condenser of capacity C is connected by leads of resistance r so as to be in parallel with a coil of self-induction L, the resistance of the coil and its leads being R. If this arrangement forms part of a circuit in which there is an e.m.f. of period $2\pi/p$, show that it can be replaced by a wire without self-inductance if

$$(R^2-L/C) = p^2LC(r^2-L/C),$$

and that the resistance of this equivalent wire must be

$$(Rr+L/C)/(R+r).$$

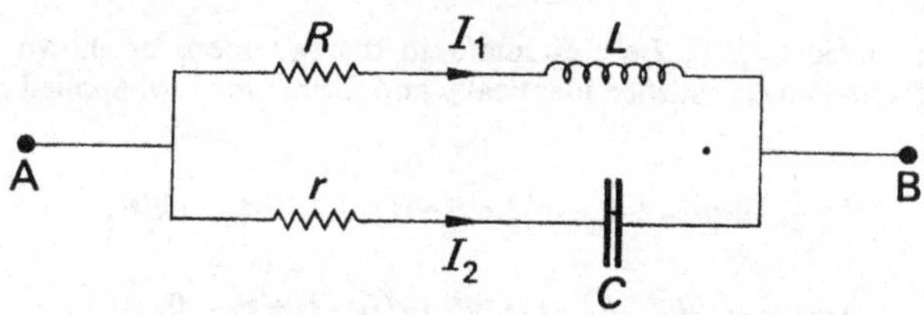

FIG. 9.26

The given arrangement is shown in Fig. 9.26 and the applied complex voltage is V_0e^{ipt}. The complex impedances in the upper and lower branches are respectively

$$R+ipL, \quad r+1/(ipC).$$

The complex impedance of the two branches together is Z, where

$$\frac{1}{Z} = \frac{1}{R+\mathrm{i}Lp}+\frac{1}{r+1/(\mathrm{i}Cp)},$$

i.e.

$$\frac{1}{Z} = \frac{R-\mathrm{i}Lp}{R^2+L^2p^2}+\frac{r+\mathrm{i}/(Cp)}{R^2+1/(Cp)^2}. \tag{1}$$

The arrangement can be replaced by a wire without self-inductance or capacitance (i.e. by a purely ohmic resistance) provided Z is real, i.e. provided $1/Z$ is real. This is true if the coefficient of i in (1) vanishes, i.e. if

$$\frac{1/(Cp)}{r^2+1/(Cp)^2}-\frac{Lp}{R^2+L^2p^2} = 0, \qquad (2$$

which reduces to

$$(R^2-L/C) = p^2LC(r^2-L/C). \tag{2a}$$

Note that this result implies that there is one frequency only for which the arrangement acts purely as an ohmic resistance. Further, for real p, L/C must lie outside the range r^2 to R^2.

The (ohmic) resistance of the equivalent wire is given by eqn. (1) when p satisfies eqn. (2) [or (2a)]. It is R_0, where

$$\frac{1}{R_0} = \frac{R}{(R^2+L^2p^2)}+\frac{r}{r^2+1/(Cp)^2} = \frac{R+rLCp^2}{R^2+L^2p^2}.$$

Then, using eqn. (2a) to eliminate p^2, we obtain

$$R_0 = (Rr+L/C)/(R+r).$$

Example 5. A circuit contains a condenser of capacity C_1, a resistance R_1, and has self-induction L_1. The corresponding quantities for a second circuit are C_2, R_2, L_2, and a part r of R_1, R_2 is common to both circuits. The mutual induction is assumed to be negligible. If the first circuit contains a periodic electromotive force $E \sin pt$, find the amplitude of the current in the second circuit.

If L_1, L_2 are adjustable, prove that this amplitude is greatest if $L_1 = 1/(C_1p^2)$, $L_2 = 1/(C_2p^2)$ and that its value is then $rE(R_1R_2-r^2)^{-1}$. Find also the current in the first circuit when these conditions are satisfied.

Let complex currents $I_1e^{\mathrm{i}pt}$, $I_2e^{\mathrm{i}pt}$ circulate in the two loops as shown in Fig. 9.27. Then Kirchhoff's first law is satisfied identically and the second law, applied to the circuits in turn, gives

$$\left(\frac{1}{\mathrm{i}pC_1}+\mathrm{i}pL_1+R_1-r\right)I_1e^{\mathrm{i}pt}+r(I_1-I_2)e^{\mathrm{i}pt} = Ee^{\mathrm{i}pt},$$

$$\left(\frac{1}{\mathrm{i}pC_2}+\mathrm{i}pL_2+R_2-r\right)I_2e^{\mathrm{i}pt}+r(I_2-I_1)e^{\mathrm{i}pt} = 0.$$

$$\therefore \quad \left(\frac{1}{\mathrm{i}pC_1}+\mathrm{i}pL_1+R_1\right)I_1-rI_2 = E, \tag{1}$$

$$-rI_1+\left(\frac{1}{\mathrm{i}pC_2}+\mathrm{i}pL_2+R_2\right)I_2 = 0. \tag{2}$$

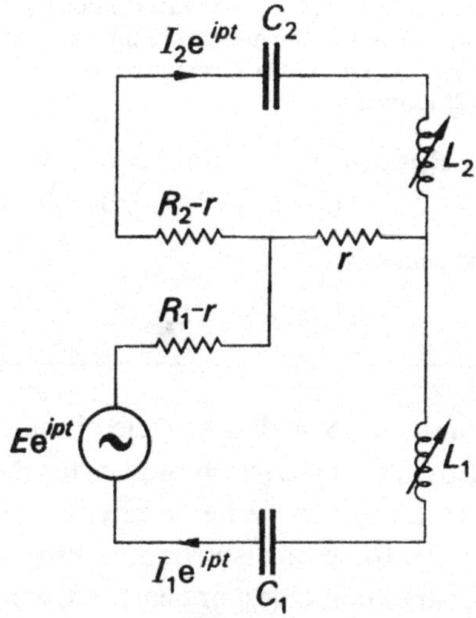

FIG. 9.27

Solving for I_2, we find

$$I_2 = \frac{Er}{\left(\frac{1}{ipC_1}+ipL_1+R_1\right)\left(\frac{1}{ipC_2}+ipL_2+R_2\right)-r^2}$$

$$= \frac{Er}{\left\{R_1+ip\left(L_1-\frac{1}{C_1p^2}\right)\right\}\left\{R_2+ip\left(L_2-\frac{1}{C_2p^2}\right)\right\}-r^2}$$

$$= \frac{Er}{\left\{R_1R_2-r^2-p^2\left(L_1-\frac{1}{C_1p^2}\right)\left(L_2-\frac{1}{C_2p^2}\right)+ipR_1\left(L_2-\frac{1}{C_2p^2}\right)+ipR_2\left(L_1-\frac{1}{C_1p^2}\right)\right\}}.$$

$$\therefore \quad |I_2| = Er/|P|,$$

where

$$P^2 = \left\{p^2\left(L_1-\frac{1}{C_1p^2}\right)\left(L_2-\frac{1}{C_2p^2}\right)+r^2-R_1R_2\right\}^2+p^2\left\{R_1\left(L_2-\frac{1}{C_2p^2}\right)+R_2\left(L_1-\frac{1}{C_1p^2}\right)\right\}^2$$

and $|I_2|$ gives the amplitude of the current in the second circuit.

If L_1, L_2 are adjustable (this is symbolized by the arrows in Fig. 9.27), all other quantities being fixed, $|I_2|$ is greatest when P^2 is least. But $\partial P^2/\partial L_1 = 0 = \partial P^2/\partial L_2$ when

$$L_1 = \frac{1}{C_1p^2}, \quad L_2 = \frac{1}{C_2p^2}$$

and in this case

$$\frac{\partial^2 P^2}{\partial L_1^2} = 2p^2R_1^2, \quad \frac{\partial^2 P^2}{\partial L_2^2} = 2p^2R_2^2,$$

$$\frac{\partial^2 P^2}{\partial L_1\,\partial L_2} = 2p^2(r^2-R_1R_2)+2p^2R_1R_2 = 2p^2r^2.$$

$$\therefore \quad \left(\frac{\partial^2 P^2}{\partial L_1\,\partial L_2}\right)^2-\frac{\partial^2 P^2}{\partial L_1^2}\frac{\partial^2 P^2}{\partial L_2^2} = 4p^4(r^4-R_1^2R_2^2) < 0$$

since $r < R_1, R_2$. Therefore P is least and $|I_2|$ is greatest when L_1, L_2 take the stated values. The least value of P is $R_1R_2 - r^2$ and so the greatest amplitude of the current in the second circuit is $rE/(R_1R_2 - r^2)$ when the impedances are purely resistive.

In this case eqns. (1), (2) reduce to

$$R_1I_1 - rI_2 = E, \qquad -rI_1 + R_2I_2 = 0.$$

$$\therefore \quad I_1 = ER_2/(R_1R_2 - r^2)$$

and the current in the first circuit is

$$(ER_2 \sin pt)/(R_1R_2 - r^2).$$

We saw in the previous section that electric circuits and networks have frequencies of (free) oscillation corresponding to the normal modes of oscillation of a mechanical system. When a mechanical system is subjected to the action of a periodic force or disturbance, resonance occurs when the frequency of the disturbance coincides, or nearly so, with a natural frequency of the system. The discussion of the following two circuits illustrates the electrical case.

Example 1. Series circuit resonance

The impedance between A and B in the circuit of Fig. 9.28 is

$$Z = R + \mathrm{i}\left(\omega L - \frac{1}{\omega C}\right) = Z_0 e^{\mathrm{i}\alpha},$$

where $Z_0^2 = R^2 + \{\omega L - 1/(\omega C)\}^2$. Hence Z_0 has a minimum value, equal to R, when $\omega^2 = \omega_0^2 = 1/(LC)$.

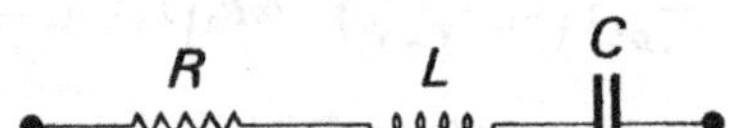

FIG. 9.28

At this frequency an e.m.f. $E_0 e^{\mathrm{i}\omega t}$ applied between A and B produces a current of larger amplitude than for any other frequency. The voltages across L and C in this case are

$$V_L = \mathrm{i}\omega_0 L \frac{E_0}{Z_0} = \mathrm{i}\sqrt{\left(\frac{L}{C}\right)}\frac{E_0}{R}, \quad V_C = \frac{1}{\mathrm{i}\omega_0 C}\frac{E_0}{Z_0} = -\mathrm{i}\sqrt{\left(\frac{L}{C}\right)}\frac{E_0}{R}. \tag{1}$$

A graph of Z_0 against ω shows a peak for $\omega = \omega_0$ and the smaller the value of R the sharper is this peak. Equation (1) shows that, when R is small, the voltages across L and C are equal and opposite but are large, and differ in phase by $\pm\pi/2$ from the applied voltage E_0. The large amplitude of the voltage V_L or V_C for the frequency $\omega_0/(2\pi)$ is an example of resonance. This resonant frequency is also the natural frequency of oscillation when a capacitance C is connected to an inductance L; the resistance R damps the motion but the smaller the value of R the sharper is the resonance.

Example 2. Parallel circuit resonance

The impedance Z between A and B for the circuit of Fig. 9.29 is given by

$$\frac{1}{Z} = i\omega C + \frac{1}{R + i\omega L}.$$

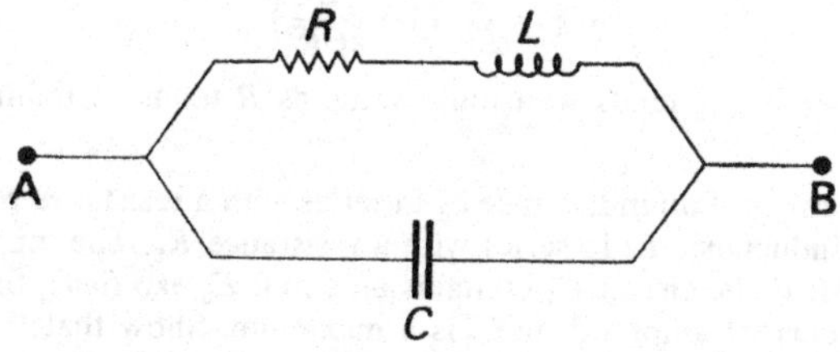

FIG. 9.29

We look more closely at this for small values of R. Then

$$\frac{1}{Z} = i\omega C + \frac{1}{i\omega L}\left(1 - \frac{iR}{\omega L}\right)^{-1} \approx i\left(\omega C - \frac{1}{\omega L}\right) + \frac{R}{\omega^2 L^2}.$$

Here again if $\omega^2 = \omega_0^2 = 1/(LC)$, Z is real and, since R is small, $|Z|$ is large; in fact, $|Z| \to \infty$ as $R \to 0$.

Therefore the circuit between A and B *obstructs* the passage of a current of the resonant frequency. The relation between Z_0 and ω is more complicated than in the previous example, but the effect is that of a filter, the nearer the frequency is to $\omega_0/(2\pi)$ the greater is Z_0. [If R is not small the maximum value of Z_0 does not occur exactly at the frequency $\omega_0/(2\pi)$, but at a value close to this.]

Exercises 9.7

1. Prove that the impedance of the system shown in Fig. Ex. 9.7.1 is purely resistive if the frequency is

$$\frac{1}{2\pi}\left(\frac{L_1 + L_2}{CL_1L_2} - \frac{1}{C^2R^2}\right)^{1/2}.$$

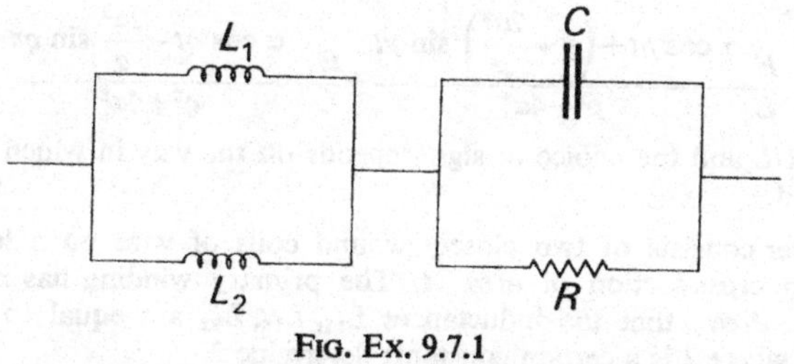

FIG. EX. 9.7.1

2. A potential difference $E_0 \cos pt$ is maintained between two points A and B which are connected through two circuits in parallel. The first consists of a condenser of capacity C and leads of total resistance r, the second of a coil of self-inductance L and resistance R. Show that, if $R = r$ and $LCp^2 = 1$, the currents in the two circuits are equal in amplitude and opposite in phase.

3. A condenser of capacity C is connected through a self-inductance L with a motor of e.m.f. $E \cos \omega t$ and negligible resistance. The plates of the condenser are also connected by a wire of resistance R and no induction. Show that, if $2CR^2 > L$, the value ω_0 of ω for which the amplitude, j_0, of j, the current in R, attains its maximum value is given by

$$\omega_0^2 = \frac{1}{LC}\left(1-\frac{L}{2CR^2}\right).$$

Show that, if $\omega = \omega_0$, j_0 tends to a finite value as R tends to infinity, and interpret this result.

4. A circuit C_1 consists of an inductance L_1 in series with a resistance R_1 and a circuit C_2 consists of an inductance L_2 in series with a resistance R_2. The mutual inductance of C_1 and C_2 is M. If the circuit C_1 contains an e.m.f. $E_0 \exp(i\omega t)$, find the value of ω for which the current amplitude in C_2 is a maximum. Show that this current cannot be greater than $E_0/2(R_1R_2)^{1/2}$ for any admissible values of L_1, L_2 and M.

5. A network contains three branches connected in parallel: the first contains two condensers in series of capacity C_1 and C_2 respectively, the second two inductances L_1 and L_2 in series, and the third contains a generator of alternating e.m.f. A point A between the condensers is connected to a point B between the inductances by a branch of resistance R. Show that the currents in the inductances can be in phase only if these currents are identical, which will occur when $C_1L_1 = C_2L_2$.

6. Terminals A and B are connected by a coil of inductance L while B is connected to a terminal D by a condenser of capacitance C. Wires each of resistance R connect A, B and D to a terminal F. An electromotive force $E_0 \cos(\omega t)$ is applied between A and D. Prove that the amplitude of the current flowing in BF is

$$E_0(1+\omega^2LC)/\sqrt{\{4\omega^2L^2+9R^2(1-\omega^2LC)^2\}},$$

and show that the difference in phase between this current and the applied electromotive force is

$$\cot^{-1}\left\{\frac{3}{2}R\left(\omega C-\frac{1}{\omega L}\right)\right\}.$$

7. A circuit consists of an e.m.f. $E \cos pt$, resistance R and inductance L. Another e.m.f. $E \cos qt$ and an inductance L in series with it are connected in parallel across the resistance R. There is no mutual inductance between the two parts of the circuit. Show that the current through the e.m.f. of period $2\pi/p$ is

$$\frac{E}{L}\,\frac{\alpha \cos pt+\left(p+\dfrac{2\alpha^2}{p}\right)\sin pt}{p^2+4\alpha^2} \pm \frac{E}{L}\cdot\frac{\alpha \cos qt-\dfrac{2\alpha^2}{q}\sin qt}{q^2+4\alpha^2}$$

where $\alpha = R/L$ and the choice of sign depends on the way in which the two e.m.f.s are connected.

8. A transformer consists of two closely wound coils of wire on a long cylinder of length l and cross-section of area A. The primary winding has m turns and the secondary n. Prove that the inductances L_{11}, L_{12}, L_{22} are equal to λm^2, λmn, λn^2 respectively, where λ is a certain constant; determine λ.

If an e.m.f. $\mathcal{E} \cos \omega t$ is applied to the primary coil, and a load consisting of a pure resistance R is connected to the secondary, determine the primary and secondary currents, neglecting the transient currents and the resistances in the transformer. Verify that the mean rate of working of the applied e.m.f. and the mean rate of dissipation of energy in the load are equal. Explain (without making any calculations) why the instantaneous rates are in general unequal.

9. A closed circuit *ABDFGA* consists of two resistances *AB, BD* each of magnitude R, a coil *DF* of self-inductance L, a condenser *FG* of capacity C and a coil *GA* of self-inductance L. The mutual inductance of the two coils is M. An alternating e.m.f. of period $2\pi/p$ is maintained across *BF*. Show that there is no value of p for which the phase difference between the currents in the two branches is either 0 or π but that there is one value of p for which this difference is $\pi/2$ if $R^2C < L-M$.

10. A circuit of self-inductance L_1 contains a condenser of capacity C_1 and a resistance R_1 in series. The corresponding quantities for a second circuit are L_2, C_2, R_2. The first circuit contains also a periodic electromotive force $E\cos\omega t$ and the mutual inductance is M. Show that the current j_2 in the second circuit is the real part of

$$\frac{-Z_{12}Ee^{i\omega t}}{Z_{11}Z_{22}-Z_{12}^2},$$

where $$Z_{11} = R_1+iX_1, \quad Z_{12} = i\omega M, \quad Z_{22} = R_2+iX_2$$

and $$X_k = \omega L_k - \frac{1}{\omega C_k} \qquad (k = 1, 2).$$

If the reactance X_2 of the second circuit is adjustable and the other quantities are fixed, show that the value of X_2 that gives the greatest amplitude for j_2 is

$$\frac{\omega^2M^2X_1}{R_1^2+X_1^2}.$$

11. An e.m.f. equal to the real part of $\mathcal{E}e^{i\omega t}$ is applied to a resistor, an inductor and a capacitor connected in series. Show that the current is equal to the real part of $\mathcal{E}e^{i\omega t}/Z$, where $Z = R+i[L\omega-1/(C\omega)]$.

In a transmission line $A_0A_1A_2 \ldots A_{n+1}$ each of the elements $A_1A_2, \ldots, A_{n-1}A_n$ has inductance L, and each of the junctions $A_1, A_2, \ldots, A_n$ is connected to earth through a capacitor of capacitance C. Resistance is negligible, and the impedance of the elements A_0A_1 and A_nA_{n+1} is zero. A (complex) alternating current x_0 is fed in at the terminal A_0, and x_r is the current along A_rA_{r+1}; show that, provided $LC\omega^2 < 4$,

$$x_r-2x_{r+1}\cos\gamma+x_{r+2} = 0,$$

where $$\cos\gamma = 1-\tfrac{1}{2}LC\omega^2.$$

Show that if the terminal A_{n+1} is connected to earth through a receiver of resistance R and inductance $\frac{1}{2}L$, and if $RC\omega = \pm\sin\gamma$, then the current x_n through the receiver is $x_0e^{\mp in\gamma}$.

9.8 Magnetic energy and forces

In § 6.2 we introduced the "work-function", $W = -IN$, from which we can obtain the force $X = -\partial W/\partial x$ tending to increase a coordinate x fixing the position of a circuit. We used the name "work-function" because the differentiation $\partial W/\partial x$ had to be made subject to the proviso that the currents remain constant during an arbitrary displacement δx. The relation $X = -\partial V/\partial x$ for the force acting in a system whose potential energy is V is obtained from considerations of conservation of energy; but the proviso of constant currents requires the expenditure of an (unknown) amount of energy during the displacement δx. We are now in a position to take account of this energy.

The law of conservation of energy, or the First Law of Thermodynamics, is usually expressed in the form

$$dU + dW = dQ.$$

Here dU stands for the change in the internal energy of the system, dW for the mechanical work done *by* the system, and dQ for the energy imparted to the system by external agents, during an arbitrary displacement.

Whereas in the electrostatic field the procedure of "collecting charges from infinite dispersion" does represent a theoretical process similar to the physical process of charging a system, "collecting current filaments from infinite dispersion" bears no resemblance to the process of "switching on" currents and so establishing a magnetic field. In the process of "switching on" the currents are allowed to increase from zero to their final values; the circuits remain throughout in their final positions. Hence for this process $dW = 0$, and dQ represents the energy supplied from the batteries which maintain the currents. In the following, instead of the general symbol U, we shall use T to denote the (internal) energy of the magnetic field.

At any stage of the "switching on" process the current and electric and magnetic fields are related by

$$\mathbf{J} = \sigma(\mathbf{E}+\mathbf{E}'), \quad \mathbf{curl}\, \mathbf{E} = -\frac{\partial \mathbf{B}}{\partial t} \quad \text{or} \quad \mathbf{E} = -\mathbf{grad}\, V - \frac{\partial \mathbf{A}}{\partial t}.$$

The rate at which the batteries, represented by the field $\mathbf{E}'$, are expending energy is

$$\mathbf{E}' \cdot \mathbf{J} = J^2/\sigma - \mathbf{J} \cdot \mathbf{E}.$$

Hence we conclude that, apart from the Joule heat J^2/σ dissipated in the resistances, the rate of supply of energy to the magnetic field is $-\iiint \mathbf{J} \cdot \mathbf{E}\, d\tau$. Hence, since $dW = 0$ in the equation of conservation of energy,

$$\frac{dT}{dt} = -\iiint \mathbf{J} \cdot \mathbf{E}\, d\tau = \iiint \mathbf{J} \cdot \left(\mathbf{grad}\, V + \frac{\partial \mathbf{A}}{\partial t}\right) d\tau$$

$$= \iiint \left\{ \operatorname{div}(V\mathbf{J}) - V \operatorname{div} \mathbf{J} + \mathbf{J} \cdot \frac{\partial \mathbf{A}}{\partial t} \right\} d\tau.$$

The current satisfies the condition $\operatorname{div} \mathbf{J} = 0$ and the integral of $\operatorname{div}(V\mathbf{J})$ becomes a surface integral over the boundaries of the system and must vanish because of the boundary conditions upon $\mathbf{J}$. Hence we obtain

$$\frac{dT}{dt} = \iiint \mathbf{J} \cdot \frac{\partial \mathbf{A}}{\partial t}\, d\tau. \tag{9.38}$$

More generally, we can say that whenever a change takes place in the system the batteries supply an amount of energy

$$\delta Q = \iiint \boldsymbol{J}\boldsymbol{\cdot}\delta\boldsymbol{A}\,\mathrm{d}\tau. \tag{9.39}$$

We have assumed a continuous distribution of current. If there are surface and linear currents in the system, then

$$\frac{\mathrm{d}T}{\mathrm{d}t} = \iiint \boldsymbol{J}\boldsymbol{\cdot}\frac{\partial \boldsymbol{A}}{\partial t}\,\mathrm{d}\tau + \iint \boldsymbol{K}\boldsymbol{\cdot}\frac{\partial \boldsymbol{A}}{\partial t}\,\mathrm{d}S + \Sigma I \oint \frac{\partial \boldsymbol{A}}{\partial t}\boldsymbol{\cdot}\mathbf{d}\boldsymbol{s} \tag{9.38a}$$

and

$$\delta Q = \iiint \boldsymbol{J}\boldsymbol{\cdot}\delta\boldsymbol{A}\,\mathrm{d}\tau + \iint \boldsymbol{K}\boldsymbol{\cdot}\delta\boldsymbol{A}\,\mathrm{d}S + \Sigma I \oint \delta\boldsymbol{A}\boldsymbol{\cdot}\mathbf{d}\boldsymbol{s}. \tag{9.39a}$$

We can obtain the integral of (9.38) or (9.38a) for the case of a field in free space. Since **curl** $\boldsymbol{H} = \boldsymbol{J}$,

$$\begin{aligned}\iiint \boldsymbol{J}\boldsymbol{\cdot}\frac{\partial \boldsymbol{A}}{\partial t}\,\mathrm{d}\tau &= \iiint \frac{\partial \boldsymbol{A}}{\partial t}\boldsymbol{\cdot}\mathbf{curl}\,\boldsymbol{H}\,\mathrm{d}\tau \\ &= \iiint \boldsymbol{H}\boldsymbol{\cdot}\mathbf{curl}\,\frac{\partial \boldsymbol{A}}{\partial t}\,\mathrm{d}\tau - \iiint \mathrm{div}\left(\frac{\partial \boldsymbol{A}}{\partial t}\times\boldsymbol{H}\right)\mathrm{d}\tau \\ &= \iiint \boldsymbol{H}\boldsymbol{\cdot}\frac{\partial \boldsymbol{B}}{\partial t}\,\mathrm{d}\tau - \iiint \left(\frac{\partial \boldsymbol{A}}{\partial t}\times\boldsymbol{H}\right)\boldsymbol{\cdot}\hat{\boldsymbol{\nu}}\,\mathrm{d}S,\end{aligned}$$

where the surface integral is taken over the boundaries of the field. If there are surface currents or linear currents present these are excluded from the region of integration by enveloping surfaces which, in the limit as they coincide with the singularities, give (see Example 1 below)

$$\iiint \boldsymbol{J}\boldsymbol{\cdot}\frac{\partial \boldsymbol{A}}{\partial t}\,\mathrm{d}\tau = \iiint \boldsymbol{H}\boldsymbol{\cdot}\frac{\partial \boldsymbol{B}}{\partial t}\,\mathrm{d}\tau - \iint \boldsymbol{K}\boldsymbol{\cdot}\frac{\partial \boldsymbol{A}}{\partial t}\,\mathrm{d}S - \Sigma I \oint \frac{\partial \boldsymbol{A}}{\partial t}\boldsymbol{\cdot}\mathbf{d}\boldsymbol{s}.$$

$$\therefore\quad \frac{\mathrm{d}T}{\mathrm{d}t} = \iiint \boldsymbol{H}\boldsymbol{\cdot}\frac{\partial \boldsymbol{B}}{\partial t}\,\mathrm{d}\tau = \frac{\mathrm{d}}{\mathrm{d}t}\iiint \frac{\boldsymbol{B}^2}{2\mu_0}\,\mathrm{d}\tau.$$

Since the field is zero when the currents are zero we integrate this last equation to give

$$T = \iiint \frac{\boldsymbol{B}^2}{2\mu_0}\,\mathrm{d}\tau. \tag{9.40}$$

We can also proceed from (9.38) or (9.38a) thus.

The vector potential at any arbitrary point depends linearly on the currents, i.e.

$$A = \frac{\mu_0}{4\pi}\iiint \frac{J'\,d\tau'}{r} + \frac{\mu_0}{4\pi}\iint \frac{K'\,dS'}{r} + \frac{\mu_0}{4\pi}\Sigma I' \oint \frac{\mathbf{ds}'}{r},$$

where r is the distance between the element of integration $d\tau'$, dS or $\mathbf{ds}'$ to the point of evaluation of A. Hence

$$\frac{dT}{dt} = \frac{\mu_0}{4\pi}\iiint d\tau \iiint \frac{J}{r}\cdot\frac{\partial J'}{\partial t}\,d\tau' + \frac{\mu_0}{4\pi}\iiint d\tau \iint \frac{J}{r}\cdot\frac{\partial K'}{dt}\,dS' + \ldots$$

If we invert the order of integration w.r. to the primed and unprimed elements of integration, we obtain

$$\frac{dT}{dt} = \iiint A'\cdot\frac{\partial J'}{\partial t}\,d\tau' + \iint A'\cdot\frac{\partial K'}{\partial t}\,dS' + \sum\cdot\frac{dI'}{dt}\oint A'\cdot\mathbf{ds}'. \quad (9.41)$$

Since the integration in (9.41) is taken over the same region as that of (9.38a) we write

$$\frac{dT}{dt} = \frac{1}{2}\iiint\left\{A\cdot\frac{\partial J}{\partial t} + J\cdot\frac{\partial A}{\partial t}\right\}d\tau + \frac{1}{2}\iint\left\{A\cdot\frac{\partial K}{\partial t} + K\cdot\frac{\partial A}{\partial t}\right\}dS$$
$$+\frac{1}{2}\Sigma\oint\left\{\frac{dI}{dt}A + I\frac{\partial A}{\partial t}\right\}\cdot\mathbf{ds}.$$

We can integrate this to obtain

$$T = \tfrac{1}{2}\iiint A\cdot J\,d\tau + \tfrac{1}{2}\iint A\cdot K\,dS + \tfrac{1}{2}\sum I\oint A\cdot\mathbf{ds}. \quad (9.42)$$

This expression corresponds to (3.24), (3.25) for the electrostatic field. We also interpret (9.40) as indicating that the energy is distributed in the field with a density $B^2/(2\mu_0)$.

Example 1. We give the details of the reduction of the integral $\iiint J\cdot(\partial A/\partial t)\,d\tau$ when surface and linear currents are present.

We surround all singularities by a surface S_0, which will eventually tend to infinity; we enclose surface currents in envelopes such as L, and we enclose linear currents in tubes such as F, see Fig. 9.30. Then we consider

$$\iint\left(\frac{\partial A}{\partial t}\times H\right)\cdot\hat{n}\,dS$$

over the boundary of the space between S_0, L, F. On S_0 the standard boundary conditions $|A| = O(1/R)$, $|H| = O(1/R^2)$, imply that the integral tends to zero.

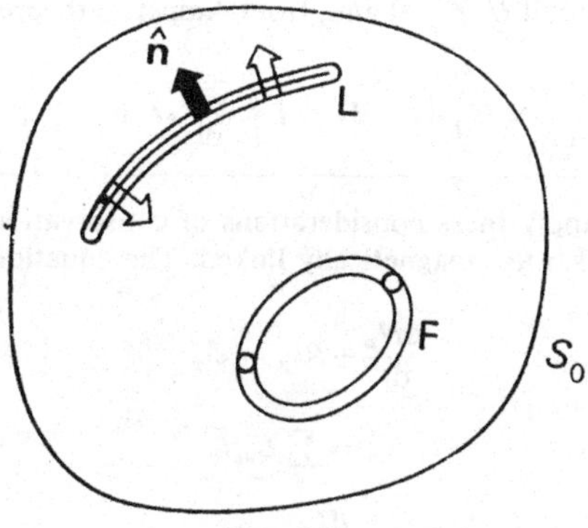

FIG. 9.30

On L the direction of $\hat{\boldsymbol{\nu}}$ is that of $\hat{\boldsymbol{n}}$ on the negative side and of $-\hat{\boldsymbol{n}}$ on the positive side Since $\boldsymbol{A}$ is continuous across a surface current $\boldsymbol{K}$ (see p. 217), the integral becomes

$$\iint_K \frac{\partial \boldsymbol{A}}{\partial t}\cdot\{(\boldsymbol{H}_1-\boldsymbol{H}_0)\times\hat{\boldsymbol{n}}\}\,\mathrm{d}S = -\iint_K \frac{\partial \boldsymbol{A}}{\partial t}\cdot\boldsymbol{K}\,\mathrm{d}S$$

by virtue of the continuity condition on $\boldsymbol{H}$.

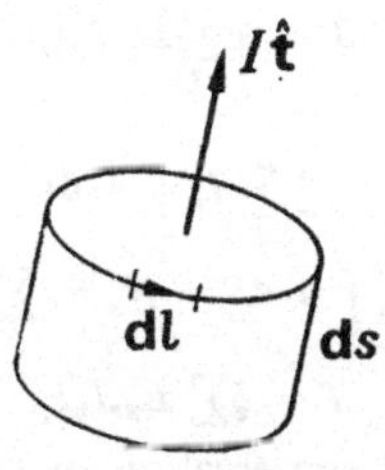

FIG. 9.31

We consider an element of F of length ds, see Fig. 9.31, and denote the direction of the linear current at this point by the unit vector $\hat{\boldsymbol{\imath}}$. We also take $\mathbf{d}\boldsymbol{l}$ to be an element of arc of the perimeter of the cross-section of F. The element of surface area, directed out of the volume between S_0, L, F, is

$$\hat{\boldsymbol{\nu}}\,\mathrm{d}S = (\hat{\boldsymbol{\imath}}\times\mathbf{d}\boldsymbol{l})\,\mathrm{d}s.$$

$$\therefore\quad \iint_F \left(\frac{\partial \boldsymbol{A}}{\partial t}\times\boldsymbol{H}\cdot\hat{\boldsymbol{\nu}}\right)\mathrm{d}S = \oint \mathrm{d}s \oint \left(\frac{\partial \boldsymbol{A}}{\partial t}\times\boldsymbol{H}\right)\cdot(\hat{\boldsymbol{\imath}}\times\mathbf{d}\boldsymbol{l}),$$

where $\oint \ldots \mathrm{d}s$ is taken around the linear circuit and $\oint \ldots \cdot\mathbf{d}\boldsymbol{l}$ is taken around the perimeter of a cross-section of F. But since F has a very small cross-section we regard $\partial\boldsymbol{A}/\partial t$ as constant for all points on the perimeter and so

$$\oint\left(\frac{\partial \boldsymbol{A}}{\partial t}\times\boldsymbol{H}\right)\cdot(\hat{\boldsymbol{\imath}}\times\mathbf{d}\boldsymbol{l}) = \frac{\partial \boldsymbol{A}}{\partial t}\cdot\oint \boldsymbol{H}\times(\hat{\boldsymbol{\imath}}\times\mathbf{d}\boldsymbol{l}) = \frac{\partial \boldsymbol{A}}{\partial t}\cdot\oint\{\hat{\boldsymbol{\imath}}(\boldsymbol{H}\cdot\mathbf{d}\boldsymbol{l})-\mathbf{d}\boldsymbol{l}(\boldsymbol{H}\cdot\hat{\boldsymbol{\imath}})\}.$$

But, close to the linear circuit $\boldsymbol{H}\cdot\hat{\boldsymbol{t}} = 0$ and, from Ampère's theorem, $\oint \boldsymbol{H}\cdot\mathbf{d}\boldsymbol{l} = I$. Therefore

$$\oint \mathrm{d}s \oint \left(\frac{\partial \boldsymbol{A}}{\partial t}\times\boldsymbol{H}\right)\cdot(\hat{\boldsymbol{t}}\times\mathbf{d}\boldsymbol{l}) = I\oint \frac{\partial \boldsymbol{A}}{\partial t}\cdot\hat{\boldsymbol{t}}\,\mathrm{d}s = I\oint \frac{\partial \boldsymbol{A}}{\partial t}\cdot\mathbf{ds}.$$

Example 2. We can apply these considerations of conservation of energy direct to a set of linear circuits which are magnetically linked. The equation for circuit p is, from eqn. (9.31),

$$\frac{\mathrm{d}N_p}{\mathrm{d}t}+RI_p = \mathscr{E}_p$$

where

$$N_p = \sum_q L_{pq}I_q.$$

$$\therefore\ \sum L_{pq}\frac{\mathrm{d}I_q}{\mathrm{d}t}+RI_p = \mathscr{E}_p.$$

We multiply each of these equations by I_p and add to obtain

$$\sum_{pq} L_{pq}\frac{\mathrm{d}I_q}{\mathrm{d}t}I_p+\sum_p RI_p^2 = \sum_p \mathscr{E}_pI_p.$$

Now $\sum \mathscr{E}_pI_p$ is the rate of supply of energy to the system from the batteries and $\sum_p RI_p^2$ is the rate of dissipation of heat and so we deduce that

$$\frac{\mathrm{d}T}{\mathrm{d}t} = \sum_{pq} L_{pq}\frac{\mathrm{d}I_q}{\mathrm{d}t}I_p = \sum_{pq} L_{pq}\frac{\mathrm{d}I_p}{\mathrm{d}t}I_q = \frac{1}{2}\frac{\mathrm{d}}{\mathrm{d}t}\sum_{pq} L_{pq}I_pI_q.$$

$$\therefore\ T = \tfrac{1}{2}\sum_{pq} L_{pq}I_pI_q.$$

But from eqn. (9.42) we see that

$$T = \tfrac{1}{2}\sum_p I_p\oint \boldsymbol{A}\cdot\mathbf{ds}.$$

But

$$\oint \boldsymbol{A}\cdot\mathbf{ds} = \iint \boldsymbol{B}\cdot\mathbf{dS} = N_p = \sum_q L_{pq}I_q.$$

$$\therefore\ T = \tfrac{1}{2}\sum L_{pq}I_pI_q.$$

This shows that the direct use of conservation of energy agrees with the result in eqn. (9.42).

We consider now the expression for the force X exerted by the field tending to increase the coordinate x, where x is a typical (generalized) coordinate specifying the positions of the conductors, etc., in the system. The conservation of energy becomes

$$\delta T+\sum X\,\delta x = \delta Q$$

for any change. We consider, in order to avoid complications, a system containing only volume currents, and make a displacement δx, *keeping the current constant*. From eqns. (9.39), (9.42)

$$\delta Q = \iiint \boldsymbol{J}\cdot\boldsymbol{\delta A}\,\mathrm{d}\tau, \quad \delta T = \tfrac{1}{2}\iiint \boldsymbol{J}\cdot\boldsymbol{\delta A}\,\mathrm{d}\tau.$$

$$\therefore\ \delta T+\sum X\,\delta x = 2\,\delta T, \quad \text{i.e.} \quad X = \left(\frac{\partial T}{\partial x}\right)_I.$$

This confirms our use of the force function and the expression $X = -\partial W/\partial x$ with $W = -T$.

If magnetization of the medium can occur, either permanent $\boldsymbol{M}$, or induced $\boldsymbol{M}_1$, we can adopt the analysis leading to eqn. (9.40) with very little modification. When permanent magnetization is introduced into a field we must regard this as "transported from infinity" and not "switched on". Hence if the permanent magnetization of a volume element $d\tau$ is increased by $\boldsymbol{\delta M}$, then the energy required for this is $-\iiint \boldsymbol{B}\cdot\boldsymbol{\delta M}\, d\tau$. Since the equations $\boldsymbol{J}=\sigma(\boldsymbol{E}+\boldsymbol{E}')$ and $\textbf{curl}\, \boldsymbol{E} = -\partial \boldsymbol{B}/\partial t$ must apply in the presence of magnetization just as in free space, the equation of conservation of energy becomes

$$\delta T = \iiint \boldsymbol{J}\cdot\boldsymbol{\delta A}\, d\tau - \iiint \boldsymbol{B}\cdot\boldsymbol{\delta M}\, d\tau.$$

Here the change in the magnetic field arising from the introduction of $\boldsymbol{\delta M}$ over and above any increment in the currents is included in the increment $\boldsymbol{\delta A}$. But the equation $\textbf{curl}\, \boldsymbol{H} = \boldsymbol{J}$ also holds in the presence of magnetization, so we deduce as before

$$\delta T = \iiint (\boldsymbol{H}\cdot\boldsymbol{\delta B} - \boldsymbol{B}\cdot\boldsymbol{\delta M})\, d\tau. \tag{9.43}$$

The general relation for $\boldsymbol{H}$ when both permanent and induced magnetization may be present is $\boldsymbol{H} = \boldsymbol{B}/\mu_0 - \boldsymbol{M}_1 - \boldsymbol{M}$.

$$\begin{aligned} \therefore\ \boldsymbol{H}\cdot\boldsymbol{\delta B} - \boldsymbol{B}\cdot\boldsymbol{\delta M} &= \boldsymbol{\delta B}\cdot(\boldsymbol{B}/\mu_0 - \boldsymbol{M}_1 - \boldsymbol{M}) - \boldsymbol{B}\cdot\boldsymbol{\delta M} \\ &= \delta\{B^2/(2\mu_0) - \boldsymbol{B}\cdot\boldsymbol{M}\} - \boldsymbol{\delta B}\cdot\boldsymbol{M}_1. \end{aligned}$$

Now if the relation between $\boldsymbol{B}$, $\boldsymbol{M}_1$ is linear (the definition of permeability μ is not clear when both permanent and induced magnetization are present), then $\boldsymbol{\delta B}\cdot\boldsymbol{M}_1 = \frac{1}{2}\delta(\boldsymbol{B}\cdot\boldsymbol{M}_1)$.

$$\therefore\ \boldsymbol{H}\cdot\boldsymbol{\delta B} - \boldsymbol{B}\cdot\boldsymbol{\delta M} = \delta\{B/(2\mu_0) - \tfrac{1}{2}\boldsymbol{B}\cdot\boldsymbol{M}_1 - \boldsymbol{B}\cdot\boldsymbol{M}\}.$$

Hence we may "integrate w.r. to δ" and obtain the expression for the magnetic energy density

$$B^2/(2\mu_0) - \tfrac{1}{2}\boldsymbol{B}\cdot\boldsymbol{M}_1 - \boldsymbol{B}\cdot\boldsymbol{M}. \tag{9.44}$$

We have already obtained the special case for the energy distribution in a vacuum. The other special case of a (linear) medium of constant permeability μ is obtained either from (9.44) with $\boldsymbol{M} = \boldsymbol{0}$, i.e.

$$B^2/(2\mu_0) - \tfrac{1}{2}\boldsymbol{B}\cdot\boldsymbol{M}_1 = \tfrac{1}{2}\boldsymbol{B}\cdot(\boldsymbol{B}/\mu_0 - \boldsymbol{M}_1) = \tfrac{1}{2}\boldsymbol{B}\cdot\boldsymbol{H}, \tag{9.45}$$

or it may also be obtained from (9.43) with $\boldsymbol{\delta M} = \boldsymbol{0}$, and $\boldsymbol{H} = \boldsymbol{B}/\mu$. Then

$$\boldsymbol{H}\cdot\boldsymbol{\delta B} = (\boldsymbol{B}\cdot\boldsymbol{\delta B})/\mu = \delta\{B^2/(2\mu)\} = \delta\{\tfrac{1}{2}\boldsymbol{B}\cdot\boldsymbol{H}\}. \tag{9.45a}$$

9.9 Dissipation of energy by quasi-steady currents

When the instantaneous value of the current in a resistor is I the rate of dissipation of energy in that resistor is RI^2 at that instant. For an alternating current $I = I_0 \cos(\omega t+\delta)$ the dissipation has the instantaneous value $H = RI_0^2 \cos^2(\omega t+\delta)$ and the mean value, taken over one period, is

$$\overline{H} = \tfrac{1}{2}RI_0^2.$$

Similarly, if V is the instantaneous potential difference between the ends of the resistor

$$H = V^2/R, \qquad \overline{H} = V_0^2/(2R),$$

where $\overline{H}$ is the mean value corresponding to an alternating voltage $V_0 \cos(\omega t+\delta)$. Here the values V_0, I_0 are the *peak values* (or amplitudes) of the alternating quantities. If we write $I_e = I_0/\sqrt{2}$, $V_e = V_0/\sqrt{2}$, then the values I_e, V_e are the *effective values* and the mean rate of dissipation of energy is

$$\overline{H} = I_eV_e = I_e^2R = V_e^2/R.$$

This mean rate of dissipation of energy is related to effective values of the voltage and current by formulae identical to those of direct current quantities. It is easy to see that for sinusoidal currents I_e and V_e are the r.m.s. (root mean square) values of I, V.

If we use complex numbers to represent the currents and voltages, e.g. $I_0e^{i\omega t}$ where I_0 is a complex number, the instantaneous rate of production of heat (or dissipation of energy) is NOT Re $I_0^2e^{2i\omega t}$. We can obtain the *mean values* of products and squares of alternating quantities by a suitable combination of complex numbers, but we cannot (conveniently) obtain instantaneous values of such products or squares. Suppose two alternating quantities x, y are given by

$$x = x_0 \cos \omega t = \text{Re } x_0e^{i\omega t},$$
$$y = y_0 \cos(\omega t+\delta) = \text{Re } y_0e^{i\delta}e^{i\omega t}.$$

Then $$\overline{xy} = \frac{\omega}{2\pi}\int_0^{2\pi/\omega} x_0y_0 \cos \omega t \cos(\omega t+\delta)\, dt = \frac{1}{2}x_0y_0 \cos\delta.$$

But $$\text{Re } \tfrac{1}{2}xy^* = \text{Re } \tfrac{1}{2}x^*y = \text{Re } \tfrac{1}{2}x_0y_0e^{\pm i\delta} = \tfrac{1}{2}x_0y_0 \cos\delta = \overline{xy}. \quad (9.46)$$

(We use asterisks * to denote complex conjugate to avoid confusion with the bar denoting mean value.)

If a current (in complex notation) $I_0 e^{i(\omega t+\delta)}$ flows between points at a potential difference V_0 the mean rate of dissipation of energy between these points is Re $\frac{1}{2} I_0 V_0 e^{i\delta} = \frac{1}{2} I_0 V_0 \cos\delta$. The factor $\cos\delta$ is called the *power factor* because the power dissipated is reduced by this factor from the product of the effective values $(V_0/\sqrt{2})(I_0/\sqrt{2})$.

These ideas are of importance when it is necessary to extract the greatest possible power from a generator (such as a battery, a dynamo or the output of a radio receiver). Consider first the simple case of a torch battery producing heat by discharging across a resistance r (Fig. 9.32). Even if $r = 0$ (so that the battery is short-circuited), only a finite current flows, and this is determined by the *internal resistance* R of the battery. The current flowing when r is in circuit is then $E/(R+r)$ and the e.m.f. across r is $Er/(R+r)$. Consequently the power developed is

$$\frac{E^2 r}{(R+r^2)} = \frac{E^2}{R\left(\sqrt{\frac{R}{r}}+\sqrt{\frac{r}{R}}\right)^2}.$$

Since $x+1/x \geqslant 2$, for $x > 0$, and equality holds only if $x = 1$, the maximum power is $E^2/(4R)$ when $r = R$, that is, when the battery discharges into a circuit whose resistance *matches* its own.

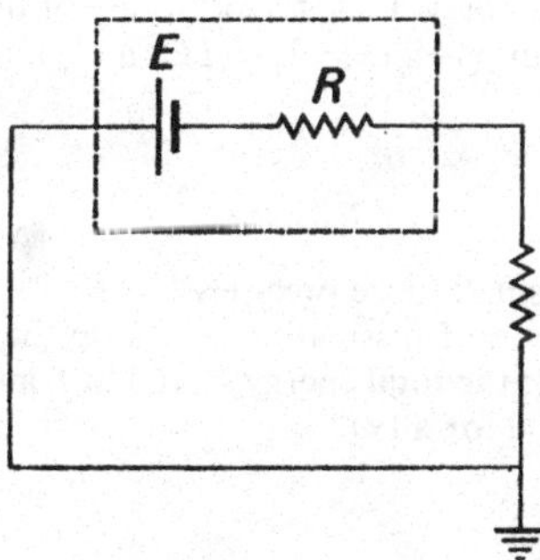

FIG. 9.32

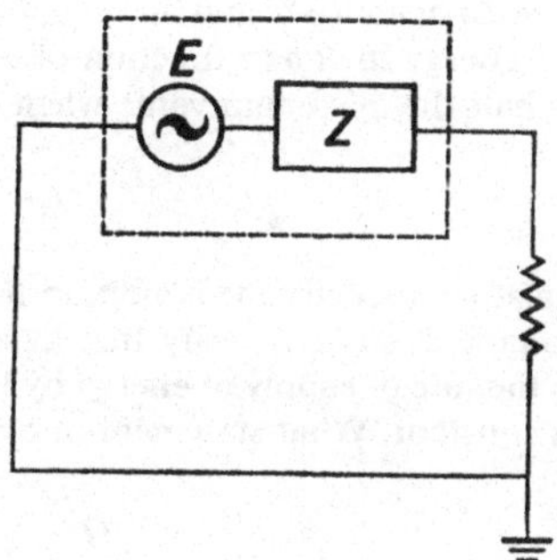

FIG. 9.33

Turning now to the case of alternating currents, we consider a generator or a radio circuit with an internal impedance Z (see Fig. 9.33). If the system is to generate power by discharging across an external circuit, it is clear from what has been said already that it is only the *resistive* part of the external circuit which can generate heat or otherwise extract power from the electrical circuit. So it is most useful to consider the case of a purely resistive load.

The current is now given by $I = E/(Z+r)$, the e.m.f. by $V = Er/(Z+r)$, so that

$$\text{Re}\,\frac{1}{2}IV^* = \frac{\frac{1}{2}E^2r}{(Z+r)(Z^*+r)} = \frac{\frac{1}{2}E^2}{\dfrac{ZZ^*}{r}+r+Z+Z^*}$$

which has a maximum value, when $r^2 = ZZ^* = |Z|^2$, of

$$\tfrac{1}{2}E^2/(Z+Z^*+2\sqrt{ZZ^*}) = \tfrac{1}{2}E^2/(\sqrt{Z}+\sqrt{Z^*})^2.$$

However, since $\sqrt{(Z^*)} = (\sqrt{Z})^*$, this may be written $\frac{1}{4}E^2/\{(\text{Re}\sqrt{Z})^2\}$. Of course, if Z is not purely resistive, it is to be expected that it will depend on frequency in some way, so that the load will not be properly matched except at certain frequencies. But it is usually possible to achieve approximate matching over the *range of frequencies* required.

Example 1. When a current I_0 (in complex notation) flows through an inductance the potential difference between the terminals is $V_0 = I_0Z = i\omega LZ_0$. Hence the power dissipated is $\text{Re}\,\frac{1}{2}I_0V_0^* = \text{Re}\,(-i\omega LI_0^2) = 0$. This shows that *no* power is dissipated in the inductance. (The power factor is $\cos\pi/2 = 0$.)

The mean energy of the magnetic field arising from the current is the mean value of $\frac{1}{2}LI^2$.

$$\therefore\ \bar{T} = \tfrac{1}{2}\,\text{Re}\,(\tfrac{1}{2}LI_0I_0^*) = \tfrac{1}{4}L\,|I_0|^2.$$

Example 2. An e.m.f. $E\cos\omega t$ drives a current through a circuit consisting of a resistance R, a capacitance C and an inductance L in series. Show that the mean rate of dissipation of energy in R as a function of ω is a maximum when $\omega = 1/\sqrt{(LC)}$ and, if R is small, is half the maximum value when

$$\omega = \frac{1}{\sqrt{(LC)}} \pm \frac{R}{2L}$$

approximately. (Specify the condition that R is to be small more precisely.)

When $\omega = 1/\sqrt{(LC)}$, verify that at all times the rate of dissipation of energy in R is equal to the rate of supply of energy by the e.m.f., while the total energy stored in L and C remains constant. What statement of this nature is true for all ω?

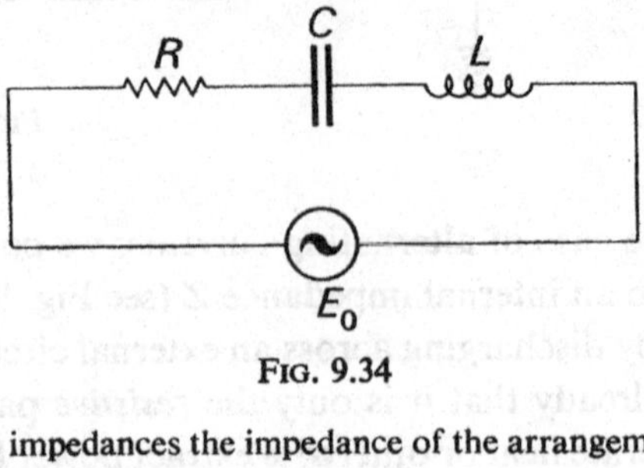

FIG. 9.34

In terms of complex impedances the impedance of the arrangement in Fig. 9.34 is

$$Z = R+i\{\omega L-1/(\omega C)\} = Z_0e^{i\alpha},\quad Z_0^2 = R^2+\{\omega L-1/(\omega C)\}^2.$$

The complex current $I = I_0 e^{i\omega t}$ is given by

$$I_0 Z = E_0, \quad I_0 = (E_0/Z)e^{-i\alpha}.$$

Hence the mean rate of dissipation in R is

$$\bar{H} = \text{Re}\,\tfrac{1}{2}RI_0 I_0^* = \tfrac{1}{2}R\,|I_0|^2 = RE_0^2/(2Z_0^2).$$

When $\bar{H}$ is a maximum Z_0 is a minimum, i.e. when $\omega L = 1/(\omega C)$.

$$\therefore\ \omega^2 = 1/\sqrt{(LC)}, \quad \bar{H}_{\max} = E_0^2/(2R).$$

When $$\omega = \frac{1}{\sqrt{(LC)}} \pm \frac{R}{2L}, \quad \omega L = \sqrt{\left(\frac{L}{C}\right)} \pm \frac{R}{2},$$

$$\frac{1}{\omega C} = \sqrt{\left(\frac{L}{C}\right)}\left\{1 \mp \frac{R}{2}\sqrt{\left(\frac{C}{L}\right)}\right\} + O(R^2).$$

$$\therefore\ \omega L - \frac{1}{\omega C} = \sqrt{\left(\frac{L}{C}\right)} \pm \frac{R}{2} - \sqrt{\left(\frac{L}{C}\right)}\left\{1 \mp \frac{R}{2}\sqrt{\left(\frac{C}{L}\right)} + O(R^2)\right\} = \pm R + O(R^2).$$

$$\therefore\ Z_0^2 = R^2 + \{\omega L - 1/(\omega C)\}^2 = 2R^2 + O(R^3).$$

$$\therefore\ \bar{H} = RE_0^2/(4R^2) + O(1) \approx E_0^2/(4R) = \tfrac{1}{2}\bar{H}_{\max}.$$

Since R is small we neglect the term $O(1)$ in comparison with E_0^2/R. The condition that R is small is to be interpreted as the ratio $R/(L/C)^{1/2}$ is small, R and $\sqrt{(L/C)}$ having the same dimensions.

The remainder of the question refers to instantaneous values calculated at an arbitrary instant. When $\omega = 1/\sqrt{(LC)}$, $Z = R$, $\alpha = 0$, $I_0 = E_0/R$ so that the instantaneous values of current and e.m.f. are

$$I = (E_0/R)\cos\omega t, \quad \mathscr{E} = E_0 \cos\omega t.$$

The rate of dissipation in R is

$$RI^2 = (E_0^2/R)\cos^2\omega t.$$

The rate of supply by $\mathscr{E}$ is

$$\mathscr{E}I = (E_0^2/R)\cos^2\omega t,$$

showing the required equality.

The energy "stored in L" is $\tfrac{1}{2}LI^2 = \tfrac{1}{2}LI_0^2\cos^2\omega t$.

The complex expression for the charge on C is $q_0 e^{i\omega t}$ where $q_0 = I_0/(i\omega)$. Consequently the instantaneous value of q is $(I_0/\omega)\sin\omega t$. The energy stored in C is therefore

$$q^2/(2C) = \{I_0^2/(2C\omega^2)\}\sin^2\omega t = \tfrac{1}{2}LI_0^2\sin^2\omega t.$$

The sum of these two energies is

$$\tfrac{1}{2}LI_0^2(\cos^2\omega t + \sin^2\omega t) = \tfrac{1}{2}LI_0^2,$$

which is constant as required. This result means that, when the transients have been damped out, in addition to the dissipation of energy as heat in R there is a continual passage of energy from C to L and back again without loss. At certain times all this energy is stored in the condenser as electrostatic energy and at others it is all magnetic energy in the inductance, the condenser being uncharged at that instant.

Example 3. An electric circuit is as shown in Fig. 9.35, with the source supplying an e.m.f. equal to $V_0 \cos\omega t$. Find the mean rate of dissipation of energy in r. Show that if both C and r are adjusted so as to maximize this rate of dissipation, then its maximum is the same as it would be if r and R were in series with no other circuit elements present and r were adjusted to maximize the mean rate of dissipation.

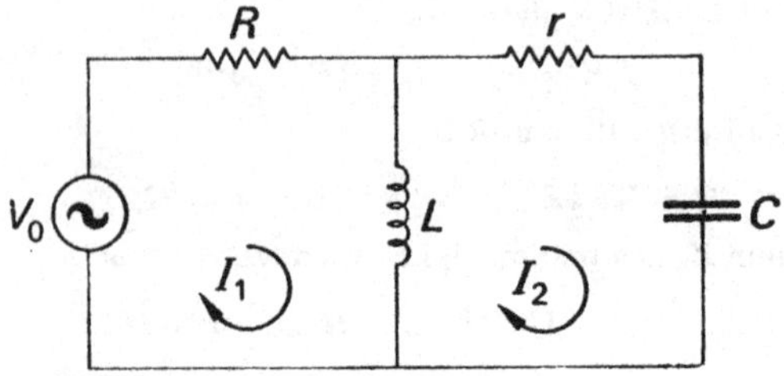

FIG. 9.35

Using complex circulating currents (see Fig. 9.35), the two equations derived from Kirchhoff's second law are

$$RI_1+iL\omega(I_1-I_2) = V_0,$$

$$rI_2+iL\omega(I_2-I_1)+\frac{1}{iC\omega}I_2 = 0.$$

Solving for I_2, we find

$$I_2 = \frac{iL\omega V_0}{\left(Rr+\frac{L}{C}\right)+i\omega\left\{L(R+r)-\frac{R}{C\omega^2}\right\}}.$$

Therefore
$$|I_2| = \frac{L\omega V_0}{\sqrt{\left[\left(Rr+\frac{L}{C}\right)^2+\omega^2\left\{L(R+r)-\frac{R}{C\omega^2}\right\}^2\right]}}. \quad (1)$$

From eqn. (9.46) the mean rate of dissipation of energy in the resistance r is $\bar{H} = \text{Re}\{\frac{1}{2}rI_2I_2^*\} = \frac{1}{2}r\,|I_2|^2$, i.e.

$$\bar{H} = \frac{\frac{1}{2}L^2\omega^2V_0^2r}{\left(Rr+\frac{L}{C}\right)^2+\omega^2\left\{L(R+r)-\frac{R}{C\omega^2}\right\}^2}.$$

Since C occurs in the denominator of $\bar{H}$ only, $\bar{H}$ considered as a function of C will be greatest when this denominator is least, i.e. when (differentiating w.r. to $1/C$)

$$L\left(Rr+\frac{L}{C}\right)-R\left\{L(R+r)-\frac{R}{C\omega^2}\right\} = 0,$$

i.e. when
$$\frac{1}{C}\left(L^2+\frac{R^2}{\omega^2}\right) = R^2L.$$

In this case

$$\bar{H}_{\max} = \frac{\frac{1}{2}L^2\omega^2V_0^2r\left(1+\frac{\omega^2L^2}{R^2}\right)}{\left\{Rr\left(1+\frac{\omega^2L^2}{R^2}\right)+L^2\omega^2\right\}^2}.$$

$\bar{H}_{\max}$, considered as a function of r, is itself a maximum when

$$r = L^2\omega^2\Big/\left\{R\left(1+\frac{\omega^2L^2}{R^2}\right)\right\}$$

and in this case
$$\bar{H}_{\max} = \tfrac{1}{8}V_0^2/R,$$

which is the mean rate of dissipation of energy in R when L, C are absent and r is adjusted so as to make H a maximum. [When R, r only are present, $H = \frac{1}{2}V_0^2/(R+r)$.]

Miscellaneous Exercises IX

1. The plates A and B of a condenser of capacity C initially carry charges Q and $-Q$ respectively, where $Q > 0$, and are suddenly connected by a uniform wire of total resistance R_1 and self-inductance L_1. Show that in the resulting discharge of the condenser the flow of current in the wire is always in the same direction if $CR_1^2 > 4L_1$ but that it oscillates if $CR_1^2 < 4L_1$.

Show further that, if the system is such that the discharge is oscillatory, the greatest positive charge on plate B during the subsequent flow of current is $Qe^{-\pi/n}$, where

$$n = \sqrt{(4L_1/CR_1^2-1)}.$$

A secondary coil of resistance R_2 and self-inductance L_2 is inductively coupled with the primary circuit, the coefficient of mutual inductance being M. Prove that, if the condenser in the primary circuit is now discharged, the discharge will only be oscillatory if the cubic equation in x,

$$(L_1L_2-M^2)x^3+(L_1R_2+L_2R_1)x^2+(R_1R_2+L_2/C)x+R_2/C = 0,$$

has a pair of complex roots.

2. A generator with e.m.f. of amount $E\cos\omega t$ is connected through a circuit with a condenser of capacity C, a coil of inductance L and resistance r, and a resistance R. The ends of the resistance R are connected through a second circuit that is shielded from the first and also contains a condenser of capacity C and a coil of inductance L and resistance r. Prove that if the capacity C of the condensers can be varied, then the amplitude of the current in the second circuit is greatest when $\omega^2LC = 1$, and prove that its value is then

$$\frac{ER}{(2Rr+r^2)}.$$

3. Four points $\alpha, \beta, \gamma, \delta$ are connected by the following impedances:

(*a*) a resistance R and a capacitance C in parallel between α and β,

(*b*) a resistance R and an inductance L in series between γ and δ,

(*c*) three resistances each of amount R joining the three pairs of points $\beta\gamma$, $\delta\alpha$, and $\beta\delta$.

A harmonic e.m.f. of period $2\pi/\omega$ is maintained between α and γ. Assuming that there are no transient currents so that the different currents in the members of the system also vary harmonically with time, show that the current in $\beta\delta$ is always zero if $L = CR^2$, and that it is of maximum amplitude when $\omega\sqrt{(LC)} = 2$ if $L \neq CR^2$ and ω is considered as the only variable parameter.

4. An alternating electromotive force $E\cos pt$ acts in a circuit, of self-inductance L and negligible resistance, which is inductively coupled with a second circuit of self-inductance L', resistance R and capacity C, the coefficient of mutual inductance of the circuits being M ($< L^{1/2}L'^{1/2}$). Show that the value of p which gives the greatest amplitude to the oscillations of current induced in the second circuit is $\{L/C(LL'-M^2)\}^{1/2}$, and that the greatest amplitude is $EM/(RL)$.

5. A coil is subject to a small periodic deformation such that its self-inductance at time t is the real part of $L_0+L_1e^{i\omega t}$, where L_0, L_1 and ω are real constants and L_1 is small compared with L_0. The ends of the coil are connected through a constant resistance R to a battery of constant electromotive force E. Show that, neglecting transient currents, the current in this circuit will be given approximately by the real part of

$$\frac{E}{R}+\frac{\omega EL_1}{R(R^2+\omega^2L_0^2)^{1/2}}\,e^{i(\omega t+\delta)},$$

and determine δ.

6. Four points A_1, A_2, A_3, A_4 are connected by six branches A_1A_2, A_1A_3, A_2A_3, A_2A_4, A_3A_4, A_1A_4, containing respectively a capacitance C and resistance $2R$ in series, a capacitance C and resistance $2R$ in series, a capacitance C, a resistance R, a resistance R, resistance zero. The initial currents are all zero and the charges on the respective condensers are Q, $2Q$, 0 in such a way that the negatively charged plate of one is connected to the positively charged plate of the other. Show that at time t the charge on the third condenser is

$$Q(e^{-t/RC}-e^{-t/4RC}).$$

7. A circuit consists of two inductors, each of inductance L, in series with capacitance C. A battery of e.m.f. E and resistance R is connected in parallel with the capacitance. Each inductor is coupled by a mutual inductance M to a circuit of inductance L and capacitance C in series. After a steady state has been attained the battery connexion is broken. Show that at time t later the current in the first circuit is

$$\frac{1}{2}\frac{E}{R}\{(1-LY)\cos\omega_1 t+(1+LY)\cos\omega_2 t\}$$

and in each of the other two circuits is

$$\frac{EMY}{R}(\cos\omega_1 t-\cos\omega_2 t),$$

where $$\omega_{1,2}^2=\frac{3L\pm Y^{-1}}{4C(L^2-M^2)} \quad \text{and} \quad Y=(L^2+8M^2)^{-1/2}.$$

8. A primary circuit contains a resistance R_1, an inductance L_1 and a capacitance C_1 in series with an e.m.f. $E\sin\omega t$. A secondary circuit contains a resistance R_2, an inductance L_2, and a capacitance C_2 also in series. There is a mutual inductance M between the two circuits.

If C_1, C_2 are varied but R_1, L_1, R_2, L_2, E, ω are kept constant show that the amplitude of the steady current in the secondary is stationary either when

$$\omega^2L_1C_1=\omega^2L_2C_2=1$$

or when

$$\frac{R_1}{R_2}\left(\omega L_1-\frac{1}{\omega C_1}\right)^2=\frac{R_2}{R_1}\left(\omega L_2-\frac{1}{\omega C_2}\right)^2=M^2\omega^2-R_1R_2.$$

Hence show that if $R_1R_2 < M^2\omega^2$ the greatest current in the secondary is $E/2(R_1R_2)^{1/2}$.

9. Current flows in a closed circuit. Prove that if the circuit is displaced in a static magnetic field, the current being maintained at a constant value J, the mechanical work performed on the circuit is the product of J with the change in the flux (of the static field) linking the circuit.

Two rigid circuits, whose mutual inductance is M, carry currents J_1 and J_2. The first circuit is free to move without rotation parallel to an axis Ox, M being a function of the displacement x. Prove that the component in the direction Ox of the force acting on it is $J_1J_2\,\partial M/\partial x$.

The first circuit has resistance R_1, self-inductance L_1 and negligible mass. It is acted upon by a light spring which exerts a restoring force $-\lambda x$. The second circuit has resistance R_2, self-inductance L_2 and contains a battery generating an e.m.f. E. For small displacements the mutual inductance follows the law

$$M=N+kx,$$

N and k being constants. Prove that small departures from the equilibrium conditions of position and current vary with time as the sum of two terms, the one varying as

$e^{-p_1 t}$ and the other as $e^{-p_2 t}$, where p_1 and p_2 are the roots of the equation

$$(L_1' L_2 - N^2)p^2 - (R_1 L_2 + R_2 L_1')p + R_1 R_2 = 0,$$

in which L_1' is defined by

$$L_1' = L_1 + \frac{k^2 E^2}{\lambda R_2^2}.$$

10. A periodic e.m.f. of rectangular wave form, such that

$$V = V_0 \quad \text{for} \quad 2n\pi/\omega < t < (2n+1)\pi/\omega,$$

and

$$V = 0 \quad \text{for} \quad (2n+1)\pi/\omega < t < (2n+2)\pi/\omega,$$

where n is an integer, is applied to a circuit containing a condenser of capacity C_1 and inductance L_1, but without resistance, coupled inductively with a second circuit containing C_2, L_2 and without resistance. The coefficient of mutual inductance between the two circuits is M. Show that, in the steady state, the current in the second circuit is

$$\frac{2V_0 M C_1 C_2}{\pi} \sum_{m=0}^{\infty} \frac{(2m+1)^2 \omega^3 \cos(2m+1)\omega t}{1 - A(2m+1)^2\omega^2 + B(2m+1)^4\omega^4}$$

where $A = L_1 C_1 + L_2 C_2$ and $B = C_1 C_2 (L_1 L_2 - M^2)$.

11. An electrical generator of resistance R is connected in series with a capacitance C, a switch and a generator which develops a potential difference $E = E_0 |\sin \omega t|$. If C is uncharged and the switch is closed at $t = 0$, show that the current I at time t is

$$I = E_0 \frac{\alpha}{R} \frac{8}{\pi} \sum_{n=1}^{\infty} \frac{n}{4n^2 - 1} \frac{\sin(2\omega n t - \tan^{-1} 2n\alpha)}{(1 + 4n^2\alpha^2)^{1/2}}$$

$$- E_0 \frac{\alpha}{R} \frac{e^{-\omega t/\alpha}(1 + e^{\pi/\alpha})}{(1+\alpha^2)(1 - e^{\pi/\alpha})},$$

where $\alpha = \omega CR$.

12. A wave filter consists of a large number of identical units. Each series element consists of a capacitance C and inductance L in parallel, and each shunt element of the same two impedances but arranged in series. Show that alternating current of frequency $p/2\pi$ will be attenuated if p lies between the values

$$(\sqrt{17} \pm 1)/\{4(LC)^{1/2}\}.$$

13. A wave filter consists of a number of identical T elements with the horizontal arms joined in series. A typical horizontal arm $A_r B_r A_{r+1}$ contains an impedance Z_1 in each of $A_r B_r$ and $B_r A_{r+1}$ and the vertical arm from B_r to earth contains an impedance Z_2. The total number of elements is n and the potential of A_0 is $v(t)$ while that of A_n is zero. Show that the Laplace transform of the current i_r in $B_{r-1} A_r B_r$ is

$$\frac{V \cosh (n-r)\theta}{Z_2 \sinh \theta \sinh n\theta}$$

where $\cosh \theta = 1 + Z_1/Z_2$ and V is the Laplace transform of $v(t)$.

Determine the current i_r when Z_2 is a resistance R, and $2Z_1$ is a series combination of R' and C, $v = 0$ for $t < 0$ and $v = E$ (constant) for $t > 0$.

14. A particle satisfies the equation of motion

$$\frac{d}{ds}\left(m \frac{dx_i}{ds}\right) = F_i \qquad (i = 1, \ldots, 4),$$

where ds is an increment of proper time and F_i is the gradient of a scalar field,

$$F_i = \phi_i.$$

Prove that m cannot be a constant of the motion (unless ϕ is also) and deduce that, if m is a function of position only, then

$$m = m_0 + \phi/c^2,$$

where m_0 is constant.

If ϕ is independent of t in some Lorentz frame, deduce the energy integral

$$\frac{m_0 c^2 + \phi}{\left(1 - \frac{v^2}{c^2}\right)^{1/2}} = E,$$

where E is constant, and interpret this for weak fields and slow motions.

15. Verify that the Lorentz-force equation of motion for a particle of charge e and mass m, moving in an electromagnetic field with 4-potential $(\boldsymbol{A}, \phi/c)$, can be derived from the relativistic Lagrangian

$$L = -mc^2\sqrt{\left(1 - \frac{v^2}{c^2}\right)} + e\,\boldsymbol{v}\cdot\boldsymbol{A} - e\phi.$$

A particle of rest mass m and charge e is projected from the point $(a, 0, 0)$ with velocity $(0, v_0, 0)$ in a uniform field $\boldsymbol{E} = (0, 0, E)$, $\boldsymbol{B} = (0, 0, B)$. Show that the rectangular coordinates of the particle at time t are

$$x_1 = a + \frac{\gamma_0 v_0}{\Omega}(1 - \cos \Omega\tau),$$

$$x_2 = \frac{\gamma_0 v_0}{\Omega} \sin \Omega\tau,$$

$$x_3 = \frac{\gamma_0 mc^2}{eE}\left\{\cosh\left(\frac{eE}{mc}\tau\right) - 1\right\},$$

$$t = \frac{\gamma_0 mc}{eE} \sinh\left(\frac{eE}{mc}\tau\right),$$

where $\gamma_0 = \dfrac{1}{\sqrt{\left(1 - \frac{v_0^2}{c^2}\right)}}$, $\Omega = \dfrac{eB}{m}$ and τ is the proper time.

16. By considering the collision of two identical perfectly elastic particles, or otherwise, derive a formula for the relative mass of a particle of proper mass m moving with relative speed v.

Obtain a formula for the change in wavelength of a photon scattered through an angle θ by a free electron of mass m, initially at rest.

17. A source of light has equation of motion $\boldsymbol{x} = \boldsymbol{x}(t)$ and an observer moves with $\boldsymbol{x} = \boldsymbol{x}'(t)$. Show that if the light is emitted with frequency ν, the frequency of reception ν' is given by

$$\frac{\nu'}{\nu} = \frac{\gamma'}{\gamma}\frac{(1 - \boldsymbol{n}\cdot\boldsymbol{v}'/c)}{(1 - \boldsymbol{n}\cdot\boldsymbol{v}/c)}$$

where $\boldsymbol{v}$ is the velocity of the source at the retarded time, $\boldsymbol{n}$ the unit vector from source to observer, and γ and γ' are the Lorentz factors associated with $\boldsymbol{v}$ and $\boldsymbol{v}'$ respectively.

Consider two observers P_1 and P_2 such that P_1 is inertial and P_2 moves in a circle with velocity v about P_1. If P_1 emits light of frequency ν_1 which is received by P_2 at frequency ν_1', and P_2 emits light of frequency ν_2 which is received by P_1 at frequency ν_2', show that

$$\frac{\nu_2'}{\nu_2} = \left(\frac{\nu_1'}{\nu_1}\right)^{-1} = \sqrt{\left(1 - \frac{v^2}{c^2}\right)}.$$

18. State the relativistic conservation laws for a system of interacting particles in the absence of external forces.

A particle of rest-energy W_0, having energy W and linear momentum p, disintegrates spontaneously into two photons (particles of zero rest-energy). Show that if the direction of motion of one of the photons makes an angle θ with the direction of the original particle, then it has energy

$$E(\theta) = \frac{\frac{1}{2}W_0^2}{W - cp\cos\theta}.$$

Show also that if the second photon is emitted at an angle ϕ to the direction of the original particle, then

$$W(\sin\theta - \sin\phi) = cp\sin(\theta - \phi).$$

19. Frames of reference S and $\bar{S}$ are related by the Lorentz transformation

$$\bar{x} = \beta(x - vt), \quad \bar{t} = \beta(t - vx/c^2),$$
$$\bar{y} = y, \quad \bar{z} = z, \quad \beta = (1 - v^2/c^2)^{-1/2}.$$

Establish the following:

(i) The Lorentz contraction factor for a moving rigid rod.
(ii) The time difference recorded by clocks attached to S and $\bar{S}$.
(iii) The relativistic Doppler equation

$$\nu = \beta\bar{\nu}(1 + v/\bar{u})$$

for a plane wave of frequencies ν and $\bar{\nu}$ travelling with speeds u and $\bar{u}$ relative to S and $\bar{S}$ in the direction of the velocity $\mathbf{v}$ of $\bar{S}$ relative to S. Interpret the cases $c = \infty$ and $c = \bar{u}$.

20. By consideration of the invariance of phase under Lorentz transformations, or otherwise, derive the Doppler formula

$$\omega = \omega_0 \frac{1 - (\mathbf{v}\cdot\mathbf{e})/c}{\sqrt{(1 - v^2/c^2)}}$$

for the proper frequency ω of a light wave in terms of ω_0, the frequency as observed by an observer with respect to whom the light source has a velocity $\mathbf{v}$, $\mathbf{e}$ being the wave-normal of the light wave in the system of the observer.

Derive also the aberration formula

$$\tan\theta = \frac{\sqrt{(1 - v^2/c^2)}\sin\theta_0}{\cos\theta_0 - v/c},$$

θ_0 being the angle between the wave-normal and $\mathbf{v}$ in the observer's system, and θ the corresponding angle in the system in which the light source is at rest.

21. A particle (A) whose velocity components are $(u_1, 0, 0)$ in a frame S collides with another particle (B) of equal mass which is at rest relative to S. If the momenta $\mathbf{p}'_A$, $\mathbf{p}'_B$ after the collision make angles θ and ϕ respectively with the x-axis, show that

$$\tan\theta\tan\phi = \frac{2}{1+\gamma},$$

where $\gamma = (1 - u_1^2/c^2)^{-1/2}$.

22. Two perfectly synchronized clocks C_1, C_2 are initially at rest in an inertial frame S. The clock C_2 then accelerates away from C_1 with constant acceleration g relative to its instantaneous rest frame until it reaches a velocity $\mathbf{v}$. It then decelerates at the same rate until coming to rest, accelerates back towards C_1 at the same rate until reaching velocity $-\mathbf{v}$ and then decelerates at the same rate, arriving back at C_1 at rest in the inertial

frame S. C_1 has remained at rest in S for the whole of C_2's journey. Show that C_1's clock is ahead of C_2's by an amount

$$\frac{4v}{g}\left[\frac{1}{\sqrt{(1-v^2/c^2)}}-\frac{c}{v}\sinh^{-1}\left(\frac{v}{\sqrt{(c^2-v^2)}}\right)\right].$$

State what assumptions you have made in deriving this result.

23. Derive the law of velocity transformation under a Lorentz transformation between inertial frames S, $\bar{S}$ in the form

$$u = \frac{\bar{u}+V}{1+\bar{u}V/c^2}, \quad v = \frac{\bar{v}}{\beta(1+\bar{u}V/c^2)}, \quad w = \frac{\bar{w}}{\beta(1+\bar{u}V/c^2)},$$

where $\beta = (1-V^2/c^2)^{-1/2}$.

If $\bar{u}_1 = -\bar{W}\cos\bar{\alpha}$, $\bar{v}_1 = \bar{W}\sin\bar{\alpha}$, $\bar{w}_1 = 0$ in $\bar{S}$ transforms to $u_1 = -W_1\cos\alpha_1$, $v_1 = W_1\sin\alpha_1$, $w_1 = 0$ in S, show that

$$1+VW_1\cos\alpha_1/c^2 = (1-V^2/c^2)(1-V\bar{W}\cos\bar{\alpha}/c^2)^{-1},$$

$$1-W_1^2/c^2 = (1-V^2/c^2)(1-\bar{W}^2/c^2)(1-V\bar{W}\cos\bar{\alpha}/c^2)^{-2}.$$

Also if $\bar{u}_2 = \bar{W}\cos\bar{\alpha}$, $\bar{v}_2 = \bar{W}\sin\bar{\alpha}$, $\bar{w} = 0$ transforms to $u_2 = W_2\cos\alpha_2$, $v_2 = W_2\sin\alpha_2$, $w_2 = 0$, show that

$$1-VW_2\cos\alpha_2/c^2 = (1-V^2/c^2)(1+V\bar{W}\cos\bar{\alpha}/c^2)^{-1},$$

$$1-W_2^2/c^2 = (1-V^2/c^2)(1-\bar{W}^2/c^2)(1+V\bar{W}\cos\bar{\alpha}/c^2)^{-2}.$$

The velocities $(\bar{u}_1, \bar{v}_1)$, $(\bar{u}_2, \bar{v}_2)$ are taken to be those of a particle before and after making an elastic collision with a rigid boundary $\bar{x} = 0$; the velocities (u_1, v_1), $(u_2, v_2$ are those relative to an inertial frame S in which the boundary moves with uniform) velocity V normal to itself. In the limiting case in which the particle becomes a photon of frequencies ν_1, ν_2 in S before and after reflexion at a moving mirror, deduce that

$$(1+V\cos\alpha_1/c)\nu_1 = (1-V\cos\alpha_2/c)\nu_2.$$

CHAPTER 10

THE DISPLACEMENT CURRENT AND MAXWELL'S EQUATIONS

10.1 The displacement current

We have developed the theory of electromagnetism by successive removals of restriction on the sources (the charges and currents) of the fields. By considering non-steady currents we were able to incorporate electromagnetic induction into the theory; the field equations which govern the system, so far, have been obtained as

$$\mathbf{curl}\ \boldsymbol{E} = -\partial \boldsymbol{B}/\partial t, \qquad \mathbf{curl}\ \boldsymbol{H} = \boldsymbol{J} \tag{10.1}$$

together with div $\boldsymbol{B} = 0$ and div $\boldsymbol{D} = \varrho$.

We have also retained the condition div $\boldsymbol{J} = 0$; because div **curl** $\boldsymbol{H} \equiv 0$, the condition div $\boldsymbol{J} = 0$ is necessary if the second of equations (10.1) is to be consistent.

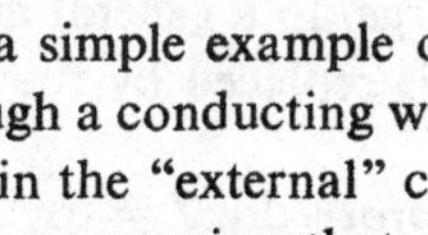

FIG. 10.1

Nevertheless, situations exist for which div $\boldsymbol{J} \neq 0$. We now generalize the theory to include this case, a simple example of which is found in the discharge of a condenser through a conducting wire, see Fig. 10.1. During the discharge a current I flows in the "external" circuit, and the charge q decreases. Conservation of charge requires that

$$I = -\mathrm{d}q/\mathrm{d}t.$$

Although this current is not steady it produces a magnetic field—as could easily be shown by experiment—but we cannot apply eqns. (10.1) to this case since I does not flow in a closed circuit, i.e. div $\boldsymbol{J} \neq 0$ everywhere.

During the process of discharge the electric field between the plates varies in strength, as does the displacement vector $\boldsymbol{D}$ also. We introduced the vector $\boldsymbol{D}$ as the measure of the "displacement" of charge across the space between the plates; this idea is easily extended, when $\boldsymbol{D}$ varies with time, to suggest the idea of a "rate of displacement of charge" between the plates, or a *displacement current.*

At any instant the surface density of charge on the plates is $\pm q/\alpha$, where α is the area of each plate and

$$|\boldsymbol{D}| = \sigma = q/\alpha.$$

Therefore

$$\frac{\mathrm{d}}{\mathrm{d}t}|\boldsymbol{D}| = -\frac{I}{\alpha}. \tag{10.2}$$

The vector $\boldsymbol{D}$ is directed from A towards B in Fig. 10.1 so that the "displacement current", $\mathrm{d}\boldsymbol{D}/\mathrm{d}t$ in this case, should be directed from A to B also. The minus sign in eqn. (10.2) therefore denotes a current, strictly a current density, directed from B towards A, which is in the same sense as I. Further, since this current (density) corresponds to unit area of the plate, the total current strength arising from the change of displacement is $\alpha(I/\alpha) = I$. This current then "completes the circuit" and there is a current strength I flowing in a closed circuit. Maxwell's contribution was his assertion that this displacement current and the conduction current both contribute to the magnetic field. He therefore suggested that the second of eqns. (10.1) should be written

$$\mathbf{curl}\, \boldsymbol{H} = \boldsymbol{J} + \partial \boldsymbol{D}/\partial t. \tag{10.3}$$

The name *displacement current* (density) is given to the term $\partial \boldsymbol{D}/\partial t$. The results obtained in Chapter 9 are correct, in so far as magnetic effects are concerned, as long as the displacement current is negligible. This applies when the changes implied in $\partial/\partial t$ take place slowly, at low frequencies, or the electric field is small.

The equality between the two currents in the circuit of Fig. 10.1 arose from the conservation of charge. In a general motion of charge the "steady state" restriction div $\boldsymbol{J} = 0$ is replaced by the equation of continuity (see Vol. 1, p. 150).

$$\partial \varrho/\partial t + \operatorname{div} \boldsymbol{J} = 0.$$

Bearing in mind that div $\boldsymbol{D} = \varrho$, we can write the equation of continuity in the form

$$\operatorname{div}(\partial \boldsymbol{D}/\partial t + \boldsymbol{J}) = 0,$$

a condition which is necessary if eqn. (10.3) is to be satisfied in general.

The field equations, when all restrictions on the charges and currents giving rise to these fields have been lifted, are now to be written

$$\mathbf{curl}\,\boldsymbol{E} = -\partial \boldsymbol{B}/\partial t, \qquad \mathbf{curl}\,\boldsymbol{H} = \partial \boldsymbol{D}/\partial t + \boldsymbol{J}, \tag{10.4, 5}$$

$$\text{div}\,\boldsymbol{D} = \varrho, \qquad \text{div}\,\boldsymbol{B} = 0. \tag{10.6, 7}$$

These are known as Maxwell's equations; they are supplemented by the relations

$$\boldsymbol{D} = \varepsilon \boldsymbol{E} = \varepsilon_0(\boldsymbol{E}+\boldsymbol{P}), \tag{10.8}$$

$$\boldsymbol{H} = \boldsymbol{B}/\mu = \boldsymbol{B}/\mu_0 - \boldsymbol{M}, \tag{10.9}$$

$$\boldsymbol{J} = \sigma(\boldsymbol{E}+\boldsymbol{E}'), \tag{10.10}$$

which, because they express the properties of the medium in which the field is established, are often called the *constitutive relations*; the last is the field form of Ohm's Law. (In this equation σ stands for electrical conductivity, and not surface density of charge.)

10.2 Electromagnetic radiation

In a sense the magnetic effect of the displacement current is a counterpart to electromagnetic induction. In the first a variable electric field produces a magnetic field in the immediate neighbourhood: in the second, a variable magnetic field produces an electric field in its neighbourhood. Both these effects take place in the absence of conductors and can occur in any medium or in a vacuum. It is the combination of these two effects which produces electromagnetic radiation across a medium, or a vacuum. We can describe the process crudely as follows.

Suppose a charge (or a current) varies continuously and at a rate which varies with time, e.g. a charge oscillating in a (broadcasting) aerial. This motion produces an electric field in the neighbourhood of the aerial; because of the variations in charge this electric field itself varies and so produces a (varying) magnetic field in its neighbourhood, slightly farther out from the source. This varying magnetic field produces in its turn a (varying) electric field in its neighbourhood; the combined effect of these variations is to propagate the fields outwards from the source, and constitutes electromagnetic radiation.

We consider analytically this process in a vacuum. In the field, away from the source, Maxwell's equations together give

$$\mathbf{curl}\,\boldsymbol{E} = -\partial \boldsymbol{B}/\partial t, \qquad \mathbf{curl}\,\boldsymbol{H} = \partial \boldsymbol{D}/\partial t,$$

$$\text{div}\,\boldsymbol{D} = 0, \qquad \text{div}\,\boldsymbol{B} = 0,$$

and

$$\boldsymbol{D} = \varepsilon_0 \boldsymbol{E}, \qquad \boldsymbol{H} = \boldsymbol{B}/\mu_0 .$$

We eliminate the magnetic vectors and obtain

$$\textbf{curl curl } \boldsymbol{E} = -\frac{\partial}{\partial t}\{\textbf{curl } (\mu_0 \boldsymbol{H})\} = -\mu_0 \frac{\partial}{\partial t}\left\{\varepsilon_0 \frac{\partial \boldsymbol{E}}{\partial t}\right\}.$$

$$\therefore \ \textbf{grad} \text{ div } \boldsymbol{E} - \nabla^2 \boldsymbol{E} = -\varepsilon_0 \mu_0 \frac{\partial^2 \boldsymbol{E}}{\partial t^2}.$$

Since there is no charge density present, div $\boldsymbol{E} = 0$ and $\boldsymbol{E}$ satisfies the (vector) wave equation

$$\nabla^2 \boldsymbol{E} = \varepsilon_0 \mu_0 \frac{\partial^2 \boldsymbol{E}}{\partial t^2}. \tag{10.11}$$

Similar manipulations, eliminating $\boldsymbol{E}$, $\boldsymbol{D}$, lead to the same (vector) wave-equation for $\boldsymbol{B}$, viz.

$$\nabla^2 \boldsymbol{B} = \varepsilon_0 \mu_0 \frac{\partial^2 \boldsymbol{B}}{\partial t^2}. \tag{10.12}$$

The velocity of propagation of these waves is $(\varepsilon_0 \mu_0)^{-1/2}$. This velocity is the same as the velocity of light, as shown in eqn. (9.17). This is how Maxwell's theory predicted the existence of electromagnetic radiation, a prediction that was confirmed subsequently by Hertz's experiments [see Vol. 1, § 1.1(5)].

The field vectors for the static cases were obtained from potential fields

$$\boldsymbol{E} = -\textbf{grad } V, \quad \boldsymbol{B} = \textbf{curl } \boldsymbol{A}.$$

These can now be generalized. From Maxwell's equation div $\boldsymbol{B} = 0$ we can write

$$\boldsymbol{B} = \textbf{curl } \boldsymbol{A}, \tag{10.13}$$

where $\boldsymbol{A}$, the ***vector potential***, is arbitrary to the extent that the gradient of a scalar may be added to it. Equation (10.13) together with (10.4) gives

$$\textbf{curl}\left(\boldsymbol{E} + \frac{\partial \boldsymbol{A}}{\partial t}\right) = \boldsymbol{0}.$$

Therefore

$$\boldsymbol{E} = -\textbf{grad } V - \frac{\partial \boldsymbol{A}}{\partial t}. \tag{10.14}$$

The two potentials introduced here are the generalizations of the static potentials. One of the remaining two of Maxwell's equations (10.5) gives, when ε and μ are uniform (i.e. independent of time and position),

$$\textbf{curl } (\boldsymbol{B}/\mu) = \textbf{curl}\left(\frac{1}{\mu}\textbf{ curl } \boldsymbol{A}\right) = \varepsilon\left(-\textbf{grad } \frac{\partial V}{\partial t} - \frac{\partial^2 \boldsymbol{A}}{\partial t^2}\right) + \boldsymbol{J}.$$

$$\therefore \ \textbf{grad} \text{ div } \boldsymbol{A} - \nabla^2 \boldsymbol{A} = -\mu\varepsilon\left(\textbf{grad } \frac{\partial V}{\partial t} + \frac{\partial^2 \boldsymbol{A}}{\partial t^2}\right) + \mu \boldsymbol{J}.$$

$$\therefore \ \nabla^2 \boldsymbol{A} - \mu\varepsilon \frac{\partial^2 \boldsymbol{A}}{\partial t^2} = -\mu \boldsymbol{J} + \textbf{grad}\left(\text{div } \boldsymbol{A} + \mu\varepsilon \frac{\partial V}{\partial t}\right). \tag{10.15}$$

Further, eqns. (10.6) and (10.14) give

$$\text{div}\ \boldsymbol{D} = -\text{div}\left(\varepsilon\ \textbf{grad}\ V + \varepsilon\frac{\partial \boldsymbol{A}}{\partial t}\right) = \varrho.$$

$$\therefore\ \text{div}\ \textbf{grad}\ V = -\frac{\varrho}{\varepsilon} - \frac{\partial}{\partial t}(\text{div}\ \boldsymbol{A}).$$

Adding the term $-\mu\varepsilon\ \partial^2 V/\partial t^2$ to each side of this equation we obtain

$$\nabla^2 V - \mu\varepsilon\frac{\partial^2 V}{\partial t^2} = -\frac{\varrho}{\varepsilon} - \frac{\partial}{\partial t}\left(\text{div}\ \boldsymbol{A} + \mu\varepsilon\frac{\partial V}{\partial t}\right). \tag{10.16}$$

We resolve the indeterminacy in the vector $\boldsymbol{A}$ by imposing the condition

$$\text{div}\ \boldsymbol{A} + \mu\varepsilon\frac{\partial V}{\partial t} = 0 \tag{10.17}$$

so that eqns. (10.15), (10.16) become, respectively,

$$\nabla^2 \boldsymbol{A} - \mu\varepsilon\frac{\partial^2 \boldsymbol{A}}{\partial t^2} = -\mu \boldsymbol{J}, \quad \nabla^2 V - \mu\varepsilon\frac{\partial^2 V}{\partial t^2} = -\frac{\varrho}{\varepsilon}. \tag{10.18, 19}$$

In free space, where $\mu = \mu_0$, $\varepsilon = \varepsilon_0$, $\varrho = 0$, $\boldsymbol{J} = \boldsymbol{0}$, these latter equations show that $\boldsymbol{A}$ and V satisfy the wave-equation for propagation with velocity $(\mu_0\varepsilon_0)^{-1/2}$, similar to the eqns. (10.11), (10.12) for $\boldsymbol{E}$, $\boldsymbol{B}$.

Also, for the static case eqns. (10.18), (10.19) reduce to

$$\nabla^2 \boldsymbol{A} = -\mu \boldsymbol{J}, \quad \nabla^2 V = -\varrho/\varepsilon,$$

which were obtained previously (see § 7.2 and eqn. (4.14) when ε is uniform). The solutions of eqns. (10.18), (10.19), known as retarded potentials, are

$$\boldsymbol{A} = \frac{1}{4\pi}\iiint\frac{[\mu \boldsymbol{J}]\,d\tau}{r}, \quad V = \frac{1}{4\pi}\iiint\frac{[\varrho]\,d\tau}{\varepsilon r}, \tag{10.20}$$

where the brackets [ ] stand for the integrand at time $t - r/c$, when the potential is calculated for time t. (These retarded potentials will be discussed further in Vol. 3, Chapter 13.)

10.3 Boundary conditions at discontinuities

Boundary conditions at discontinuities were first discussed in connection with electrostatic fields in § 2.8 and for magnetostatic fields in § 6.4. There we considered the effect of the presence of surface distributions of charge on the potentials V and $\boldsymbol{A}$. From a consideration of the integrals such as $4\pi V = \iiint \varrho\, d\tau/(\varepsilon r)$ discontinuities arise because r vanishes when the field point crosses a surface distribution. We found that for simple surface distributions of charge V is continuous as the field point crosses the surface charge. The only modification necessary now is that the integral expressions should be replaced by the retarded potentials. Because the retardation r/c is zero when r vanishes the previous result must still apply for the contribution to the potential from the immediately local surface distribution. Hence we conclude that

$$V_+ - V_- = 0 \tag{10.21}$$

at surfaces carrying a surface distribution of charge.

Similar arguments based on the integrals $4\pi \boldsymbol{A} = \iiint \mu \boldsymbol{J}\, d\tau/r$ and the corresponding retarded potentials show that $\boldsymbol{A}$ is continuous at a surface carrying a surface current $\boldsymbol{K}$, i.e.

$$\boldsymbol{A}_+ - \boldsymbol{A}_- = \boldsymbol{0}. \tag{10.22}$$

In previous investigations into the discontinuities in the field vectors or the derivatives of the potentials, we used devices of evaluating line integrals or surface integrals around suitable (small) figures. Here we give a somewhat different method of obtaining the results based on the cartesian forms of the equations. We consider a "transition layer" between the regions labelled + (outside) and − (inside). This layer is of thickness δ and the coordinate axes have been chosen so that at the point under consideration the layer lies between the planes $z = 0$, $z = \delta$ for a small area surrounding the

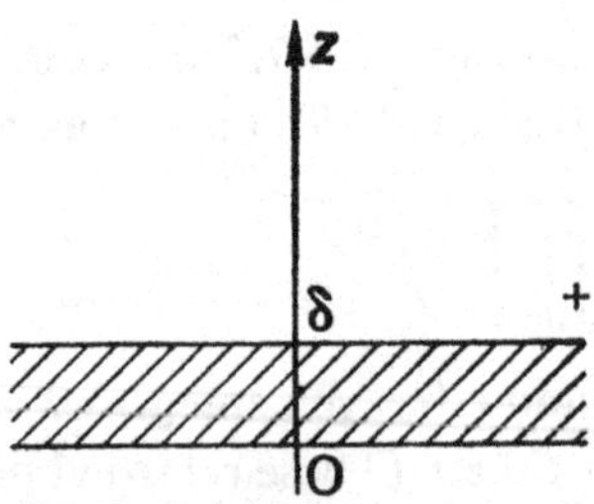

FIG. 10.2

origin, Fig. 10.2. Within the layer a charge density ϱ is "compressed", when δ is decreased, to a surface charge σ, where

$$\sigma = \lim_{\delta \to 0} (\varrho\delta)$$

and ϱ becomes infinite. Similarly a volume current $\boldsymbol{J}$ is "compressed" into a surface current $\boldsymbol{K}$ where

$$\boldsymbol{K} = \lim_{\delta \to 0} \delta \boldsymbol{J} \quad (\boldsymbol{J}\cdot\boldsymbol{k} = 0 \quad \text{for} \quad z = 0, z = \delta).$$

We further assume that at all stages of the process of compression the field quantities and their derivatives are finite, and the change in ϱ or J is continuous across the layer.

To find the discontinuity in V we note that V satisfies the equation div (ε **grad** V) $= -\varrho$, i.e.

$$\frac{\partial}{\partial x}\left(\varepsilon\frac{\partial V}{\partial x}\right)+\frac{\partial}{\partial y}\left(\varepsilon\frac{\partial V}{\partial y}\right)+\frac{\partial}{\partial z}\left(\varepsilon\frac{\partial V}{\partial z}\right)=-\varrho.$$

We integrate this equation w.r. to z (along the z-axis) through the layer obtaining

$$\int_0^\delta\left\{\frac{\partial}{\partial x}\left(\varepsilon\frac{\partial V}{\partial x}\right)+\frac{\partial}{\partial y}\left(\varepsilon\frac{\partial V}{\partial y}\right)\right\}dz+\left[\varepsilon\frac{\partial V}{\partial z}\right]_0^\delta=-\int_0^\delta\varrho\,dz.$$

The integral on the left-hand side has a value $F\delta$ where F is a finite quantity, being the value of the integrand (which remains finite in the layer) at some internal point of the layer (by the mean value theorem). The integral on the right-hand side is in the layer above unit area of the surface at $z = 0$. Therefore, in the limit $\delta \to 0$,

$$\left(\varepsilon\frac{\partial V}{\partial z}\right)_+-\left(\varepsilon\frac{\partial V}{\partial z}\right)_-=-\sigma. \qquad (10.23)$$

The vector potential A satisfies the equation

$$\nabla^2 A=-\mu J,$$

which in cartesians is

$$\frac{\partial^2 A}{\partial x^2}+\frac{\partial^2 A}{\partial y^2}+\frac{\partial^2 A}{\partial z^2}=-\mu J,$$

where μ is uniform. Integrating this equation across the transition layer gives

$$\left[\frac{1}{\mu}\frac{\partial A}{\partial z}\right]_0^\delta=-\int_0^\delta J\,dz.$$

$$\therefore\quad\left(\frac{1}{\mu}\frac{\partial A}{\partial z}\right)_+-\left(\frac{1}{\mu}\frac{\partial A}{\partial z}\right)_-=-K. \qquad (10.24)$$

Since Oz is in the direction of the normal the two results (10.23) and (10.24) can be written, without reference to any particular frame of axes, as

$$\left(\varepsilon\frac{\partial V}{\partial n}\right)_+-\left(\varepsilon\frac{\partial V}{\partial n}\right)_-=-\sigma,\qquad\left(\frac{1}{\mu}\frac{\partial A}{\partial n}\right)_+-\left(\frac{1}{\mu}\frac{\partial A}{\partial n}\right)_-=-J, \qquad (10.25, 26)$$

where $\partial/\partial n$ denotes differentiation along the normal.

We saw in the case of static fields that there is a surface equation corresponding to each of the field equations. The results for the general case are the same as those obtained earlier, but we give here the derivation of only two of them by means of the method we have just used.

To find the boundary conditions on the tangential components of $\boldsymbol{H}$ we consider the equation

$$\operatorname{curl} \boldsymbol{H} = \boldsymbol{J} + \frac{\partial \boldsymbol{D}}{\partial t}$$

written out in component form thus:

$$\frac{\partial H_z}{\partial y} - \frac{\partial H_y}{\partial z} = J_x + \frac{\partial D_x}{\partial t}, \quad \frac{\partial H_x}{\partial z} - \frac{\partial H_z}{\partial x} = J_y + \frac{\partial D_y}{\partial t},$$

$$\frac{\partial H_y}{\partial x} - \frac{\partial H_x}{\partial y} = J_z + \frac{\partial D_z}{\partial t}.$$

We integrate each of these w.r. to z across the transition layer. The left-hand sides give

$$\begin{aligned}
\int_0^\delta \left(\frac{\partial H_z}{\partial y} - \frac{\partial H_y}{\partial z}\right) dz &= \left(\frac{\partial H_z}{\partial y}\right)_m \delta - \Big[H_y\Big]_0^z, \\
\int_0^\delta \left(\frac{\partial H_x}{\partial z} - \frac{\partial H_z}{\partial x}\right) dz &= \Big[H_x\Big]_0^\delta - \left(\frac{\partial H_z}{\partial x}\right)_m \delta, \qquad (10.27) \\
\int_0^\delta \left(\frac{\partial H_y}{\partial x} - \frac{\partial H_x}{\partial y}\right) dz &= \left(\frac{\partial H_y}{\partial x} - \frac{\partial H_x}{\partial y}\right)_m \delta.
\end{aligned}$$

The notation $(\ )_m$ denotes an appropriate mean value of the integrand. The integration of the right-hand sides gives

$$\begin{aligned}
\int_0^\delta \left(J_x + \frac{\partial D_x}{\partial t}\right) dz &= K_x + \left(\frac{\partial D_x}{\partial t}\right)_m \delta, \\
\int_0^\delta \left(J_y + \frac{\partial D_y}{\partial t}\right) dz &= K_y + \left(\frac{\partial D_y}{\partial t}\right)_m \delta, \qquad (10.28) \\
\int_0^\delta \left(J_z + \frac{\partial D_z}{\partial t}\right) dz &= \left(\frac{\partial D_z}{\partial t}\right)_m \delta.
\end{aligned}$$

The third of these integrals gives only the second contribution because $\boldsymbol{J}$ is parallel to the surfaces of the layer, i.e. $J_z = 0$ at these points. The "compression" of $\boldsymbol{J}$ leads to the components K_x, K_y of the surface current. The equations (10.27), (10.28) give the limiting values

$$H_{y+} - H_{y-} = -K_x, \quad H_{x+} - H_{x-} = K_y, \quad 0 = 0,$$
$$\boldsymbol{k} \times (\boldsymbol{H}_+ - \boldsymbol{H}_-) = \boldsymbol{K}. \tag{10.29}$$

Next we derive the boundary conditions satisfied by D_n by considering the equation div $\boldsymbol{D} = \varrho$, i.e.

$$\frac{\partial D_x}{\partial x} + \frac{\partial D_y}{\partial y} + \frac{\partial D_z}{\partial z} = \varrho.$$

Then
$$\int_0^\delta \left(\frac{\partial D_x}{\partial x} + \frac{\partial D_y}{\partial y} \right) dz + \Big[D_z \Big]_0^\delta = \int_0^\delta \varrho \, dz.$$

Since we assume, as above, that $\partial D_x / \partial x$, $\partial D_y / \partial y$ are finite, the limiting value is

$$D_{z+} - D_{z-} = \sigma \quad \text{or} \quad \boldsymbol{k} \cdot (\boldsymbol{D}_+ - \boldsymbol{D}_-) = \sigma. \tag{10.30}$$

A similar treatment starting from:

$$\text{curl}\, \boldsymbol{E} = -\frac{\partial \boldsymbol{B}}{\partial t} \quad \text{leads to} \quad \boldsymbol{k} \times (\boldsymbol{E}_+ - \boldsymbol{E}_-) = \boldsymbol{0}; \tag{10.31}$$

and
$$\text{div}\, \boldsymbol{B} = 0 \quad \text{leads to} \quad \boldsymbol{k} \cdot (\boldsymbol{B}_+ - \boldsymbol{B}_-) = 0. \tag{10.32}$$

Since $\boldsymbol{k}$ is the unit vector perpendicular to the surface, we can rewrite (10.29, 30, 31, 32) with no reference to a frame of axes as follows:

$$\begin{aligned}
\text{curl}\, \boldsymbol{E} &= -\frac{\partial \boldsymbol{B}}{\partial t} && \text{corresponds to} \quad \hat{\boldsymbol{\nu}} \times (\boldsymbol{E}_+ - \boldsymbol{E}_-) = \boldsymbol{0}; \\
\text{curl}\, \boldsymbol{H} &= \boldsymbol{J} + \frac{\partial \boldsymbol{D}}{\partial t} && \text{corresponds to} \quad \hat{\boldsymbol{\nu}} \times (\boldsymbol{H}_+ - \boldsymbol{H}_-) = \boldsymbol{K}; \\
\text{div}\, \boldsymbol{D} &= \varrho && \text{corresponds to} \quad \hat{\boldsymbol{\nu}} \cdot (\boldsymbol{D}_+ - \boldsymbol{D}_-) = \sigma; \\
\text{div}\, \boldsymbol{B} &= 0 && \text{corresponds to} \quad \hat{\boldsymbol{\nu}} \cdot (\boldsymbol{B}_+ - \boldsymbol{B}_-) = 0,
\end{aligned} \tag{10.33}$$

where $\hat{\boldsymbol{\nu}}$ is the unit normal directed from side 1 into side 2 at the surface of discontinuity.

10.4 Poynting's vector

An electromagnetic field arises from a distribution of charges and currents (which may be varying). If the dielectrics, conductors and other bodies in the field are stationary, the forces acting on them, to hold them in position, do no work and the complete system receives no energy from them. The only source of energy for the system is the "electromotive field" $\boldsymbol{E}'$, i.e. batteries, which produce the electric currents. These batteries provide energy at the rate $\boldsymbol{J}\cdot\boldsymbol{E}'\,d\tau$ in a volume element $d\tau$ of the region.

The state of the system can be described at any instant by the values of the field strength, current strengths, charge density, etc., which satisfy Maxwell's equations (10.4)–(10.7). From these we can deduce that

$$\boldsymbol{H}\cdot\text{curl}\,\boldsymbol{E}-\boldsymbol{E}\cdot\text{curl}\,\boldsymbol{H} = -\boldsymbol{H}\cdot\frac{\partial \boldsymbol{B}}{\partial t}-\boldsymbol{E}\cdot\frac{\partial \boldsymbol{D}}{\partial t}-\boldsymbol{J}\cdot\boldsymbol{E}.$$

But, since $\boldsymbol{J} = \sigma(\boldsymbol{E}+\boldsymbol{E}')$, it follows that

$$\boldsymbol{J}\cdot\boldsymbol{E} = J^2/\sigma-\boldsymbol{J}\cdot\boldsymbol{E}'$$

and

$$\boldsymbol{J}\cdot\boldsymbol{E}' = \frac{J^2}{\sigma}+\boldsymbol{H}\cdot\frac{\partial \boldsymbol{B}}{\partial t}+\boldsymbol{E}\cdot\frac{\partial \boldsymbol{D}}{\partial t}+\boldsymbol{H}\cdot\text{curl}\,\boldsymbol{E}-\boldsymbol{E}\cdot\text{curl}\,\boldsymbol{H}$$

$$= \frac{J^2}{\sigma}+\boldsymbol{H}\cdot\frac{\partial \boldsymbol{B}}{\partial t}+\boldsymbol{E}\cdot\frac{\partial \boldsymbol{D}}{\partial t}+\text{div}\,(\boldsymbol{E}\times\boldsymbol{H}).$$

The integral of this relation is taken throughout the interior, V, of an arbitrary surface S, so that

$$\iiint\limits_V \boldsymbol{J}\cdot\boldsymbol{E}'\,d\tau = \iiint\limits_V \frac{J^2}{\sigma}\,d\tau+\iiint\limits_V\left(\boldsymbol{H}\cdot\frac{\partial \boldsymbol{B}}{\partial t}+\boldsymbol{E}\cdot\frac{\partial \boldsymbol{D}}{\partial t}\right)d\tau+\oint\!\!\oint\limits_S \boldsymbol{E}\times\boldsymbol{H}\cdot\mathbf{d}\boldsymbol{S}. \tag{10.34}$$

We can give a physical interpretation to each of the volume integrals as follows:

$\iiint\limits_V \boldsymbol{J}\cdot\boldsymbol{E}'\,d\tau$ is rate of supply of energy from batteries,

$\iiint\limits_V \frac{J^2}{\sigma}\,d\tau$ is rate of production of heat,

$\iiint\limits_V \boldsymbol{H}\cdot\frac{\partial \boldsymbol{B}}{\partial t}\,d\tau$ is the rate of increase of magnetic energy [see eqn. (9.40)],

$\iiint\limits_V \boldsymbol{E}\cdot\frac{\partial \boldsymbol{D}}{\partial t}\,d\tau$ is the rate of increase of electric energy [see eqn. (4.23)]

taking place inside S. We therefore interpret $\oint\!\!\oint\limits_S \boldsymbol{E}\times\boldsymbol{H}\cdot\mathbf{d}\boldsymbol{S}$ as the rate at which energy is transported outwards across the surface S. Strictly we cannot deduce that the "density of energy transport", i.e. the energy transported across unit area (cf. current density) is $\boldsymbol{E}\times\boldsymbol{H}$ at every point. This rate of transport could be augmented by any vector whose divergence vanishes; however, the assumption that $\boldsymbol{E}\times\boldsymbol{H}$ is the actual density of energy flow must give the correct result when integrated over any closed surface. We make the assumption that this is the correct expression for the flow density; this is *Poynting's theorem*. It is confirmed in a variety of special cases, some of which will be discussed later. The vector $\boldsymbol{E}\times\boldsymbol{H}$ is called the *Poynting vector*.

Example 1. We consider a section of length l of a current of strength I uniformly distributed in a long straight conducting wire with circular cross-section of radius a, the resistance of the length l being R.

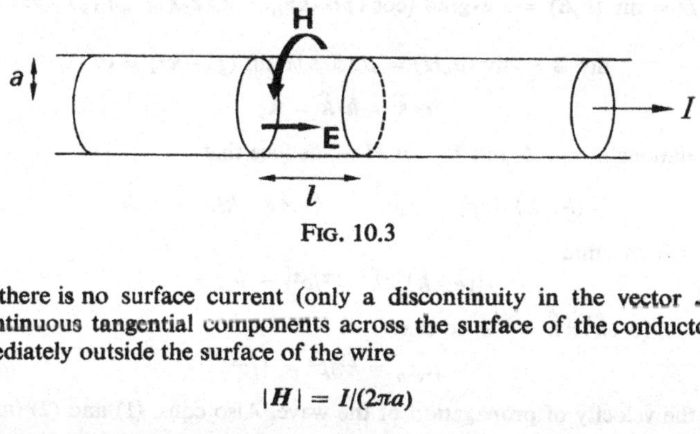

FIG. 10.3

Because there is no surface current (only a discontinuity in the vector $\boldsymbol{J}$) both $\boldsymbol{E}$, $\boldsymbol{H}$ have continuous tangential components across the surface of the conductor (see Fig. 10.3). Immediately outside the surface of the wire

$$|\boldsymbol{H}| = I/(2\pi a)$$

and is directed around the circumference of the wire. Since the current is uniform the electric field, or potential gradient, is given by $|\boldsymbol{E}| = V/l$, where V is the potential difference between the ends of the section. Since $\boldsymbol{E}$ and $\boldsymbol{H}$ are perpendicular to each other,

$$|\boldsymbol{E}\times\boldsymbol{H}| = \frac{VI}{2\pi al}, \quad \text{where} \quad V = RI.$$

$$\therefore \quad |\boldsymbol{E}\times\boldsymbol{H}| = RI^2/(2\pi al).$$

The vector $\boldsymbol{E}$ is parallel to the direction of I so that the vector product $\boldsymbol{E}\times\boldsymbol{H}$ is directed into the wire. Poynting's theorem implies that the energy crossing the surface of the wire from the field into the metal is equal to the Joule heat developed inside the section.

Example 2. We investigate *the propagation of a plane wave in free space.*

We represent the wave by the field vectors

$$\boldsymbol{E} = \boldsymbol{e}\cos(pt-kz), \quad \boldsymbol{H} = \boldsymbol{h}\cos(pt-kz),$$

where $\boldsymbol{e}$, $\boldsymbol{h}$ are constant vectors. These give a plane wave propagated in the z-direction, i.e. in the direction of the unit vector $\hat{\boldsymbol{k}}$, with speed p/k. These vectors must satisfy Maxwell's equations. Then

$$-\frac{\partial \boldsymbol{B}}{\partial t} = \textbf{curl}\,\boldsymbol{E} = -\boldsymbol{e}\times\textbf{grad}\,\{\cos(pt-kz)\}$$
$$= -(\boldsymbol{e}\times\hat{\boldsymbol{k}})k\sin(pt-kz).$$

Integration w.r. to t now gives

$$\boldsymbol{B} = -(\boldsymbol{e}\times\hat{\boldsymbol{k}})(k/p)\cos(pt-kz).$$

Similarly,

$$\frac{\partial \boldsymbol{D}}{\partial t} = \textbf{curl}\,\boldsymbol{H} = -\boldsymbol{h}\times\textbf{grad}\,\{\cos(pt-kz)\},$$
$$= -(\boldsymbol{h}\times\hat{\boldsymbol{k}})k\sin(pt-kz).$$
$$\therefore\quad \boldsymbol{D} = (\boldsymbol{h}\times\hat{\boldsymbol{k}})(k/p)\cos(pt-kz).$$

The remaining equations give

$$\operatorname{div}\boldsymbol{D} = \operatorname{div}(\varepsilon_0\boldsymbol{E}) = \varepsilon_0\boldsymbol{e}\cdot\textbf{grad}\,\{\cos(pt-kz)\} = \varepsilon_0(\boldsymbol{e}\cdot\hat{\boldsymbol{k}})k\sin(pt-kz) = 0$$

and

$$\operatorname{div}\boldsymbol{B} = \operatorname{div}(\mu_0\boldsymbol{H}) = \mu_0(\boldsymbol{h}\cdot\hat{\boldsymbol{k}})k\sin(pt-kz) = 0.$$

Hence

$$\boldsymbol{e}\cdot\hat{\boldsymbol{k}} = \boldsymbol{h}\cdot\hat{\boldsymbol{k}} = 0. \tag{1}$$

From the relations $\boldsymbol{D} = \varepsilon_0\boldsymbol{E}$ and $\boldsymbol{B} = \mu_0\boldsymbol{H}$ we deduce that

$$(\boldsymbol{h}\times\hat{\boldsymbol{k}})(k/p) = \varepsilon_0\boldsymbol{e},\qquad -(\boldsymbol{e}\times\hat{\boldsymbol{k}})(k/p) = \mu_0\boldsymbol{h}. \tag{2}$$

We eliminate $\boldsymbol{h}$ and find

$$-[(\boldsymbol{e}\times\hat{\boldsymbol{k}})\times\hat{\boldsymbol{k}}](k^2/p^2) = \mu_0\varepsilon_0\boldsymbol{e}.$$

From eqn. (1), $(\boldsymbol{e}\times\hat{\boldsymbol{k}})\times\hat{\boldsymbol{k}} = -\boldsymbol{e}$, so that

$$\mu_0\varepsilon_0 = k^2/p^2 = 1/c^2,$$

where c is the velocity of propagation of the wave. Also eqns. (1) and (2) further imply that $\boldsymbol{e}\cdot\boldsymbol{h} = 0$; this means that $\boldsymbol{e}$, $\boldsymbol{h}$, $\hat{\boldsymbol{k}}$ form a right-handed set of mutually perpendicular directions. These waves are therefore transverse waves and the amplitudes are related by

$$|\boldsymbol{h}| = (\varepsilon_0 p/k)|\boldsymbol{e}| = \varepsilon_0 c|\boldsymbol{e}|,\qquad |\boldsymbol{e}| = \mu_0 c|\boldsymbol{h}|.$$

The Poynting vector for this field is

$$\boldsymbol{E}\times\boldsymbol{H} = (\boldsymbol{e}\times\boldsymbol{h})\cos^2(pt-kz).$$

But

$$\boldsymbol{e}\times\boldsymbol{h} = (\boldsymbol{h}\times\hat{\boldsymbol{k}})\times\boldsymbol{h}/(\varepsilon_0 c) = (\boldsymbol{e}\times\hat{\boldsymbol{k}})\times\boldsymbol{e}/(\mu_0 c)$$
$$= |\boldsymbol{h}|^2\hat{\boldsymbol{k}}/(\varepsilon_0 c) = |\boldsymbol{e}|^2\hat{\boldsymbol{k}}/(\mu_0 c).$$

Thus the wave discussed corresponds to a flow of energy in the positive z-direction, $\hat{\boldsymbol{k}}$, the intensity of the flow having a mean value

$$|\boldsymbol{h}|^2/(2\varepsilon_0 c) = |\boldsymbol{e}|^2/(2\mu_0 c).$$

Example 3. If the electric-intensity $\boldsymbol{E}$ and magnetic-intensity $\boldsymbol{B}$ are combined into a single vector $\boldsymbol{M} = \boldsymbol{E}+\mathrm{i}\boldsymbol{B}$, show that Maxwell's equations for free space are satisfied if

$$\boldsymbol{M} = \textbf{curl}\left\{\frac{\mathrm{i}}{c^2}\frac{\partial \boldsymbol{S}}{\partial t}+\textbf{curl}\,\boldsymbol{S}\right\},$$

where $\boldsymbol{S}$ is a vector function (regarded as referred to rectangular coordinates) such that

$$\nabla^2 \boldsymbol{S} = \frac{1}{c^2}\frac{\partial^2 \boldsymbol{S}}{\partial t^2}.$$

Show that a possible $\boldsymbol{M}$ can be generated in this way from

$$\boldsymbol{S} = \mathrm{i}C\boldsymbol{k}r^{-1}\sin\frac{\omega r}{c}\cos\omega t,$$

where C is a real constant, $\boldsymbol{k}$ a constant unit vector, and r distance from the origin. Obtain the resulting expression for $\boldsymbol{E}$.

The equations to be satisfied are

$$\mathbf{curl}\,\boldsymbol{E} = -\partial\boldsymbol{B}/\partial t, \quad \mathbf{curl}\,\boldsymbol{H} = \partial\boldsymbol{D}/\partial t, \quad \mathrm{div}\,\boldsymbol{D} = 0, \quad \mathrm{div}\,\boldsymbol{B} = 0$$

with $\boldsymbol{D} = \varepsilon_0\boldsymbol{E}$, and $\boldsymbol{B} = \mu_0\boldsymbol{H}$.

The "potentials" which satisfy the equations are $\boldsymbol{A}$ and V, where

$$\boldsymbol{E} = -\mathbf{grad}\,V - \partial\boldsymbol{A}/\partial t, \quad \boldsymbol{B} = \mathbf{curl}\,\boldsymbol{A}, \quad \text{with} \quad \mathrm{div}\,\boldsymbol{A} + \varepsilon_0\mu_0\,\partial V/\partial t = 0.$$

If we choose $\boldsymbol{A} = \varepsilon_0\mu_0\,\partial\boldsymbol{S}/\partial t$, $V = -\mathrm{div}\,\boldsymbol{S}$, the last equation is satisfied identically and

$$\boldsymbol{E} = \mathbf{grad}\,\mathrm{div}\,\boldsymbol{S} - \varepsilon_0\mu_0\,\partial^2\boldsymbol{S}/\partial t^2 = \mathbf{curl}\,\mathbf{curl}\,\boldsymbol{S} + \nabla^2\boldsymbol{S} - \varepsilon_0\mu_0\partial^2\boldsymbol{S}/\partial t^2,$$

$$\boldsymbol{B} = \varepsilon_0\mu_0\,\mathbf{curl}\,(\partial\boldsymbol{S}/\partial t).$$

$$\therefore \quad \boldsymbol{M} = \boldsymbol{E} + \mathrm{i}\boldsymbol{B} = \mathbf{curl}\,\{\mathbf{curl}\,\boldsymbol{S} + \mathrm{i}\varepsilon_0\mu_0\,\partial\boldsymbol{S}/\partial t\} + \nabla^2\boldsymbol{S} - \varepsilon_0\mu_0\,\partial^2\boldsymbol{S}/\partial t^2.$$

Since $\varepsilon_0\mu_0 = 1/c^2$ the required form of solution is obtained.

Since the given vector $\boldsymbol{S}$ has only a z-component the wave-equation becomes

$$\left(\frac{\partial^2}{\partial x^2} + \frac{\partial^2}{\partial y^2} + \frac{\partial^2}{\partial z^2}\right)S_z = \frac{1}{c^2}\frac{\partial^2 S_z}{\partial t^2}.$$

Because S_z depends on the coordinates only through r the left-hand side becomes

$$\frac{1}{r^2}\frac{\mathrm{d}}{\mathrm{d}r}\left(r^2\frac{\mathrm{d}S_z}{\mathrm{d}r}\right) = \frac{1}{r^2}\frac{\mathrm{d}}{\mathrm{d}r}\left(-\sin\frac{\omega r}{c} + \frac{\omega r}{c}\cos\frac{\omega r}{c}\right)\mathrm{i}C\cos\omega t$$

$$= \frac{1}{r^2}\left(-\frac{\omega}{c}\cos\frac{\omega r}{c} + \frac{\omega}{c}\cos\frac{\omega r}{c} - \frac{\omega^2}{c^2}\,r\sin\omega r\right)\mathrm{i}C\cos\omega t$$

$$= -\frac{\omega^2\,\mathrm{i}C}{c^2 r}\sin\omega r\cos\omega t$$

and the right-hand side is

$$\frac{1}{c^2}\frac{\partial^2 S_z}{\partial t^2} = -\frac{\omega^2\,\mathrm{i}C}{c^2 r}\sin\frac{\omega r}{c}\cos\omega t.$$

Hence this vector $\boldsymbol{S}$ satisfies the wave-equation.

The vector $\boldsymbol{S}$ is purely imaginary, and $\boldsymbol{E}$ is the real part of the expression for $\boldsymbol{M}$, hence

$$\boldsymbol{E} = \frac{\mathrm{i}}{c^2}\,\mathbf{curl}\,\frac{\partial\boldsymbol{S}}{\partial t} = -\frac{C}{c^2}\,\mathbf{curl}\left(\frac{\boldsymbol{k}}{r}\sin\frac{\omega r}{c}\cos\omega t\right)$$

$$= \frac{C}{c^2}\,\boldsymbol{k}\times\mathbf{grad}\left(\frac{1}{r}\sin\frac{\omega r}{c}\right)\cos\omega t$$

$$= \frac{C}{c^2}\frac{\boldsymbol{k}\times\boldsymbol{r}}{r}\left(-\frac{1}{r^2}\sin\frac{\omega r}{c^2} + \frac{\omega}{cr}\cos\frac{\omega r}{c}\right)\cos\omega t.$$

10.5 Maxwell's stress tensor and electromagnetic momentum

In § 4.6 we discussed the ponderomotive forces exerted by the electrostatic field in terms of a stress tensor T_{ij}. This discussion assumed that the forces exerted on bodies on side 2 (the inside of the surface G in Fig. 10.4)

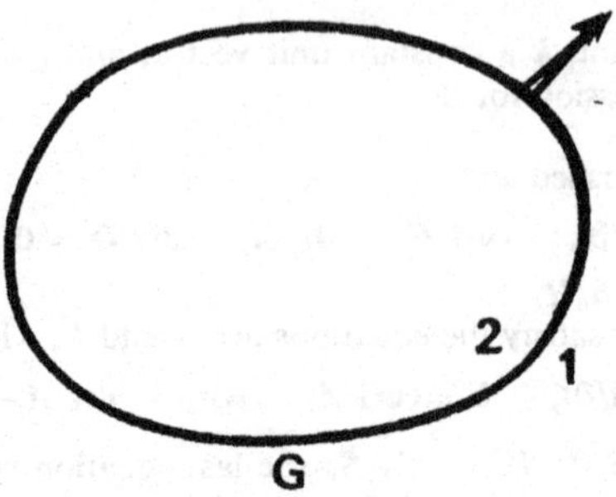

FIG. 10.4

were exactly equivalent to the forces from the stress system acting across G from side 1 onto side 2. Because the system considered in § 4.6 was stationary we wrote

$$\oint T_{ij}n_j \, dS = \iiint \varrho E_i \, d\tau$$

$$\oint e_{ijk}x_jT_{kl}n_l \, dS = \iiint e_{ijk}\varrho x_jE_k \, d\tau$$

[see eqn. (4.32)].

In the present case where the system is altering with time **curl** $\boldsymbol{E} \neq 0$, and we must consider the possibility of a change of the momentum inside G in addition to the ponderomotive forces exerted inside G. We therefore supplement the equations of § 4.6 by the addition of a term corresponding to a momentum density g_i at each point inside G. The supplemented equations are

$$\oint\!\!\!\oint T_{ij}n_j \, dS = \iiint F_i \, d\tau + \iiint \frac{\partial g_i}{\partial t} \, d\tau \tag{10.35}$$

and

$$\oint\!\!\!\oint e_{ijk}x_jT_{kl}n_l \, dS = \iiint e_{ijk}x_jF_k \, d\tau + \iiint e_{ijk}x_j \frac{\partial g_k}{\partial t} \, d\tau. \tag{10.36}$$

Roughly, eqns. (10.35), (10.36) state that the forces exerted by the stress system on G are exactly equivalent to the ponderomotive forces on bodies (carrying currents or charges) inside G together with the rate of change of any additional momentum inside G. Equation (10.35) states the equivalence between the forces, and equation (10.36) states the equivalence of moments about the (arbitrary) origin.

In the present case the force F_i is given by

$$\boldsymbol{F} = \varrho \boldsymbol{E} + \boldsymbol{J} \times \boldsymbol{B}; \quad F_i = \varrho E_i + e_{ilm} J_l B_m.$$

As in § 4.6, we deduce that

$$\frac{\partial T_{ij}}{\partial x_j} - \frac{\partial g_i}{\partial t} = F_i = \varrho E_i + e_{ilm} J_l B_m \tag{10.37}$$

and

$$e_{ijk} T_{kj} = 0 \tag{10.38}$$

corresponding to (4.33), (4.34). The second of these equations implies that T_{ij} is symmetric in the suffixes.

In order to avoid the complications of forces exerted on non-uniformities of permittivity or permeability, we consider only the case of a vacuum (with ε_0, μ_0) in which charge ϱ and current density $\boldsymbol{J}$ are the only elements subject to ponderomotive forces.

Manipulations similar to but more complicated than those in § 4.6 show that

$$\begin{aligned} F_i &= \varepsilon_0 E_i \frac{\partial E_j}{\partial x_j} + e_{ilm} B_m \left(\mu_0^{-1} e_{lrs} \frac{\partial B_s}{\partial x_r} - \varepsilon_0 \frac{\partial E_l}{\partial t} \right) \\ &= \frac{\partial}{\partial x_j} \left(\varepsilon_0 E_i E_j - \frac{1}{2} \varepsilon_0 \delta_{ij} E_k E_k \right) + \frac{\partial}{\partial x_j} \left(\mu_0^{-1} B_i B_j - \frac{1}{2} \mu_0^{-1} \delta_{ij} B_k B_k \right) \\ &\quad - \frac{\partial}{\partial t} (\varepsilon_0 e_{ijk} E_j B_k). \end{aligned} \tag{10.39}$$

(The manipulations to establish this result are given in the example below.)

Comparison with eqn. (10.37) suggests the identification

$$T_{ij} = \varepsilon_0 (E_i E_j - \tfrac{1}{2} \delta_{ij} E_k^2) + \mu_0^{-1} (B_i B_j - \tfrac{1}{2} \delta_{ij} B_k^2) \tag{10.40}$$

and

$$g_i = \varepsilon_0 e_{ijk} E_j B_k, \quad \text{i.e.} \quad \boldsymbol{g} = \varepsilon_0 (\boldsymbol{E} \times \boldsymbol{B}).$$

Therefore

$$\boldsymbol{g} = \varepsilon_0 \mu_0 (\boldsymbol{E} \times \boldsymbol{H}) = \boldsymbol{P}/c^2$$

where $\boldsymbol{P}$ is the Poynting vector.

The stress tensor in the electrical components is identical with that found for the electrostatic case [eqn. (4.35)], and the magnetic part has an (exactly) equivalent form. Hence the interpretation in terms of forces in various special configurations is the same as in § 4.6 for electric components, and is similar for magnetic components.

The new feature is the momentum density $\boldsymbol{g}$. This means that, even in a vacuum, when an electromagnetic field corresponds to a flux of energy $\boldsymbol{P}$ at any point there is also an associated momentum $\boldsymbol{P}/c^2$ at that point.

Example. The manipulations which establish (10.39) use an important formula involving the alternating symbol e_{ijk} and the change of dummy suffixes to other letters. The relation is

$$e_{ijk}e_{ilm} = \delta_{jl}\delta_{km} - \delta_{jm}\delta_{kl}$$

which we use on the term

$$e_{ilm}B_m e_{lrs}\frac{\partial B_s}{\partial x_r} = e_{lmi}e_{lrs}B_m\frac{\partial B_s}{\partial x_r} = (\delta_{mr}\delta_{is} - \delta_{ms}\delta_{ir})B_m\frac{\partial B_s}{\partial x_r}$$

$$= B_m\frac{\partial B_i}{\partial x_m} - B_m\frac{\partial B_m}{\partial x_i}. \tag{1}$$

Next we turn the first term of eqn. (10.39) into the derivative of a product, and use a similar operation on (1) thus:

$$E_i\frac{\partial E_j}{\partial x_j} = \frac{\partial}{\partial x_j}(E_iE_j) - E_j\frac{\partial E_i}{\partial x_j};$$

$$B_m\frac{\partial B_i}{\partial x_m} = B_j\frac{\partial B_i}{\partial x_j} = \frac{\partial}{\partial x_j}(B_iB_j) - B_i\frac{\partial B_j}{\partial x_j}.$$

The last term of the above is zero because div $\boldsymbol{B} = \partial B_j/\partial x_j = 0$. We can write (10.39) now in the form

$$F_i = \frac{\partial}{\partial x_j}(\varepsilon_0 E_iE_j) - \varepsilon_0 E_j\frac{\partial E_i}{\partial x_j} + \frac{\partial}{\partial x_j}(\mu_0^{-1}B_iB_j)$$

$$-\mu_0^{-1}B_m\frac{\partial B_m}{\partial x_i} - e_{ilm}\left(\varepsilon_0 B_m\frac{\partial E_l}{\partial t}\right). \tag{2}$$

One more manipulation can be applied to the second term of (2), viz.

$$E_j\frac{\partial E_i}{\partial x_j} = E_j\left(\frac{\partial E_i}{\partial x_j} - \frac{\partial E_j}{\partial x_i}\right) + E_j\frac{\partial E_j}{\partial x_i}.$$

The first term on the right-hand side here is similar to the result (1), so that we can write

$$E_j\frac{\partial E_i}{\partial x_j} = E_j\frac{\partial E_j}{\partial x_i} + e_{kij}e_{klm}E_j\frac{\partial E_l}{\partial x_m} = E_j\frac{\partial E_j}{\partial x_i} + e_{ijk}E_j\left(e_{klm}\frac{\partial E_l}{\partial x_m}\right).$$

The last bracket above is the k-component of **curl** $\boldsymbol{E}$ so that, from Maxwell's equation,

$$E_j\frac{\partial E_i}{\partial x_j} = E_j\frac{\partial E_j}{\partial x_i} + e_{ijk}E_j\frac{\partial B_k}{\partial t}.$$

Hence

$$F_i = \frac{\partial}{\partial x_j}(\varepsilon_0 E_iE_j + \mu_0^{-1}B_iB_j) - \varepsilon_0 E_j\frac{\partial E_j}{\partial x_i} - \mu_0^{-1}B_j\frac{\partial B_j}{\partial x_i}$$

$$-e_{ilm}\varepsilon_0 B_m\frac{\partial E_l}{\partial t} - e_{ijk}\varepsilon_0 E_j\frac{\partial B_k}{\partial t}.$$

Since

$$E_j\frac{\partial E_j}{\partial x_i} = \frac{\partial}{\partial x_i}\left(\frac{1}{2}E_kE_k\right) = \delta_{ij}\frac{\partial}{\partial x_j}\left(\frac{1}{2}E_k^2\right)$$

with a similar result for the term $B_j(\partial B_j/\partial x_i)$, we can write all the differentiations in one bracket thus:

$$F_i = \frac{\partial}{\partial x_i}\left(\varepsilon_0 E_iE_j - \frac{1}{2}\varepsilon_0\delta_{ij}E_k^2 + \mu_0^{-1}B_iB_j - \frac{1}{2}\mu_0^{-1}\delta_{ij}B_k^2\right)$$

$$-e_{ijk}\varepsilon_0\left(E_j\frac{\partial B_k}{\partial t} + \frac{\partial E_j}{\partial t}B_k\right).$$

This is equivalent to eqn. (10.39).

Miscellaneous Exercises X

1. State Maxwell's equations for free space and prove that they are satisfied by

$$\boldsymbol{E} = -\frac{\partial \boldsymbol{A}}{\partial t}, \qquad \boldsymbol{H} = \operatorname{curl} \boldsymbol{A},$$

provided that

$$\operatorname{div} \boldsymbol{A} = 0, \qquad \nabla^2 \boldsymbol{A} = \frac{1}{c^2}\frac{\partial^2 \boldsymbol{A}}{\partial t^2}.$$

Derive $\boldsymbol{E}$ and $\boldsymbol{H}$ when

$$\boldsymbol{A} = \{a \cos 2\pi k(z-ct)\ b \sin 2\pi k(z-ct)\ 0\}.$$

Verify that $\boldsymbol{E}$ and $\boldsymbol{H}$ are perpendicular and that their directions rotate about the z-axis with frequency kc.

2. Verify that a possible form of the electric intensity vector $\boldsymbol{E}$ in the free space between two perfectly conducting planes $z = 0$ and $z = a$ is

$$E_x = E \cos\alpha \sin(kz\cos\alpha)\sin\{k(x\sin\alpha - ct)\},$$
$$E_y = 0,$$
$$E_z = E\sin\alpha\cos(kz\cos\alpha)\cos\{k(x\sin\alpha - ct)\},$$

where E and α ($\neq \frac{1}{2}\pi$) are constants. Show that k must satisfy a certain equation. Find the corresponding magnetic intensity vector $\boldsymbol{H}$ containing only periodic terms. Prove that the average energy flux is entirely in the x-direction and that the average flux across the area bounded by $y = 0$, $y = b$, $z = 0$ and $z = a$ is $(abE^2 \sin\alpha)/(4\mu_0 c)$.

3. A perfectly conducting medium occupies the half-space $x \leqslant 0$ while $x > 0$ is empty. On the boundary $x = 0$ there is a surface current density $\boldsymbol{J}$ and a surface charge density σ. Assuming that the electromagnetic field vanishes in $x < 0$ derive the boundary conditions for the plane $x = 0$.

Show that the magnetic field

$$\boldsymbol{H} = \{0\ \ \cos\alpha x\cos(\omega t - \gamma z)\ \ 0\}$$

in Cartesian coordinates is a solution of Maxwell's equations in free space provided that the constants α, γ, ω satisfy a certain relation. Find the corresponding form for the electric field $\boldsymbol{E}$. Show also that for certain values of α this solution is appropriate to the region between two perfect conductors defined by $x \leqslant 0$ and $x \geqslant a$. Calculate the charge and current densities induced on the plane $x = 0$.

4. A condenser formed by two parallel plates of area A and distance d apart ($d^2 \ll A$) has the space between the plates filled with a slightly conducting medium in which the conductivity σ and the dielectric constant K are functions of the distance x from one plate. At time $t = 0$ a charge Q_0 is given to the plate at $x = 0$ whilst the plate at $x = d$ is earthed. Neglecting edge effects, show that at time t the charge on the plate at $x = 0$ has become

$$Q = Q_0 \exp\left[-\frac{\sigma t}{\varepsilon_0 K}\right]$$

evaluated at $x = 0$.

Find the electric intensity at a point in the dielectric, and show that the charge density ϱ is given by

$$\varrho = -\frac{Q_0}{\varepsilon_0 A}\left\{\frac{\mathrm{d}}{\mathrm{d}x}\left(\frac{\sigma}{K}\right)\right\} t \exp\left\{-\left(\frac{\sigma t}{\varepsilon_0 K}\right)\right\}.$$

5. An electromagnetic wave in free space is represented by the equations

$$\boldsymbol{E} = \boldsymbol{E}_0 e^{i(\omega t - kz)}, \quad \boldsymbol{B} = \boldsymbol{B}_0 e^{i(\omega t - kz)}$$

where $\boldsymbol{E}_0$ and $\boldsymbol{B}_0$ are independent of z and t. If the component E_{0y} of $\boldsymbol{E}_0$ is zero and E_{0x} and E_{0z} are functions of x only, show by using Maxwell's equations that

$$B_x = B_z = 0, \quad E_x = -(ik/\beta^2)\frac{\partial E_z}{\partial x},$$

where $\beta^2 = (\omega^2/c^2) - k^2$ and c is the velocity of light. Hence or otherwise show that

$$\frac{\partial^2 E_z}{\partial x^2} + \beta^2 E_z = 0.$$

Find the appropriate solution for E_z if the space is bounded by the perfectly conducting plane walls $x = 0$ and $x = a$, and show that the least value of ω for wave propagation is $\pi c/a$.

ANSWERS TO THE EXERCISES

Exercises 6.3 (p. 215)

4. If $GP = x$, the field is

$$\frac{I}{2\pi}\left\{\frac{3}{x}+\frac{\sqrt{3}}{a-x\sqrt{3}}-\frac{(a+2x\sqrt{3})}{[a^2/4+\{x+a/(2\sqrt{3})\}^2]\sqrt{3}}\right\}$$

perpendicular to GA.

8. $4(m/a+n/b)I/(5\sqrt{5})$ along the axis.

Exercises 6.7 (p. 238)

1. $\dfrac{2\mu_0 a}{\pi}\left\{\ln\left(\dfrac{2+\sqrt{2}}{1+\sqrt{3}}\right)+\sqrt{3}+1-2\sqrt{2}\right\}.$

Miscellaneous Exercises VI (p. 239)

3. $\dfrac{q\omega a^3\cos\theta}{12\pi R^2}.$

5. $\{0 \quad 0 \quad \tfrac{1}{4}\mu_0 j(a^2-r^2)-\tfrac{1}{2}\mu_0 ja^2 \ln a\}$; inside $\{0 \quad \tfrac{1}{2}\mu_0 jr^2 \quad 0\}$, outside $\{0 \quad \mu_0 ja^2/(2r) \quad 0\}$.

Exercises 8.1 (p. 268)

1. Use $\displaystyle\iiint\limits_{\text{Volume}} (\mathbf{grad}\ V)^2\, d\tau = \iint\limits_{\Sigma} V\,\mathbf{grad}\ V\cdot d\mathbf{S} = V_0 \iint\limits_{\Sigma} \mathbf{grad}\ V\cdot d\mathbf{S}$

$$= V_0 \iiint\limits_{\text{Volume}} \nabla^2 V\, d\tau = 0.$$

2 If P is a point on the sphere, q_i is the charge at a point A_i inside the sphere and q_0 is the charge at a point A_0 outside, then

$$V_P = \frac{1}{4\pi\varepsilon_0}\left(\frac{q_i}{A_iP}+\frac{q_0}{A_0P}\right).$$

The mean value over the sphere of radius a is

$$\frac{1}{4\pi a^2}\oiint\limits_{\Sigma} V_P\, dS = \frac{1}{16\pi a^2\varepsilon_0}\left\{q_i\oiint\limits_{\Sigma}\frac{dS}{A_iP}+q_0\oiint\limits_{\Sigma}\frac{dS}{A_0P}\right\}.$$

But $\frac{1}{4\pi\varepsilon_0}\oiint_{\Sigma}\frac{dS}{A_tP}$ is the potential at A_t due to a unit surface layer of charge spread over the surface of the sphere and is therefore equal to a/ε_0 [see Example 4, p. 43]. Similarly, $\frac{1}{4\pi\varepsilon_0}\oiint_{\Sigma}\frac{dS}{A_0P}$ is $a^2/(\varepsilon_0 r)$. The required result follows by addition over all charges in the field.

3. Use the mean value theorem as in Example 4, note 5 on p. 268.
4. Consider V_1-V_2 and use the results of exercises 1 and 3.
5. Consider $V = V_1-V_2$ outside Σ and apply Green's theorem

$$\oiint_{\Sigma} V\frac{\partial V}{\partial n}\,dS = \iiint_{\text{all space outside}} \{V\nabla^2V+(\nabla V)^2\}\,d\tau$$

with $V = 0$ on Σ and $\nabla^2V = 0$ outside Σ.

6. $4\pi R^2\varepsilon_0 E$.
8. $3\varepsilon_0 F\cos\theta$.
10. $A = -F,\ B = a^2F,\ C = Q$; Area $2\pi\left(a^2+\frac{Q}{3F}\right)$, charge $\frac{3}{4}Fa^2+\frac{Q}{2}+\frac{Q^2}{12Fa^2}$.

Exercises 8.2 (p. 283)

1. Point charges $-Q_1$, $-Q_2$ at (optical) images of A, B respectively in the plane; (i) $3Q_1^2/(32\pi\varepsilon_0 a^2\sqrt{2})$; (ii) $Q_1^2(2\sqrt{2}-1)/(32\pi\varepsilon_0 a^2)$.
5. When the dipole is parallel to the plane.
6. $-Q/2$.
7. An additional repulsion on $+Q$ and an additional attraction on $-Q$, each being of magnitude $QQ'/(4\pi\varepsilon_0 d^2)$.
8. A repulsion (from the centre of the sphere) of magnitude $Q^2ac/\{4\pi\varepsilon_0(a^2-c^2)^2\}$.
11. Components of attraction:

towards dielectric $\frac{Q^2}{16\pi\varepsilon_0}\left\{\frac{1}{d^2}+\frac{d}{(a^2+d^2)^{3/2}}\right\}\left(\frac{K-1}{K+1}\right)$,

towards the other charge $\frac{Q^2}{16\pi\varepsilon_0}\left\{\frac{a}{(a^2+d^2)^{3/2}}\left(\frac{K-1}{K+1}\right)-\frac{1}{d^2}\right\}$.

Miscellaneous Exercises VIII (p. 327)

1. $-Q\left(\frac{a}{r}+\frac{\lambda a^3P_2}{r^3}\right)$.
10. $\frac{Q}{4\pi\varepsilon_0 a}\left\{\frac{\pi}{2}-\sum_{n=0}^{\infty}\frac{r^{2n+1}P_{2n+1}}{(2n+1)a^{2n+1}}\right\}$.
11. $\frac{Q}{4\pi\varepsilon_0 r}\sum_{n=0}^{\infty}\left(\frac{c}{r}\right)^n P_n(\cos\theta)P_n(\cos\alpha)$.

Exercises 9.4 (p. 364)

5. $x = \frac{\varepsilon_0}{eE}\cosh\left(\frac{eEy}{cp_0}\right)$.

Exercises 9.5 (p. 373)

2. $J = -\omega nBA(R\cos\omega t + L\omega\sin\omega t)/(R^2+L^2\omega^2)$; $\frac{1}{2}\omega^2n^2B^2A^2R/(R^2+L^2\omega^2)$.
3. $-4a^2\mu_0\omega j(R\cos\omega t+L\omega\sin\omega t)/\{2\pi d(R^2+L^2\omega^2)\}$.

Exercises 9.6 (p. 385)

8. $q_1 = \frac{1}{2}Q(\cos nt+\cos mt)$, $q_2 = \frac{1}{2}Q(\cos mt-\cos nt)$, where
$n^2 = 1/\{C(L+M)\}$, $m^2 = 2/\{C(L-M)\}$.
9. $CR^2 < 4L$.
10. $LRC\dfrac{d^2J}{dt^2}+R\dfrac{dJ}{dt}+LJ = V(t)$.

Exercises 9.7 (p. 397)

4. $\omega^2_{max} = R_1R_2(L_1L_2-M^2)$.
8. $\lambda = \mu_0/l$.

Miscellaneous Exercises X (p. 433)

1. $E = 2\pi kc\,\{a\sin[2\pi k(z-ct)] \quad -b\cos[2\pi k(z-ct)] \quad 0\}$,
$H = 2\pi k\,\{-b\cos[2\pi k(z-ct)] \quad -a\sin[2\pi k(z-ct)] \quad 0\}$.
2. $H = ka\cos\alpha = n\pi$; $H = \dfrac{1}{\mu_0 c}\{0 \quad E\cos(kz\cos\alpha)\cos[k(x\sin\alpha-ct)] \quad 0\}$.
3. $E = \dfrac{1}{\varepsilon_0\omega}\{\gamma\cos\alpha x\cos(\omega t-\gamma z) \quad 0 \quad -\alpha\sin\alpha x\sin(\omega t-\gamma z)\}$;
$\alpha = n\pi/a$; $J = \{0 \quad 0 \quad \cos\alpha x\cos(\omega t-\gamma z)\}$;
$\sigma = (\gamma/\omega)\cos\alpha x\cos(\omega t-\gamma z)$.
4. $|E| = \dfrac{Q_0}{\varepsilon_0 AK}\exp\left(-\dfrac{\sigma t}{\varepsilon_0 K}\right)$.
5. $E_z = B_n\sin(n\pi x/a)$.

INDEX

Aberration 344
Alternating currents 387, 406
Ampère's law 206, 257, 332
Anisotropic medium 257
Anode 281
Argand diagram 388
Atom 255

Biot–Savart law 206, 332
Blasius's theorem 309

Capacitor (condenser) 366
Capillaries 204
Cathode 281
Cauchy–Riemann conditions 305
Champion's experiment 336, 348
Circuits (magnetic) 258
Complex numbers 388, 406
Complex potential 305, 312
Complex variable 299, 305, 311
Conformal transformation 311
Conjugate functions 305
Connection (of a space) 299
Conservation of charge 211
Conservation of energy 400
Constitutive relations 419
"Corkscrew" rule 198, 199, 204, 205
Cyclic constant 222, 299
Cyclic potential 221

Diamagnetic substance 256, 258
Dielectric 255, 263, 278
Dipole (magnetic) 199, 221, 223, 224
Discontinuities 217, 247, 421
Displacement (electric) 203, 206, 418
Displacement current 417
Doppler effect 339

Earnshaw's theorem 267
Earth's magnetic field 225, 243

Effective values (of alternating quantities) 406
Einstein 337
Electrode 281
Electromagnetic induction 256, 365, 419
Electromagnetic momentum 430
Electromotive force (e.m.f.) 332, 365, 366, 388, 389
Electrostatics 263
Equivalent magnetic shell 217

Faraday's law 332, 351
Ferromagnetic substance 256, 258
Field lines 201
Field strength (magnetic) 203
Fitzgerald contraction 343
Fleming's rule 201
Flux 229, 332, 366, 367
Flux density 199, 206
Forced vibration 3, 392
Frequency 388
Fresnel's experiment 336, 337, 348

Galilean transformation 336, 342
Gauss's theorem 317
Gilbert 197
Gravitation 263, 316
Green's theorem 318

Harmonic functions 285, 299
Heaviside 356
Henry 207, 367
Hertz 420
Hysteresis 258

Images 270, 278, 302
Impedance 390, 407
Inductance 229, 365, 367

Inertial frame 336
Isotropic medium 257

Kaufmann 359
Kirchhoff's laws 197, 365, 388, 390

Laplace's equation 264, 284, 305
Larmor precession 360
Legendre polynomial 284, 322
Lenz's law 332
Linear medium 257, 405
Lodestone 243
Lorentz force 201, 356, 359
Lorentz transformation 336, 341, 349, 351, 357

MacCullagh's formula 321
Magnetic circuit 258
Magnetostatics 263, 280
Maxwell 355, 418
Molecular current 255
Molecule 255
Multiple connection 221
Mutual inductance 231

Neumann's law 332, 333
Normal modes 381, 396

Oersted's experiment 198
Ohm's law 197, 419

Paramagnetic substance 256, 258
Particle (magnetic) 199, 224, 243, 280
Permanent magnets 243
Permeability 206, 223, 255, 258
Phase 388
Photon 347
Planck's law 347
Poisson's (equivalent) currents 244, 256, 257
Potential (electrostatic) 217
Potential (magnetostatic scalar) 221, 223, 280
Potential (vector) 201, 207, 217, 243
Potential energy 202
Power factor 407
Poynting's vector 425
Pressure (of a fluid) 318
Proper 343, 357, 358

Quasi-steady currents 387, 406

Radiation (electromagnetic) 419
Reactance 390
Relative motion (uniform) 197
Relative permeability 258
Residual magnetization 259
Residues (complex variable) 310
Resonance 396
Rest mass (proper mass) 358
Retarded potential 421

Saturation 259
Scalar potential 221, 223
Self-inductance 230
Shell (magnetic) 199, 202, 223
Simultaneous events 337
Solenoid 204, 230
Solid angle theorem 271
Special relativity 347
Standard boundary conditions 263, 307, 318
Stokes's theorem 200, 202
Straight current 198, 204
Stress tensor 430
Surface current 211
Susceptibility 258
Switching 375

Tesla 199
Test circuit 198
Thomas precession 344
Time dilation 343, 346
Transformer 368
Transient 370, 388
Two-dimensional problems 298

Uniqueness theorem 264, 307

Vector potential 201, 207, 217, 243, 420
Volume current 211
Vortex filaments 198

Weber 206, 367
Wilson and Wilson's experiment 336
Work function 202, 225, 233, 372, 373, 399

Made in the USA
Monee, IL
20 October 2026

66035297R00144